PERGAMON INTERNATIONAL LIBRARY
of Science, Technology, Engineering and Social Studies
*The 1000-volume original paperback library in aid of education, industrial training and the enjoyment of leisure*
Publisher: Robert Maxwell, M.C.

# NMR Data Handbook for Biomedical Applications

# NMR Data Handbook for Biomedical Applications


PAULA T. BEALL, Ph.D.

Department of Physiology
Baylor College of Medicine
Houston, Texas 77030, USA

SHARAD R. AMTEY, Ph.D.

Department of Radiology
University of Texas Medical School
Houston, Texas 77030, USA

SITAPATI R. KASTURI, Ph.D.

Department of Physics
Tata Institute
Bombay, India



PERGAMON PRESS

New York • Oxford • Toronto • Sydney • Paris • Frankfurt

Pergamon Press Offices:

| | |
|---|---|
| **U.S.A.** | Pergamon Press Inc., Maxwell House, Fairview Park, Elmsford, New York 10523, U.S.A. |
| **U.K.** | Pergamon Press Ltd., Headington Hill Hall, Oxford OX3 0BW, England |
| **CANADA** | Pergamon Press Canada Ltd., Suite 104, 150 Consumers Road, Willowdale, Ontario M2J 1P9, Canada |
| **AUSTRALIA** | Pergamon Press (Aust.) Pty. Ltd., P.O. Box 544, Potts Point, NSW 2011, Australia |
| **FRANCE** | Pergamon Press SARL, 24 rue des Ecoles, 75240 Paris, Cedex 05, France |
| **FEDERAL REPUBLIC OF GERMANY** | Pergamon Press GmbH, Hammerweg 6, D-6242 Kronberg-Taunus, Federal Republic of Germany |



**Library of Congress Cataloging in Publication Data**
Beall, Paula T.
NMR data handbook for biomedical applications.

Includes index.
1. Nuclear magnetic resonance spectroscopy--Handbooks, manuals, etc. 2. Nuclear magnetic-resonance--Diagnostic use--Handbooks, manuals, etc. 3. Nuclear Magnetic resonance spectroscopy--Tables. 4. Medicine--Research--Handbooks, manuals, etc. 5. Biology--Research--Handbooks, manuals, etc. I. Amtey, Sharad R. II. Kasturi, Sitapati R. III. Title, IV. Title: N.M.R. data handbook for biomedical applications. V. Series.
QP519.9.N83B43 1984 610.28 83-25660
ISBN 0-08-030774-4
ISBN 0-08-030775-2 (pbk.)



***Printed in the United States of America***

# CONTENTS

# PREFACE

In the decade of the 1970s, nuclear magnetic resonance (NMR) spectroscopy came into its own as a new and exciting tool in biological research. Many scientists and physicians were introduced to the power and diversity of NMR as a diagnostic method through the use of whole body NMR scanners. As these complex and expensive instruments become more common in research and medicine in the 1980s, a new generation of investigators is entering what has been until very recently the narrow and specialized field of biological NMR. This book was written for students, clinicians, and scientists seeking a clearly written simplified text as an introduction to the biological applications of NMR.

The text is divided into 10 chapters, each of which covers a specific block of material and has its own references. The volume is meant to serve as a laboratory handbook and a desk reference, containing basic NMR theory, useful formulae and physical constants, and compiled data from the NMR literature. Chapters 1 and 2 cover the basic theory of NMR and imaging by NMR techniques. The practical applications of multinuclear NMR and the actual methods of measuring NMR parameters are discussed in Chapters 3 and 4. Chapter 5 is a detailed practical guide to the handling of biological samples for *in vitro* studies. Examples of protocols for tissue biopsies, cultured cells, and blood experiments are given to aid in the reduction of biological variation in NMR studies. In Chapter 6 the dependence of NMR relaxation times on physical variables such as frequency, temperature, and water content is discussed, and tables of experimental data are presented. Problems and solutions to the statistical handling of NMR data are illustrated using a detailed example in Chapter 7. Chapter 8 contains 30 tables of useful physical data, formulae, and constants used in NMR. Chapter 9 contains a series of 26 tables compiled from NMR data in the literature on invertebrate, mammalian, and human systems. It covers the majority of $T_1$, $T_2$, and D measurements on organisms, tissues, cells, cell organelles, and body fluids *in vitro*. This heavily referenced section provides background material for those just becoming interested in the field. Many European and eastern European sources are included. In the last section, glossaries of both physical terms for biologists and clinicians and biomedical terms for physical scientists are included to aid in the understanding of this interdisciplinary area of research.

The volume attempts to cover the development of biological NMR through several decades of *in vitro* experiments that have laid the groundwork for and pointed to profitable areas of investigation for new *in vivo* techniques. After gaining an understanding of these basic principles and some historical perspective, the reader will be better prepared to delve into highly specialized volumes on NMR imaging techniques.

The authors wish to acknowledge the assistance of Ms. Jane Lester, Ms. Ann Thompson, Ms. Anita Harris, and Ms. Marina Evagelatos in the preparation of the manuscript. The aid and support of colleagues who contributed to the completion of the book are gratefully acknowledged. Special thanks to Dr. Carlton F. Hazlewood of the Baylor College of Medicine and to Dr. R. Vijayaraghavan of the Tata Institute of Fundamental Research, India, for their constant support and encouragement. The support of our friends and families during the preparation of the book contributed greatly to its completion. Dr. Kasturi wishes to thank his wife, Lakshmi, his daughter, Niraja, and his son, Vikas, for their patience in granting him a year's leave from family responsibilities to do research in the United States.

The authors of this book were partially supported by the following grants and contracts during the writing of the manuscript: Robert Welch Foundation Q–390 and the Office of Naval Research contracts N00014-76-C-0100 and N00014-81-K-0167.

**P. T. Beall**
**MR2 Pharma Division, CIBA-Geigy,**
**444 Saw Mill River Road, Ardsley, NY 10502**

**S. R. Amtey**
**Mag Scan, Inc., Houston, TX**

**S. R. Kasturi**
**Tata Institute, Bombay, India**

# INTRODUCTION

## HISTORICAL BACKGROUND OF BIOLOGICAL NMR

Demonstrations of the nuclear magnetic resonance (NMR) phenomenon for hydrogen protons were independently published by Bloch, Hansen, and Packard[1] and Purcell, Torrey, and Pound[2] in 1946. For this work, Bloch and Purcell shared the Nobel Prize. In their experiments, atomic nuclei of odd spins (in this case hydrogen protons in water or paraffin) in a magnetic field were shown to absorb energy at a characteristic resonance frequency. The first NMR signal was produced from the hydrogen nuclei of water molecules, but little thought was given at that time to the future biomedical application of NMR in wet living systems. In the 1950s physicists utilized NMR to expand our knowledge of the structure and motion of simple molecules. Chemists began to exploit the phenomenon of chemical shift in NMR to collect characteristic absorption spectra of organic molecules that would allow their identification and structural analysis.

In the early 1950s, only a few individuals had begun to explore the potential biological and biomedical applications of NMR. T. M. Shaw and co-workers[3–5] began to use the proton signal intensity from NMR measurements to estimate the water content of plant and animal foodstuffs. The technique has since become an important tool in the food industry for determining the optimum moisture contents for storage and the conditions for freezing meat to prevent rigor or toughness.[6] However, the man who first published studies describing the use of NMR in the testing of biomedical hypotheses was Dr. Erik Odeblad of Sweden. In 1953, while at the Karolinska Institute in Stockholm, Odeblad, a young gynecologist with a strong interest in physics (Ph.D. in 1966), teamed up with Dr. Gunnar Lindstrom at the Nobel Institute of Physics. Using the first NMR spectrometer in Sweden they began investigations on biological systems which were published in 1955 under the title of "Some Preliminary Observations on the PMR [Proton Magnetic Resonance] in Biological Samples" in *Acta Radiologica*.[7] Their initial survey included human and rabbit blood fractions, yeast cells, rabbit liver, muscle, and fat, rat liver, muscle, and fat, calf cartilage, human Achilles tendon, and $D_2O$. In 1956 they used NMR methods to study the rapid exchange between $D_2O$ and $H_2O$ in human red blood cells.[8] Over the next three decades, Odeblad's research team contributed more than 40 papers to the literature on biological NMR research, including studies on human cervical mucus during the menstrual cycle,[9] tissues and fluids of the eye,[10] human vaginal cells,[11] human milk,[12] human gingival tissue,[13] human saliva,[14] human and animal uterine myometrium,[15] and tongue cells.[16] Of special significance was their discovery of cyclical hormone-controlled changes in the viscosity of cervical mucus that could be related to ovulation and have become the basis of a worldwide birth control effort.[17] Today, Dr. Odeblad is a professor at the University of Umeä, Sweden, and continues to work toward furthering our understanding of the biophysics of water in living systems. His unique contribution, as a physician and a physicist, was that he was a leader in recognizing not just the potential of NMR in biological research, but also its potential as a diagnostic tool in medicine. As early as 1961 he was advocating the inclusion of NMR spectrometers in medical research departments.[18]

Another entirely different, early biomedical application of proton NMR was its use for measuring blood flow. In 1959 J. R. Singer[19] first demonstrated a method of relating the amplitude of an NMR proton signal in the tail of a live mouse to blood flow through the tail vascular system. In a later design, the difference in signal amplitude between a transmitter coil and a receiver coil, spaced some distance apart over a flowing stream of liquid, could be related to volume flow per minute.[20] In 1970, Morse and Singer[21] demonstrated NMR blood flow measurements in living human subjects, one of whom was Singer's daughter, whose arm was thin enough to fit between the poles of their electromagnet. This approach is now being coupled with NMR imaging to measure blood flow in large arteries and veins with better accuracy than any other method.[22]

In the late 1950s and into the 1960s, numerous investigators began to realize the potential of NMR to test basic hypotheses about the role of water in biological systems. Because tissues contain 70–90% water and because of the high sensitivity of the hydrogen nucleus, it is possible to look at high-resolution NMR spectra in the proton region to thereby gain some insight into the behavior of water from these spectra. Bratton, Hopkins, and Weinberg[23] observed a broadening of the proton line width in frog gastrocnemius muscle in a homemade spectrometer in 1965. When the muscle was stimulated to tetanic isometric contraction, the proton signal was distinctly narrowed by 20 ± 5%. In seeking to understand the line broadening, these investigators utilized pulsed NMR techniques to measure $T_1$ and $T_2$ relaxation times for the protons in the muscle. Their article was probably the first to incorporate the recognition that an explanation of the NMR findings would require a model of water in the cell which would include at least a small fraction with altered physical properties relative to pure water. Upon the physiological change of contraction, they proposed that a release of a portion of water from this "bound" fraction occurred near macromolecular surfaces.

Sussman and Chin[24] confirmed the fact that muscles undergoing physiological changes could be evaluated by NMR methods. Codfish muscle, monitored from 10 min to 6 h after death, showed a proton line width narrowing consistent with the hypothesis that a loss of macromolecular structure and an increase in water mobility would occur.

From the start, biological NMR application was an international effort. Beginning in Sweden and then later in the United States, the field rapidly gained adherents in France, Russia, Mexico, and Japan. In 1966, Rybak et al.[25] coupled an electrocardiograph and an NMR spectrometer to measure proton spectra in a beating turtle heart. The turtle heart was removed from the animal, but, as is the case with this organ in the turtle, it continued to beat for hours. Under the conditions of these experiments, a broad line for proton resonance was detectable with each contraction of the ventricle. In Russia, Bruskov also published reports on the use of NMR in biological studies in 1966.[26] Cerbon, in Mexico, proposed that water could be restricted in its motion by interactions with lipid-rich membrane systems.[27,28] Using the *Nocardia asteroides* microorganism, he was able to correlate changes in water proton and lipid proton absorption peaks in cells treated with $Ca^{2+}$ salts. In Japan, Koga et al.[29] became interested in using NMR to study water in partially dried yeast. The period of 1964–1966 produced the first international burst of interest in biological applications of proton NMR.

During the 1960s and into the 1970s, the relatively small sample chambers and high costs of NMR machines designed for the chemical industry restricted the application of NMR in biology to *in vitro* studies on model systems, cells, and excised tissues. Most of these experiments are summarized in the biological data tables in this volume. So far, all the findings of proton *in vitro* studies, which have been tested, have been confirmed *in vivo*.

These publications provide a firm foundation for the interpretation of NMR imaging data and point toward many new avenues for *in vivo* investigations (see Tables 9.2–9.26 in Chapter 9).

Some investigators also began to consider the use of NMR in studying nuclei other than hydrogen protons in living systems. The greatly reduced sensitivity of other nuclei and their low concentrations in living cells made these experiments difficult. But by using isotope enrichment and signal averaging techniques, some pioneering experiments were accomplished. While Odeblad et al.[8] had first used $D_2O$ to observe exchange in red blood cells, Freeman Cope first used $D_2O$ in 1969[30] to examine the structure of water in muscle and brain. His findings suggested that multinuclear studies using $^{1}H$, $^{2}D$, and $^{17}O$ NMR in the same system would aid in the interpretation of NMR water data (see Table 9.2 in Chapter 9). In 1965 Cope[31] also published the first $^{23}Na$ NMR studies on biological tissues, which he interpreted as evidence for the complexing of $Na^+$ to cellular proteins in muscle, brain, and kidney. However $^{23}Na$ NMR data interpretations were and still remain somewhat controversial (see Table 9.22 in Chapter 9). Biological $^{17}O$ NMR studies of water were reported by Glasel[32] as early as 1966 (see Table 9.23 in Chapter 9). In 1970, Cope and Damadian[33] extended their studies to the NMR of the $^{39}K$ nucleus in bacteria. They concluded that cellular potassium could not be considered to be in a free solution. In 1979, they were able to measure $^{39}K$ NMR in live newborn mice[34] (see Table 9.21 in Chapter 9). The use of other exotic nuclei in biological NMR was explored somewhat later.

In 1973, Moon and Richards[35] were able to accumulate sufficient $^{31}P$ signals over 5.5 h to prepare a phosphorous spectrum from rabbit blood. The primary component of the spectrum was 2,3-diphosphoglycerate, from the red cells. They also originated the idea of measuring intracellular pH from the chemical shift between inorganic phosphate in the cell and the extracellular compartment (where the pH is known). This is probably the most accurate method ever invented to measure intracellular pH. In 1974 Hoult et al. demonstrated that quantitative metabolic spectra could be obtained from rabbit muscle in approximately 400 s with improved instrumentation.[36] Their work first showed the alternations in ATP and its metabolic products, ADP, AMP, and $P_i$, as a function of time in an excised rat leg muscle. $^{31}P$ NMR has continued to yield information about metabolite levels, turnover, interactions, and compartmentation in both normal and diseased tissues. Today, *in vivo* $^{31}P$ NMR is an exciting adjunct to whole body proton imaging techniques[37,38] (see Tables 9.24, 9.25, and 9.26 in Chapter 9).

In 1977 Fung, using $^{13}C$ spectra of natural and dehydrated mouse muscle, proposed that a significant amount of mobile organic molecules existed in these cells.[39] Enriched and natural abundance $^{13}C$ spectra from many types of tissues are now used in characterizing physiological states and disease processes[40,41] (see Table 9.19 in Chapter 9).

While some investigators were exploring the possibilities of multinuclear NMR experiments using biological samples, there were a few individuals who began to look into the future of NMR as a biomedical tool. They primarily concentrated on the development of *in vivo* methods to study parts of whole living organisms. The ultimate medical usefulness of NMR would depend on their contributions, which led to whole body spectra and NMR imaging in humans.

One of the most interesting references in the area of *in vivo* NMR is the master's thesis of Thomas Richard Ligon, published in 1967 at Oklahoma State University.[42] Ligon stated that, after reading the study by Bratton, Hopkins, and Weinberg[23] on proton NMR of an isolated frog muscle, he saw no reason why this could not be done *in vivo* in human arms. As a graduate student in the Department of Physics at Oklahoma State, under Dr. V. L. Pollack, he began work on a coil design for low-field NMR experiments which could be used on a human arm at frequencies of 4.26 KHz to 1.66 MHz, surprisingly close to current imaging frequencies. Recruiting fellow students as subjects and using a number of wide-gap electromagnets on the campus, he proceeded to measure proton signal, $T_1$, and $T_2$ of the human arm. He made the fundamental discoveries that $T_1$ and $T_2$ data *in vivo* were not exactly exponential, that relaxed and contracted muscle could be differentiated, that prepuberty males had muscle values different from those of postpuberty males, that values for male and female arm differed, and that the fat content of the tissue was a problem in interpreting the data.

More mature scientists had also begun to think seriously about using NMR in living animals. In 1968, J. A. Jackson published two articles[43,44] describing the design and construction of a solenoid large enough to hold a whole rat. The proton spectra he published included that of a whole chicken egg and the body of a whole live rat. These experiments certainly qualify as early *in vivo* studies, although they were not very sophisticated.

Among those scientists still examining the diagnostic potential of NMR in *in vitro* excised tissues was Raymond Damadian. In 1971, he published data supporting the possibility that NMR could distinguish normal from cancerous tissues.[45] He must also have been thinking about the future application of NMR technology in the human body because in March of 1972 he filed a United States patent application[46] entitled, "Apparatus and Method for Detecting Cancer in Tissue" (patent no. 3789832), which dealt with this idea. Weisman et al., who were familiar with the work of Damadian, published in the same year a short paper in *Science* showing how NMR signals from a tumor implanted in the tail of a live rat could be monitored from day to day by NMR methods.[47]

However, these investigators were certainly not the only ones thinking and working on the adaptation of NMR to living animals and humans. At this point, the history of the development of NMR imaging and *in vivo* applications becomes much more complex. Some individuals made theoretical contributions toward the techniques that led to imaging by NMR. Some contributed *in vitro* data which pointed the way for future use of NMR *in vivo*. Others engineered and built the large magnets and specialized coils for the first attempts, and then there were those who did the first experimental work. Every case demonstrates the tremendous teamwork among physicists, engineers, and biologists that was necessary for the accomplishment of the first image of the human body in 1977.[48] The following paragraphs mention only a few of the leaders of research teams that made significant early contributions.

Damadian's research group at Downstate Medical Center in New York City was dedicated to making a reality out of the possibility of localized NMR scanning in living organisms. In a recent book, he stated that he came up with the idea while making $^{39}K$ NMR measurements on pellets of bacteria in 1969. After the experiment in 1971[45] that showed differences in $T_1$ relaxation times between normal and cancerous rat tissue *in vitro*, his group began to work on a method for locating tumors in a living animal. The method was called "field focusing nuclear magnetic resonance" (FONAR) and used the concept of moving the sample (or body) through a point (saddle point) in a static magnetic field. Collection of data after many such movements allowed the formation of a compiled image. This technique was used to produce the first image of a live animal in 1976.[49] A similar technique was proposed in the application for a patent by Abe and Tanaka in 1973.[50]

It had been known for some time that there was another method for the spatial localization of NMR information in simple glass and liquid constructs. The imposition of a linear magnetic field gradient on the homogeneous laboratory magnetic field would produce a one-dimensional (1D) profile or projection of proton density along the direction of the gradient.[51,52] This 1D method would be used by Mansfield and Grannell in 1973[53] to investigate periodic structures by an NMR diffraction method.

However, in 1973, Lauterbur[54] conceived of an improvement in this method for the production of NMR images. His technique consisted of coupling the resonant NMR electromagnetic field with a spatially defining inhomogeneous field or static field gradient. A two-dimensional (2D) image could be made by combining several projections taken either as the object was rotated around the gradient or as the gradient was rotated about the object. Using this method he was able to produce a 2D image of two capillaries filled with water and separated by approximately 4 mm. He suggested that the term "zeugmatography," from the Greek word *zeugma* (that which joins together), be used for this process. The technique has now been extended to three-dimensional (3D) imaging produced by the adjustment of the currents in three sets of gradient coils. Information over a large spherical volume can be collected relatively rapidly, but requires complex computer processing.[55]

Some other research groups which also made significant contributions to the development of early NMR imaging techniques are mentioned below.

Hutchison and co-workers at Aberdeen began work on (ESR) spectra in biological tissues in the 1960s. They were able to transfer their knowledge to NMR, and, in 1974, Hutchison et al.[56] published a projection reconstruction image of a dead mouse that looked like a mouse and indicated differences in $T_1$ values among the organs of the animal. This group also has made contributions in producing pure $T_1$ images[57] and in a methodology called "spin-warp" imaging.[58]

Mansfield and Grannell,[53] having had previous experience with 1D NMR diffraction, expanded their method to measure and reconstruct an image of a live human finger in 1977 by selective irradiative line scanning.[59] Hinshaw et al.[60,61] developed a sensitive point and a multiple sensitive point method for imaging, using one static and two oscillating field gradients. In this technique, a sensitive line was created through the sample and more data could be collected in a shorter period of time. They were able to publish an image of the live human wrist in 1977.[62]

Other individuals and groups certainly made many contributions to theoretical and experimental improvement of NMR imaging techniques during this period. And the historical reality of this work is that it was done with little funding and despite much criticism by a few who warned that NMR imaging would never be practical. In the midst of controversy and doubt as to whether NMR imaging with good resolution would ever be achievable, Damadian et al., in 1977, published the first NMR image of a human thorax.[48] With dedication and determination, they had succeeded in demonstrating that a large enough magnet could be built and that many of the doubts about the feasibility of whole body NMR could be overcome.

From that point on, the growth of NMR imaging was phenomenal. Commercial development of whole body scanners made their use practical in medical diagnosis. Two scientific societies and a number of specialized journals are operational. The number of published articles relating to biological uses of NMR has increased from less than half a column in *Index Medicus* in 1973 to a page and a half in 1983. Hundreds of clinicians, students, and health professionals are finding that "NMR imaging" is a term they must add to their vocabularies. And research scientists, even those who were acquainted with NMR, are discovering a whole new field of specialized applications.

This volume was written to bring together the basic information and decades worth of biological data for use by those seeking an introduction to biological NMR. After reading this book, a reader will be better able to understand specialized books on NMR imaging or state of the art clinical reports. It is the authors' hope that this book will become a convenient desk reference for their colleagues in NMR.

## REFERENCES


1. Bloch, F., Hansen, W.W., and Packard, M. The nuclear induction experiment. *Physiol. Rev.* **70:**474–485, 1946.
2. Purcell, E.M., Torrey, H.C., and Pound, R.V. Resonance absorption by nuclear magnetic moments in a solid. *Physiol. Rev.* **69:**37–43, 1946.
3. Shaw, T.M., Elsken, R.H., and Kunsman, C.H. Proton magnetic resonance absorption and water content of biological materials. *Physiol. Rev.* **85:**708–711, 1953.
4. Shaw, T.M., Elsken, R.H., and Kunsman, C.H. Moisture determinations of foods by hydrogen nuclei magnetic resonance. *J. Agric. Food Chem.* **4:**162–164, 1953.
5. Shaw, T.M., and Elsken, R.H. Determination of water by nuclear magnetic absorption in potato and apple tissue. *J. Agric. Food Chem.* **4:**162–164, 1956.
6. Troller, J.A., and Christian, J.H.B. *Water Activity and Food.* Academic Press, New York, 1978.
7. Odeblad, E., and Lindstrom, G. Some preliminary observations on the PMR in biological samples. *Acta Radiol.* **43:**469–476, 1955.
8. Odeblad, E., Bahr, B.N., and Lindstrom, G. Proton

magnetic resonance of human red blood cells in heavy-water exchange experiments. *Arch. Biochem. Biophys.* **63**:221–225, 1956.
9. Odeblad, E., and Bryhn, U. Proton magnetic resonance of human cervical mucus during the menstrual cycle. *Acta Radiol.* **47**:315–320, 1957.
10. Huggert, A., and Odeblad, E. Proton magnetic resonance studies of some tissues and fluids of the eye. *Acta Radiol.* **51**:385–392, 1958.
11. Odeblad, E. Studies on vaginal contents and cells with proton magnetic resonance. *Ann. N.Y. Acad. Sci.* **82**:189–206, 1959.
12. Odeblad, E., and Westin, B. Proton magnetic resonance of human milk. *Acta Radiol.* **49**:389–392, 1958.
13. Forsslund, G., Odeblad, E., and Bergstrand, A. Proton magnetic resonance of human gingival tissue: a preliminary report. *Acta Odontol. Scand.* **20**:121–126, 1962.
14. Odeblad, E., and Soremark, R. Studies on human saliva with nuclear magnetic resonance. *Acta Odontol. Scand.* **20**:33–42, 1962.
15. Odeblad, E., and Ingelman-Sunberg, G. Proton magnetic resonance studies on the structure of water in the myometrium. *Acta Obstet. Gynecol Scand.* **44**:117–125, 1965.
16. Odeblad, E., and Forsslund, G. Mechanism of hydration in complex biological material. *Acta Isotopica* **2**:127–132, 1962.
17. Hoglund, A., and Odeblad, E. Sperm penetration in cervical mucus, a biophysical and group-theoretical approach. In: *The Uterine Cervix in Reproduction.* Insler, V., and Bettendorf, G., eds. Georg Thieme Publishers, Stuttgart, 1977, pp. 129–134.
18. Odeblad, E. A department for medical research with nuclear magnetic resonance. *Acta Isotopica* **1**:27–39, 1961.
19. Singer, J.R. Blood flow rates by nuclear magnetic resonance measurements. *Science* **130**:1652–1653, 1959.
20. Singer, J.R. R-F methods for measuring fluid velocities. *Electronics* **33**:77–79,1960.
21. Morse, O.C., and Singer, J.B. Blood velocity measurements in intact subjects. *Science* **170**:440–442, 1970.
22. Singer, J.R., and Crooks, L.E. Nuclear magnetic resonance blood flow measurements in the human brain. *Science* **221**:654–656, 1983.
23. Bratton, C.B., Hopkins, A.L., and Weinberg, J.W. NMR studies of living muscle. *Science* **147**:738–739, 1965.
24. Sussman, M.V., and Chin, L. NMR spectrum of changes accompanying rigor mortis. *Nature* **211**:414–415, 1966.
25. Rybak, B., et al. Simultaneous recording of ventricular contractions, nuclear magnetic resonance and the electrocardiogram. *C.R. Acad. Sci. (Paris)* **262**:2285–2287, 1966.
26. Bruskov, V.I. The use of the nuclear magnetic resonance method in biological studies. *Biofizika* (Russian) **11**:195–204, 1966.
27. Cerbon, J. NMR studies of water immobilization by lipid systems in vitro and in vivo. *Biochim. Biophys. Acta* **144**:1–9, 1967.
28. Cerbon, J. Variations of the lipid phase of living microorganisms during the transport process. *Biochim. Biophys. Acta* **102**:449–458, 1965.
29. Koga, S., Echigo, A., and Oki, T. NMR spectra of water in partially dried yeast cells. *Appl. Microbiol.* **14**:466–467, 1966.
30. Cope, F.W. Nuclear magnetic resonance evidence using $D_2O$ for structured water in muscle and brains. *Biophys. J.* **9**:303–319, 1969.
31. Cope, F.W. Nuclear magnetic resonance evidence for complex of sodium ions in muscle. *Proc. Natl. Acad. Sci. U.S.A.* **54**:225–227, 1965.
32. Glasel, J.A. A study of water in biological systems of 0–17 magnetic resonance spectroscopy. *Proc. Natl. Acad. Sci. U.S.A.* **55**:479–485, 1966.
33. Cope, F.W., and Damadian, R. Cell potassium by $^{39}K$ spin echo NMR. *Nature* **228**:76–79, 1970.
34. Cope, F.W., and Damadian, R. Pulsed nuclear magnetic resonance of potassium ($^{39}K$) of whole body live and dead newborn mice. Double oscillation frequencies in $T_1$ decay curves. *Physiol. Chem. Phys.* **11**:143–149, 1979.
35. Moon, R.B., and Richards, J.D. Determination of intracellular pH by $^{31}P$ magnetic resonance. *J. Biol. Chem.* **248**:7276–7278, 1973.
36. Hoult, D.I., Busby, S.J.W., Gadian, D.G., Radda, G.K., Richards, R.E., and Seeley, P.J. Observation of tissue metabolites using $^{31}P$ nuclear magnetic resonance. *Nature* **252**:285–287, 1974.
37. Ingwall, J.S. Phosphorous nuclear magnetic spectroscopy of cardiac and skeletal muscles. *Am. J. Physiol.* **242**:729–744, 1982.
38. Gadian, D.G., and Radda, G.K. NMR studies of tissue metabolism. *Annu. Rev. Biochem.* **50**:69–83, 1981.
39. Fung, B.M. Carbon-13 and proton magnetic resonance of mouse muscle. *Biophys. J.* **19**:315–319, 1977.
40. Doyle, D.D., Chalovich, J.M., and Barany, M. Natural abundance $^{13}C$ spectra of intact muscle. *F.E.B.S. Lett.* **131**:147–150, 1981.
41. Barany, M., Doyle, D.D., Graff, G., Westler, W.M., and Markley, J.L. Changes in the natural abundance $^{13}C$ NMR spectra of intact frog muscle upon storage and caffeine contracture. *J. Biol. Chem.* **257**:2741–2743, 1982.
42. Ligon, T.R. Coil design for low field NMR experiments and NMR measurements on the human arm. M.S. Thesis, Oklahoma State University, 1967.
43. Jackson, J.A. Whole body NMR spectrometer LA3966 US AEC. University of California–Los Alamos Scientific Laboratory, 1–12; 31 July 1968.
44. Jackson, J.A. Whole body spectrometer. *Rev. Sci. Instrum.* **39**:510–513, 1968.
45. Damadian, R. Tumor detection by nuclear magnetic resonance. *Science* **171**:1151–1153, 1971.
46. Damadian, R. Apparatus and method for detecting cancer in tissue. U.S. Patent no. 3789832. Filed March, 1972.
47. Weisman, I.D., Bennett, L.H., Maxwell, L.R., and Woods, M.W. Recognition of cancer in vivo by nuclear magnetic resonance. *Science* **178**:1288–1290, 1972.
48. Damadian, R., Goldsmith, M., and Minkoff, L. NMR in cancer. FONAR image of the live human body. *Physiol. Chem. Phys.* **9**:97–100, 1977.
49. Damadian, R., Minkoff, L., Goldsmith, M., Stanford, M., and Koutcher, J. Field focusing nuclear magnetic resonance (FONAR) visualization of a tumor in a live animal. *Science* **194**:1430–1432, 1976.

50. Abe, Z., and Tanaka, K. U.S. Patent no. 3932805. August 9, 1973.
51. Gabillard, R. Measurement of relaxation time $T_1$ in the presence of an inhomogeneous magnetic field. *C.R. Acad. Sci. (Paris)* **232:**1551–1553, 1951.
52. Anderson, A.G., Garvin, R.L., Hahn, E.L., Horton, J.W., Tucker, G.L., and Walker, R.M. Spin echo serial storage memory. *J. Appl. Physiol.* **26:**1324–1327, 1955.
53. Mansfield, P., and Grannell, P.K. NMR "diffraction" in solid? *J. Phys. [C]* **6:**14–22, 1973.
54. Lauterbur, P. Image formation by induced local interactions: examples employing nuclear magnetic resonance. *Nature* **242:**190–191, 1973.
55. Lai, C.M., and Lauterbur, P. True three-dimensional image construction by nuclear magnetic resonance zeugmatography. *Phys. Med. Biol.* **26:**851–856, 1981.
56. Hutchison, J.M.S., Mallard, J.R., and Goll, C.C. In-vivo imaging of body structures using proton resonance. In: *Proceedings of the 18th Ampere Congress.* Allen, P.S., Andrew, E.R., and Bates, C.A., eds. University of Nottingham, Nottingham, England, 1974, p. 283.
57. Mallard, J.R., Hutchison, J.M.S., Edelstein, W.A., Ling, C.R., Foster, M.A., and Johnson, G. In-vivo NMR imaging in medicine: the Aberdeen approach both physical and biological. *Philos. Trans. R. Soc. Lond. [Biol.]* **289:**519–533, 1980.
58. Edelstein, W.A., Hutchison, J.M.S., Johnson, G., and Redpath, T.W. Spin warp NMR imaging and applications of human whole-body imaging. *Phys. Med. Biol.* **25:**751–753, 1980.
59. Mansfield, P., and Maudsley, A.A. Medical imaging by NMR. *Br. J. Radiol.* **50:**188–194, 1977.
60. Hinshaw, W.S. Image formation by nuclear magnetic resonance: the sensitive point method. *J. Appl. Physiol.* **47:**3709–3721, 1976.
61. Hinshaw, W.S., Andrew, E.R., Bottomley, P.A., Holland, G.N., Moore, W.S., and Worthington, B.S. Internal structural mapping by nuclear magnetic resonance. *Neuroradiology* **16:**607–609, 1978.
62. Hinshaw, W.S., Bottomley, P.A., and Holland, G.N. Radiographic thin section image of the human wrist by nuclear magnetic resonance. *Nature* **270:**722–723, 1977.

# CHAPTER 1

# *STATES OF WATER IN BIOLOGY**



## 1.1. INTRODUCTION

The role of water in biological processes has been under investigation since the times of the ancient Greeks. Those scientific philosophers divided matter into "earth, fire, and water." Classical investigations dealt with the determination of the water content of plants and animals by desiccation, but could go no further until more sophisticated chemical and physical techniques became available to biologists in the 20th century.

In the first half of this century biologists and biochemists were able to define some of the functions of water in cells. Water serves as the primary biological solvent in which all other cellular components and solutes diffuse in interacting with one another. In many biochemical reactions water serves as the donor of hydrogen ions, and the dissociation of water contributes to the regulation of cellular pH. Because of its high heat capacity, water stores metabolic energy and helps to regulate an organism's temperature. Through hydrophobic and hydrophilic interactions water contributes to the stabilization of three-dimensional macromolecular structure, and even stabilizes giant cellular constructs such as lipid bilayer membranes and microtubules. Water is the lubricant of the tissues and, in combination with polysaccharides, can provide low-friction films on surfaces. The end products of the metabolic pathways are carbon dioxide and water, so cells not only use water, they also synthesize it. Therefore, a great deal of biological function pertains to the consumption, regulation, and excretion of water. But what is this water like when it is inside cells? Since about 1950, techniques from physics have been borrowed by biologists to examine the physical state and mechanical properties of water in biological systems. In this chapter, some of these findings and their significance will be discussed.

There are three possible physical states of water in cells—gas, liquid, and solid states. In the temperature and pressure ranges of terrestrial life, we see water vapor present in the external environment or in the lungs, but rarely is the gaseous phase of water considered to be important inside the cell. Therefore, investigations of the liquid and solid states of water in biology are most commonly undertaken. Cryobiology is the study of the process of conversion of biological water to crystalline solid ice or, upon occasion, the peculiar biochemistry of organisms which prevents crystalline solid formation and substitutes other solvent forms at very low temperatures. Here, discussion will be concentrated on what has been learned about water in functioning metabolizing cells in their normal viable temperature range. Such water exists in a state somewhere between the solid state of crystalline ice and the liquid, hydrogen-bonded lattice of pure water (Fig. 1.1).

---

*Chapter 1 is a revision of an article entitled "States of Water in Biological Systems" that originally appeared in *Cryobiology* 20:324–334, 1983. By permission of Academic Press, Inc.

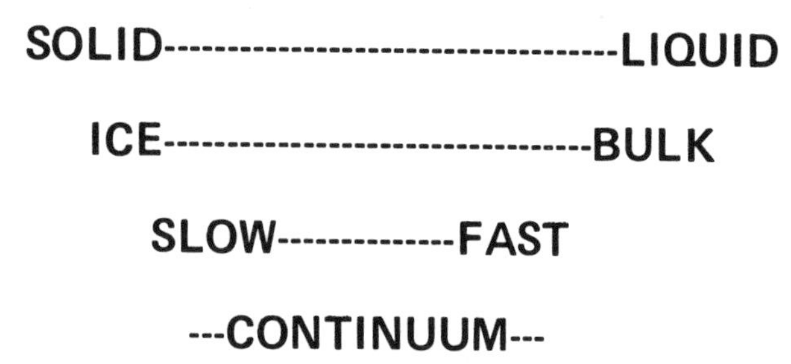


Fig. 1.1. States of water in biological systems.

## 1.2. BEHAVIOR OF PURE WATER

Excellent reviews concerning the physical findings on pure water and on water in dilute solutions of salts and polymers are available.[1–5] They point out that attempts to define the physical nature of pure liquid water have not been totally successful; therefore, it seems unlikely that we will be able to define the state of water in biological systems with any great accuracy. What we can do is make comparisons between the measurements of water in living systems and those taken on pure liquid water or ice by the same techniques.[6–9]

Numerous biophysical techniques have now been applied to living systems. Some of these are shown in Fig. 1.2 in relation to the time over which a measurement of the physical properties of water can be made. A critical number on this time scale is $\tau_{DH_2O}$ at $10^{-11}$ s. The $\tau_{DH_2O}$ is the diffusional correlation time or the time between jumps in position for water molecules in the system. In ice, $\tau_D$ is slow at $\sim 10^{-5}$ s, while in pure liquid water, $\tau_D$ is $\sim 10^{-11}$ s, or 1 million times faster. Many of the physical properties of water can be theoretically related to $\tau_D$, and the derivations of equations for biophysical techniques often include this parameter. Consequently, the ability of a biophysical technique to yield information about the physical state of water depends on how quickly a measurement can be made. On the left side of Figure 1.2 the type of structure that a technique can probe is related to the time scale. Methods such as infrared and Raman spectroscopy and inelastic neutron scattering yield information about intramolecular factors such as H–O bond lengths and hydrogen bond angles (geometric factors, Fig. 1.3). Techniques that require measurement times greater than the diffusional correlation time, however, will always yield an average over all molecules in the population, with a kinetic contribution from diffusion. They will more generally yield information about interactions between water molecules, and between

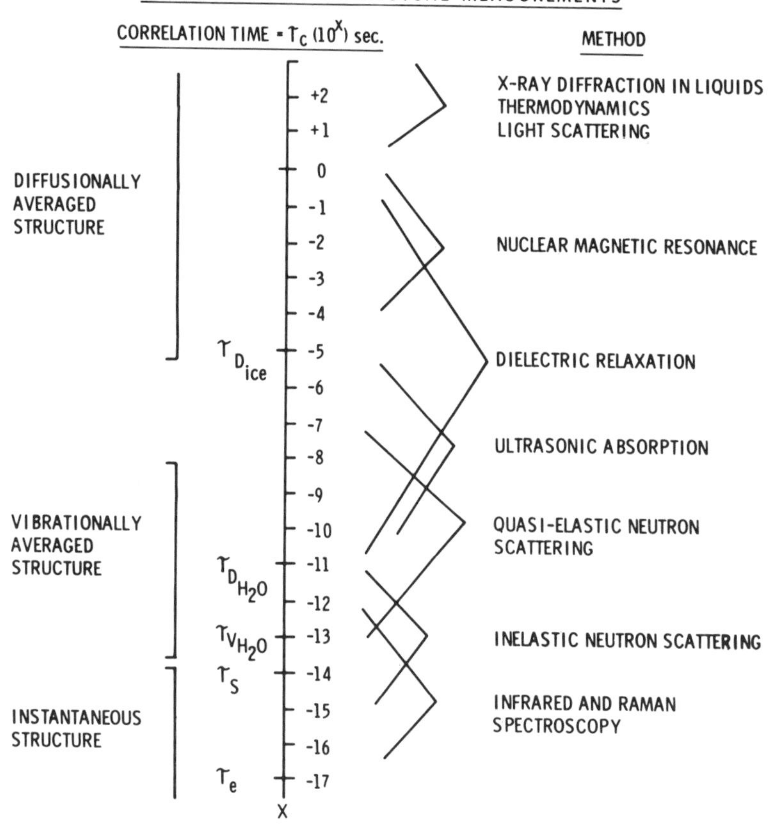


Fig. 1.2. Time scale of biophysical measurements. $\tau_D$, correlation time for water motion.

| Biophysical Methods to Monitor Water | |
|---|---|
| Geometric $< 10^{-11}$ sec | Kinetic $> 10^{-11}$ sec |
| X-ray crystallography | Thermodynamics |
| Raman spectroscopy | Nuclear Magnetic Resonance (NMR) |
| Infrared spectroscopy | |
| Neutron scattering | Dielectric relaxation |
| | Electron Spin resonance |
| | Fluorescence polarization |
| Acoustic Relaxation | |
| Quasi-elastic Neutron scattering | |

Fig. 1.3. Biophysical techniques used to monitor the behavior of water molecules.

water molecules and their environment (kinetic factors, Fig. 1.3). Some of the confusion over the physical state of water in biological systems has resulted from disparate interpretations of information derived from diverse techniques being applied to the same system. Whether water inside cells behaves exactly like pure liquid water, whether a fraction of cellular water may be relatively immobile, or whether macromolecular surfaces and charge distributions may influence water motion as a continuous gradient of forces away from surfaces, remains to be proven. Using only three examples, the state of biological water can be shown to be a still unsolved problem.

## 1.3. EXAMPLES OF WATER IN BIOLOGICAL SYSTEMS

Collagen is a protein that is the major component of the extracellular matrix of tissues; it also makes up the oriented linear fibers of the strong tendons and ligaments. X-ray diffraction has defined the spacing and geometry of protein fibers of collagen and the possible positions of water molecules, which are part of the structure.[10–14] As water is added to dry collagen, the properties of water in the system are measurable as a function of hydration. In such a well-defined biological model, it should be possible to determine the physical state of water. However, different techniques yield different information (Fig. 1.4). Walrafen[13] examined water in the interstices of collagen fibers with infrared spectroscopy, but found no differences in its behavior compared with that of bulk water. In kangaroo rat tail tendon collagen, the first few percent of added water acts as if it is in a glassy, solid-like state when examined by dielectric relaxation.[11] Techniques such as NMR, which look at average kinetic motion over all water molecules between the linear polypeptide chains, strongly support the hypothesis of reduced freedom of motion of these molecules. In a system containing 13% collagen fibers, a discernible reduction in the self-diffusion coefficient of water was measured by Hoeve and Kakivaya[11] and the thermodynamic properties of this water were changed. In a recent study on partially hydrated collagen gels, Shaw and Schy[12] reported an apparent reduction in the isotopic diffusion of water and other molecules which cannot be explained on the basis of a barrier effect. The reduced diffusion coefficient, through stop flow columns, for water and solutes can be explained only if the radius of the collagen fibers is increased from $10^{-8}$ to $10^{-6}$ cm, that is, greater by two orders of magnitude than it is according to X-ray diffraction. These results do not agree with the predictions of the Ogston[15] model for diffusion

PARTIALLY HYDRATED COLLAGEN

| TECHNIQUE | | FINDING | REF. |
|---|---|---|---|
| INFRARED SPECTROSCOPY | | NO DIFFERENCE FROM BULK | 13 |
| DIELECTRIC RELAXATION | | GLASSLIKE WATER BETWEEN FIBERS | 11 |
| NUCLEAR MAGNETIC RESONANCE ($^1$H) | | | 10 |
| | $T_1$ | REDUCED, > W. HEAT | 14 |
| | $T_2$ | REDUCED, > W. HEAT | 11 |
| | D | SLIGHTLY REDUCED 13% PROTEIN | |
| THERMODYNAMICS | | | |
| ΔH | | HIGHER THAN BULK, ONLY ONE-STATE | 11 |
| FREEZING | | NO ICE FORMATION | |
| STOP FLOW (GEL) | | REDUCED DIFFUSION | 12 |

Fig. 1.4. Summary of findings on the physical behavior of water in partially hydrated collagen.

SKELETAL MUSCLE

| TECHNIQUE | | FINDING | REF. |
|---|---|---|---|
| POSITRON ANNIHILATION | | MUSCLE $H_2O$ IS NOT POLYWATER | 72 |
| LASER RAMAN | | NO DIFFERENCE FROM BULK WATER | 57 |
| INFRARED SPECTROSCOPY | | NO DIFFERENCE FROM BULK WATER | 71 |
| DIELECTRIC RELAXATION | | 95% OF WATER LIKE BULK (9-256H$_2$) | 7 |
| NUCLEAR MAGNETIC RESONANCE | $^1H$ | $T_1$, $T_2$, REDUCED 6, 10X | 43 |
| RELAXATION | $^2H$ | $T_1$, REDUCED 6X | 17 |
| | $^{17}O$ | $T_1$, REDUCED 2X | |
| (DIFFUSION) | $D_{H_2O}$ | DIFFUSION SLOWED BY 50% | |
| EPR- SPIN-LABEL | | 5X MORE VISCOUS THAN BULK | 20 |
| FLUORESCENCE POLARIZATION | | CHANGE IN VICINAL WATER NEAR ATPASE | 22 |
| FREEZING EXPT. | | 10-20% "NONFREEZABLE WATER" | 23, 56 |

Fig. 1.5. Summary of findings on the physical behavior of water in skeletal muscle.

through a gel, in which the solvent and solutes do not interact with the protein. Shaw and Schy[12] speculate that some solvent is structured by the gel matrix so that there is no true fluid phase in the system. Their data suggest significant interactions of water with the collagen matrix out to as far as 100 Å from the protein surface. In this example, geometric techniques do not distinguish an altered physical behavior for water, but kinetic techniques suggest a reduced rate of motion, on average, for water molecules in the system.

Probably the most comprehensively studied biological sample in water biophysics is skeletal muscle because of its ordered lattice of actin and myosin filaments. Figure 1.5 summarizes some of the general findings in this system. Laser Raman spectroscopy, infrared spectroscopy, and dielectric relaxation[16] do not find intramolecular differences in hydrogen bond lengths, angles, or strengths for muscle water. However, other techniques such as NMR,[17-19] electron paramagnetic resonance (EPR),[20] fluorescence polarization,[21,22] and freezing behavior[23] suggest a restricted motion of at least a portion of cell water. Table 1.1 details some of the data on water in muscle at $-10$ to $-20$°C. A significant fraction of muscle water does not freeze, even when it is in contact with ice crystals. This is another indication of the nonideal behavior of water.

In a whole organism, such as the *Artemia salinus* (brine shrimp larvae), similar data have been recorded (Fig. 1.6). Quasi-elastic neutron scattering and NMR measurements indicate a reduced diffusion coefficient of water as a function of the hydration of the brine shrimp cysts. At the critical hydration of 0.3 g $H_2O$ per gram dry solids, physical parameters such as water diffusion, heat capacity, and free energy can be correlated with metabolic turn on, in the previously dormant dry organism.[24,25]

## 1.4. MODELS FOR NMR RELAXATION TIME DATA INTERPRETATION

At the level of the macromolecular model, in tissues and even in whole organisms, evidence exists for the reduced freedom of mobility of at least a portion of

**Table 1.1. Nonfrozen water in skeletal muscle**

| Sample | Technique | % Nonfrozen | Reference |
|---|---|---|---|
| Pig | NMR (line width) | 30 | Derbyshire and Parsons, 1972[54] |
| Frog | NMR (line width) | 20 | Belton et al., 1972[35] |
| Pig | NMR (line width) | 15 | Duff and Derbyshire, 1974[55] |
| Rat | NMR | 12 | Fung 1974[56] |
| Mouse | NMR | 9 | Fung and McGaughy, 1974[18] |
| Mouse | NMR | 10 | Rustigi et al., 1978[23] |
| Barnacle | Laser Raman | 20 | Pezolet et al., 1978[57] |
| Mouse | NMR | 10 | Peemoeller et al., 1980[58] |

**HYDRATED BRINE SHRIMP CYSTS**

| TECHNIQUE | FINDING | REF. |
|---|---|---|
| QUASI-ELASTIC NEUTRON SCATTERING-$D_{H2O}$ | REDUCED 2-10X<br>FUNC. OF HYDRATION | 50<br>61 |
| DIELECTRIC RELAXATION | REDUCED MOBILITY | 24 |
| NUCLEAR MAGNETIC RESONANCE (PROTON) $T_1$ | REDUCED, 2 FRACTIONS | 25 |
| $T_2$ | REDUCED, 2 FRACTIONS | |
| $D_{H2O}$ | REDUCED 2-10X, FUNC. OF HYDRATION | |
| THERMODYNAMICS $\triangle$H. $\triangle$S. $\triangle$G FOR H2O | FUNCTION OF HYDRATION-<br>ALL WATER IS NOT ALIKE<br>METABOLIC TURN-ON AT 0.3g $H_2O$/g. D.S. | 24 |

Fig. 1.6. Summary of findings on the physical behavior of water in brine shrimp cysts.

water molecules in each system. Controversy, however, exists over how much water is affected and to what extent[6,8,9,12,17,19,23,26–34] by the interior components of cellular life. The kinetic techniques which provide this evidence leave one without a definitive answer to this question because they measure average properties over all molecules in the system. The investigator must impose a conceptual model for further interpretation of such data. It is from model-dependent interpretation that controversy comes, not from the actual data, upon which most biophysicists do agree. Figure 1.7 is a diagram of three popular models used to interpret much of the biophysical data on water. Zimmerman and Brittin[34] first proposed that some of the biophysical data (especially from NMR) might be explained if a very small fraction of cell water (<3%) is highly immobilized on the surface of macromolecules, with correlation times ($\tau_C$) on the order of those of the $\tau_D$ of ice ($\sim 10^{-5}$ s), and if the rest of cell water is just like bulk water, with correlation times of $\sim 10^{-11}$ s (Fig. 1.7a). Rapid exchange between the two populations would yield reduced average biophysical parameters, weighted heavily by the small immobile fraction. This "two-fraction fast-exchange model" (Figure 1.7a) can be expressed for NMR relaxation times ($T_1$ and $T_2$) as

$$\frac{1}{T_a} = \frac{X}{T_s} + \frac{1 - X}{T_{H_2O}}$$

where $T_a$ is the measured relaxation time, $X$ is the fraction of cell water in the slow population, $1 - X$ is the fraction of cell water like bulk water, $T_s$ is the relaxation time of the slow fraction, and $T_{H_2O}$ is the relaxation time of bulk water.

Such a model contains many assumptions and specific predictions. Basically, we have one equation, one measured parameter ($T_a$), and three unknowns. To

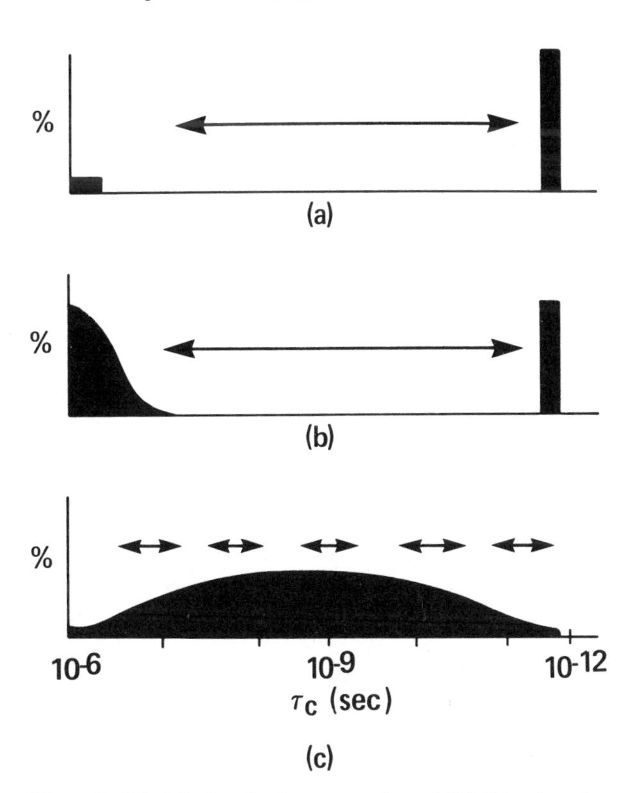


Fig. 1.7. Models for the interpretation of NMR relaxation time data in biological systems: (a) two-fraction fast-exchange model; (b) significantly large, slow-moving fraction model; and (c) continuum of correlation times model. $\tau_C$ is the motional correlation time.

solve for $X$, the relaxation times $T_s$ and $T_{H_2O}$ must be arbitrarily assigned or assumed to be those for ice and bulk water. With such an assumption, the value of $X$ is only a small fraction (<3%) of cellular water. However, the value of $X$ calculated for $T_1$ measurements cannot also be used to fit $T_2$ data for systems like muscle.[17] In addition, the nonexponential relaxation behavior for water protons suggests that the condition for "fast exchange" is not met.[28] Discussion of the failure of this model is important because this simplistic view is still widely used to interpret many types of biophysical data.

To solve for $T_s$, some other method must be used to estimate $X$. Various investigators have assumed that $X$, the fraction of slow water, may be equated to the "nonfreezable fraction" of water that does not freeze at −15°C (Table 1.1) (10–20% for skeletal muscle), or to a so-called osmotically inactive or nonsolvent fraction. Such assumptions move the interpretation to the model in Figure 1.7b. In this model a significant fraction of cell water is influenced by surfaces and charge distributions in cells, and has a distribution of $\tau_C$ near $10^{-7}$–$10^{-8}$. Water dipoles may associate with other dipoles and charges, and hydrogen bonds may form between water and other molecules such that the average overall motion of water is reduced and the residence time of a molecule at a point in space may be increased. However, model (b) (Fig. 1.7) allows for a large fraction of cellular water also to be in a bulk state in exchange with the slow fraction. By adjusting the fraction of water on the left and altering the distribution of correlation times over several orders of magnitude, this model offers ease of interpretation of data. However, simply because the model is easy to use and is the simplest case that approaches a fit to the data does not mean that it represents the true state of water in the system.

Figure 1.7c shows another popular "continuum model" in which very little water is like ice and very little water has the properties of pure liquid. Gradients of surface effects extend out into the cytoplasm from macromolecular surfaces, with many different populations mixing together. Such a distribution may not be smooth at all. It could have a hump at one end or even two humps separated by a valley of water molecules at many different correlation times. Recent data on the motion of water molecules on the surface of powdered lysozyme,[27,30,35] oriented DNA,[6,8,9] and phospholipid vesicles[36,37] support the view that at even less than one monolayer of coverage, water moves 10–100 times faster than in ice ($\tau_C$ $10^{-8}$–$10^{-9}$). This moves the left limit of the distribution toward the center and makes it difficult to propose that a tiny fraction of "bound" water with slow correlation times can weight biophysical parameters. However, neither model (b) nor (c) can completely fit $T_1$ and $T_2$ data for $^{1}H$, $^{2}D$, and $^{17}O$ NMR in muscle. Our understanding of the behavior of water in biology has not yet produced a model that can predict changes in water behavior in response to physiological changes.

However, it is possible to measure changes in the properties of water in living systems that correlate with physiological function. Alterations in water proton relaxation times have been correlated with cancer,[38–41] muscular dystrophy,[42] and developmental changes in animals.[43] Today whole-body NMR imaging systems based on differences in $T_1$ and $T_2$ relaxation times for water in tissues are being built and distributed across the United States for use with human patients[41,44] for cancer diagnosis.

While differences in the physical properties of water in biological systems may be utilized in practical applications, we still understand little about the causes of such differences. Several groups of macromolecules inside cells probably contribute to the properties of water. Alterations in the conformational state of chromatin,[45,46] in the organization and polymerization of the extensive cytoskeletal network,[16,45,47] and in the surface area of membranes inside cells[36,37] may contribute to distinguishable differences in water—macromolecular interactions measured by biophysical techniques. Until the possible contributions of these interactions are more fully understood and quantitated, we will continue to have controversy over the appropriate model for the state of water in cells.

## 1.5. SELF-DIFFUSION COEFFICIENT OF WATER IN BIOLOGICAL SYSTEMS

Controversy notwithstanding, there does exist a body of evidence in favor of reduced motion of water in biological systems which has not been disputed. NMR, quasi-elastic neutron scattering, and isotopic diffusion studies consistently find that the self-diffusion coefficient for water is reduced by at least one-half in many types of biological systems. Table 1.2 summarizes some of these findings. The major critics of these studies propose that the apparent reduction of $D_{H_2O}$ is due to low membrane permeability to water and to the obstruction effects of the membranes and macromolecular lattice. Shaw and Schy[12] have shown that reduced diffusion of water in collagen gels (which have no membrane) cannot be explained by the Ogston obstruction model,[15] unless collagen fibers are 100 times greater in radius than they have been shown to be by X-ray diffraction. Cleveland et al.[48] found, using NMR, that even parallel to the muscle lattice, $D_{H_2O}$ was still reduced by 40%. In their model of the

**Table 1.2. Self-diffusion coefficients for water ($D_{H_2O}$) in biological systems**

| System | Technique | $D_{H_2O}$ ($10^{-5}$ $cm^2/s$) | Reference |
|---|---|---|---|
| Pure water | NMR | 2.5 | Chang et al., 1973[28] |
| Bovine serum albumin 5% | NMR | 1.93 | Abetsedarskaya et al., 1968[59] |
| Agar rose gel 6% | NMR | 2.26 | Walter and Hope, 1971[60] |
| Agar rose gel 18% | Quasi-elastic neutron scattering | 1.70 | Trantham, 1980[50] |
| Polyox polymer 20% | NMR | 1.10 | Bearden (personal communication)[61] |
| Collagen gel 2% | NMR | 2.24 | Westover and Dresden, 1974[14] |
| Gelatin gel 5% | Isotope | 1.38 | Shaw and Schy, 1981[12] |
| Eggs | | | |
| Frog | NMR | 0.67 | Mild et al., 1972[62] |
| Frog | Isotope | 0.996 | Ling et al., 1967[52] |
| Chicken (yolk) | NMR | 0.60 | James and Gillen, 1972[63] |
| Skeletal muscle | | | |
| Barnacle | Isotope | 2.42 | Bunch and Kallsen, 1969[64] |
| Barnacle | Isotope | 1.35 | Ernst and Hazlewood, 1978[6] |
| Barnacle | Isotope | 1.34 | Caille and Hinke, 1974[65] |
| Frog | NMR | 1.56 | Abetsedarskaya et al., 1968[59] |
| Frog | NMR | 1.13 | Finch et al., 1971[29] |
| Frog | NMR | 1.12 | " |
| Frog | NMR | 1.41 | Yashizaki et al., 1982[33] |
| Frog | Isotope | 1.18 | Ling et al., 1967[52] |
| Frog | Isotope | 1.43 | " |
| Rat | NMR | 1.43 | Chang et al., 1973[28] |
| Rat | NMR | 1.50 | " |
| Rat | NMR | 1.09 | Finch et al., 1971[29] |
| Toad | NMR | 1.2 | Rustgi et al., 1978[23] |
| Cardiac muscle | | | |
| Rat | NMR | 0.9 | Cooper et al., 1974[66] |
| Rabbit | NMR | 0.65 | " |
| Liver | | | |
| Rat | NMR | 0.625–0.75 | " |
| Rabbit | NMR | 0.625 | " |
| Rabbit | NMR | 0.49 | " |
| Red blood cells | | | |
| Human | NMR | 0.2–0.55 | " |
| Human | NMR | 0.25–0.625 | Tanner, 1976[32] |
| Human | NMR | 1.16 | Andrasko, 1976[67] |
| Cultured cells (25°C) | | | |
| HeLa | NMR | 0.48 | Beall et al., 1982[47] |
| CHO | NMR | 0.40 | " |
| BHK | NMR | 0.50 | " |
| Lens | | | |
| Rabbit (whole) | NMR | 0.9 | Neville et al., 1974[68] |
| Rabbit (cortex) | NMR | 1.06 | " |
| Rabbit (nucleus) | NMR | 0.97 | " |
| Pea roots | | | |
| Meristematic | NMR | 0.59 | Abetsedarskaya, 1968[59] |
| Mature cells | NMR | 0.68 | " |
| Wheat endosperm | NMR | 1.2 | Callaghan et al., 1979[69] |
| Slime mold | NMR | 1.26 | Walter and Hope, 1971[60] |
| Brine shrimp cysts (50% hydration) | NMR | 0.24 | Seitz et al., 1977[70] |

actin-myosin lattice, the macromolecules would have to occupy almost the total volume of the cell to account for this finding. Tanner and co-workers[32,49] have improved NMR measurements and negated the contributions of membrane permeability by using pulsed field gradient NMR to make determinations over very short times and distances. Therefore, molecules cannot bump into membranes or barriers during this period. In three types of frog muscle, $D_{H_2O}$ is still reduced by 49%. Yashizaki et al.[33] confirmed with $^1$H- and $^{31}$P-pulsed field gradient NMR methods that both $H_2O$ and the phosphate metabolites, creatinine phosphate and inorganic phosphate, experience significant reductions in diffusive motion in muscle water. Using quasi-elastic neutron scattering methods, which also measure diffusive motion over very short times and distances ($<5$ Å), Trantham[50] confirmed that 20% agar rose gels can restrict water diffusion in ways not explained by barrier effects. Diffusion measurements are dominated by the bulk properties of cellular water and cannot be accounted for by a small "tightly bound fraction" of water. Therefore, a real 50% reduction of $D_{H_2O}$ would require that a large portion of all cellular water experience some reduction in motional freedom through interactions with charges and surfaces of other cellular components. Currently, the criticisms of diffusion measurements are not sufficient to explain the finding of a 50% reduction. The consequences of such a change in the expected properties of the biological solvent could be significant. It could explain the reduced diffusion of solutes in cells,[12] the reduced diffusion coefficient of fluorescent macromolecules in cytoplasm,[51] and possibly the reduced solvation of some compounds in cellular water.[52,53]

## 1.6. CONCLUSIONS

While much remains to be done in the study of the state of biological water, certain conclusions can be made:

(a) Water in cells does not behave as if it were *all* a dilute solution; any concept or model of the cell based on this assumption will lead to erroneous predictions if examined carefully.

(b) There is no measurable "ice-like" crystalline solid water in normal functioning cells; noncrystalline glass-like water may exist under special conditions in very small spaces.

(c) The simplified model of a small fraction of bound water and a large fraction of bulk water in cells is not supported by current biophysical evidence; while the two-fraction fast-exchange model can be used to fit one parameter, it has not been used successfully to fit a set of parameters.

(d) The most appropriate view of cell water, consistent with experimental data, is a distribution of states having many different correlation times, which is consistent with a 50% reduction of the self-diffusion coefficient. The shape of such a distribution will be a function of the type of cell, its macromolecular composition, and its physiological state.

## 1.7. REFERENCES


1. Beall, P.T. The water of life. *The Sciences* 1 Jan 1981, pp. 6–29.
2. Ben-Naim, A. *Hydrophobic Interactions.* Plenum, New York, 1980.
3. Eisenberg, D., and Kauzmann, W. *The Structure and Properties of Water.* Oxford University Press, New York, 1969.
4. Franks, F. *Water: A Comprehensive Treatise,* Vols. 1–6. Plenum Press, New York, 1972–1979.
5. Stillinger, F.H. Water revisited. *Science* **209:**451–457, 1980.
6. Ernst, E., and Hazlewood, C.F. Inorganic constituents acting in bioprocessor. I. Water. *Inorg. Persp. Biol. Med.* **2:**27–49, 1978.
7. Foster, K.R., Bidinger, J.M., and Carpenter, D.O. The electrical resistivity of cytoplasm. *Biophys. J.* **16:**991–1001, 1976.
8. Mathur-DeVre, R. The NMR studies of water in biological systems. *Prog. Biophys. Mol. Biol.* **35:**103–134, 1979.
9. Ratkovic, S. NMR studies of water in biological systems at different levels of their organization. *Sci. Yugoslav.* **7:**19–54, 1981.
10. Berendsen, H.J.C., and Michgelsen, C. Hydration structure of collagen and influence of salts. *Fed. Proc.* **25:**998–1002, 1966.
11. Hoeve, C.A.J., and Kakivaya, S.R. On the structure of water absorbed in collagen. *J. Phys. Chem.* **80:**745–749, 1976.
12. Shaw, M., and Schy, A. Diffusion coefficient measurement by the stop-flow method in a 5% collagen gel. *Biophys. J.* **34:**375–381, 1981.
13. Walrafen, G. [Personal communication], 1980.
14. Westover, C.J., and Dresden, M.H. Collagen hydration pulsed nuclear magnetic resonance studies of structural transition. *Biochim. Biophys. Acta* **365:**389–399, 1974.
15. Ogston, A.G., Preston, B.N., and Wells, J.D. On the transport of compact particles through solution of chain-polymers. *Proc. R. Soc. London* (Series A) **333:**297–307, 1973.
16. Beall, P.T. Contribution of cytoskeleton to cellular water properties in diverse functional states. *Fed. Proc.* **40:**206–213, 1981.
17. Civan, M.M., and Shporer, M. Pulsed magnetic resonance study of $^{17}$O, $^2$D and $^1$H of water in frog striated muscle. *Biophys. J.* **15:**229–306, 1975.
18. Fung, B.M., and McGaughy, T.W. The state of water in muscle as studied by pulsed NMR. *Biochim. Biophys. Acta* **343:**663–673, 1974.
19. Hazlewood, C.F., Nichols, B.L., Chang, D.C., and Brown, B. On the state of water in developing muscle. *Johns Hopkins Med. J.* **128:**117–131, 1971.
20. Belagyi, J. Water structure in striated muscle by spin

labeling technique. *Acta Biochim. Biophys. Acad. Sci. Hung.* **10:**63–70, 1975.
21. Knight, V.A., and Wiggins, P.M. A possible role for water in performance of cellular work. II. Measurements of scattering of light by actomyosin. *Bioelectrochem. Bioenerg.* **6:**135–146, 1979.
22. Wiggins, P.A., and Knight V.A. A possible role for water in the performance of cellular work. III. ATP-ase induced increases in the microviscosity of water in suspensions of sarcoplasmic reticulum vesicles and of actomyosin. *Bioelectrochem. Bioenerg.* **6:**323–335, 1979.
23. Rustigi, S.N., Peemoeller, H., Thompson, R.T., Kydon, D.W., and Pintar, M.M. A study of molecular dynamics and freezing phase transition in tissues by proton spin-relaxation. *Biophys. J.* **22:**439–452, 1978.
24. Clegg, J.S. Interrelationships between water and cellular metabolism in Artemia cysts. VIII. Sorption isotherms and derived thermodynamic quantities. *J. Cell Physiol.* **94:**123–137, 1978.
25. Seitz, P.K. Water proton magnetic resonance of metabolic and ametabolic *Artemia* embryos. Ph.D. thesis. University of Texas, 1977.
26. Bratton, C.B., Hopkins, A.L., and Weinberg, J.W. Nuclear magnetic resonance studies of living muscle. *Science* **147:**738–741, 1965.
27. Bryant, R.G., and Shirley, W.M. Dynamical deductions from nuclear magnetic resonance relaxation measurements at the water-protein interface. *Biophys J.* **34:**3–16, 1980.
28. Chang, D.C., Hazlewood, C.F., Nichols, B.L., and Rorschach, H.D. Implications of diffusion coefficient measurements for the structure of cellular water. *Ann. N.Y. Acad. Sci.* **204:**434–443, 1973.
29. Finch, E.D., Harmon, J.F., and Muller, B.H. Pulsed NMR measurements of the diffusion constant of water in muscle. *Arch. Biochem. Biophys.* **147:**299–310, 1971.
30. Koenig, S.H., Hallenga, K., and Shporer, M. Protein-water interaction studied by solvent $^{1}H$, $^{2}H$ and $^{17}O$ magnetic relaxation. *Proc. Natl. Acad. Sci. U.S.A.* **72:**2667–2671, 1975.
31. Peemoeller, H., Shenoy, R.K., and Pintar, M.M. Two-dimensional NMR time evolution correlation spectroscopy in wet lysozyme. *J. Magn. Reson.* **45:**193–204, 1981.
32. Tanner, J.E., In: *Magnetic Resonance in Colloid and Interface Science.* Reising, H.A., Wade, C.G., eds., ACS Symposium Number 34. American Chemical Society, Washington, D.C., 1976, p. 16.
33. Yashizaki, K., Yoshiteru, S., Nishikawa, H., and Morinmoto, T. Application of pulsed-gradient $^{31}P$ NMR on frog muscle to measure the diffusion rates of phosphorus compounds in cells. *Biophys. J.* **38:**209–211, 1982.
34. Zimmerman, J.R., and Brittin, W.E. Nuclear magnetic resonance studies in multiple phase systems: lifetime of a water molecule in an absorbancy phase on silica gel. *J. Phys. Chem.* **61:**1328–1333, 1957.
35. Belton, P.S., Jackson, R.R., and Packer, K.J. Pulsed NMR study of water in striated muscle. I. Transverse relaxation times and freezing effects. *Biochim. Biophys. Acta* **286:**16–27, 1972.
36. Lis, L.J., McAlister, M., and Rand, R.P. Interactions between neutral phospholipid bilayer membranes. *Biophys. J.* **37:**657–666, 1982.
37. Parsegian, V.A. Long range van der Waals forces. In: *Physical Chemistry: Enriching Topics from Colloid and Surface Science.* van Olphen, H., Mysels, K.J., eds. La Jolla, California, Theorex, 1975.
38. Beall, P. T., Asch, B.B., Chang, D.C., Medina, D., and Hazlewood, C.F. Distinction of normal preneoplastic, and neoplastic mouse mammary primary cell cultures by water NMR relaxation times. *J.N. C.I.* **64:**335–338, 1980.
39. Beall, P.T., and Hazlewood, C.F. Distinction of the normal, preneoplastic and neoplastic states by water proton NMR relaxation times. In: *NMR Imaging Techniques.* Partain, L., ed. W.B. Saunders, Philadelphia, 1983, p. 312.
40. Damadian, R. Tumor detection by nuclear magnetic resonance. *Science* **171:**1151–1153, 1971.
41. Damadian, R., Goldsmith, M., and Minkoff, L. NMR in cancer. XVI. FONAR image of the live human body. *Physiol. Chem. Phys.* **9:**97–100, 1977.
42. Chang, D.C., Misra, L.K., Beall, P.T., Fanguy, R.C., and Hazlewood, C.F. Nuclear magnetic resonance study of muscle water protons in muscular dystrophy of chickens. *J. Cell Physiol.* **107:**139–143, 1981.
43. Hazlewood, C.F., Chang, D.C., Nichols, B.L., and Woessner, D.E. Nuclear magnetic resonance transverse relaxation times of water protons in skeletal muscle. *Biophys. J.* **14:**584–606, 1974.
44. Hazlewood, C.F., Yamanashi, W.S., Rangel, R.A., and Todd, L.E. *In vivo* NMR imaging and $T_1$ measurements of water protons in the human brain. *Magn. Reson. Imaging* **1:**3–10, 1982.
45. Beall, P.T. Water-macromolecular interactions during the cell cycle. In: *Nuclear Cytoplasmic Interactions.* Whitson, G., ed. Academic Press, New York, 1980, pp. 223–249.
46. Beall, P.T., Hazlewood, C.F., and Rao, P.N. Nuclear magnetic resonance pattern of intracellular water as a function of the Hela cell cycle. *Science* **192:**904–907, 1976.
47. Beall, P.T., Hazlewood, C.F., and Chang, D.C. Microtubule organization and the self-diffusion coefficient of water in baby hamster kidney cells as a function of temperature. *J. Cell Biol.* **96:**375a, 1982.
48. Cleveland, G.G., Chang, D.C., Hazlewood, C.F., and Rorschach, H.D. Nuclear magnetic resonance measurement of skeletal muscle. Anisotrophy of the diffusion coefficient of the intracellular water. *Biophys. J.* **16:**1043–1053, 1976.
49. Tanner, J.E. Self-diffusion of water in frog muscle. *Biophys. J.* **28:**107–116, 1979.
50. Trantham, E.C. Quasi-elastic neutron scattering study of water in agarose gels. Ph.D. thesis. Rice University, Houston, Texas, 1980.
51. Wojcieszyn, J.W., Schlegel, R.A., Wu, E., and Jacobson, K.A. Diffusion of injected macromolecules within the cytoplasm of living cells. *Proc. Natl. Acad. Sci. U.S.A.* **78:**4407–4410, 1981.
52. Ling, G.N., Ochsenfeld, M.M., and Karreman, G. Is the cell membrane a universal rate limiting barrier to the movement of water between the living cell and its surrounding medium. *J. Gen. Physiol.* **50:**1807–1820, 1967.
53. Kushmerick, M.J., and Podolsky, R.J. Ionic mobility in muscle cells. *Science* **166:**1297–1298, 1969.
54. Derbyshire, W., and Parsons, S.L. NMR investigations

of frozen muscle systems. *J. Magn. Reson.* **6**:344–351, 1972.

55. Duff, I.D., and Derbyshire, W. NMR investigations of frozen porcine muscle. *J. Magn. Reson.* **15**:310–316, 1974.
56. Fung, B.M. Non-freezable water and spin-lattice relaxation time in muscle containing a growing tumor. *Biochim. Biophys. Acta* **362**:209–214, 1974.
57. Pezolet, M., Pigeon-Gosselin, M., Savoie, R., and Caille, J.P. Laser Raman investigations of intact single muscle fibers on the state of water in muscle tissue. *Biochim. Biophys. Acta* **544**:394–406, 1978.
58. Peemoeller, H., Pintar, M. M., and Kydon, D. W. Nuclear magnetic resonance analysis of water in natural and deuterated mouse muscle above and below freezing. *Biophys. J.* **29**:427–435, 1980.
59. Abetsedarskaya, L.A., Miftahutdinova, F.G., and Fedotov, V.D. State of water in live tissues. *Biofizika* **13**:750–758, 1968.
60. Walter, J.A., and Hope, A.B. Proton magnetic resonance studies of water in slime mold plasmodia. *Aust. J. Biol. Sci.* **24**:497–507, 1971.
61. Bearden, D. Unpublished results, personal communication, 1983.
62. Mild, K.H., James, T.L., and Gillen, K.T. Nuclear magnetic resonance relaxation time and self-diffusion coefficient measurements of water in frog ovarian eggs. *J. Cell Physiol.* **80**:155–158, 1972.
63. James, T., and Gillen, K.T. NMR relaxation time and self-diffusion constant of water in hen egg white and yolk. *Biochim. Biophys. Acta* **286**:10–15, 1972.
64. Bunch, W.H., and Kallsen, G. Rate of intercellular diffusion as measured in barnacle muscle. *Science* **164**:1178–1179, 1969.
65. Caille, J.P., and Hinke, J.A.M. The volume available to diffusion in the muscle fiber. *Can. J. Physiol. Pharmacol.* **52**:814–828, 1974.
66. Cooper, R.L., Chang, D.B., Young, A.C., Martin, C.J., and Johnson, B.A. Restricted diffusion in biophysical systems: experiment. *Biophys. J.* **14**:161–177, 1974.
67. Andrasko, J. Water diffusion permeability of human erythrocytes studied by pulsed gradient NMR technique. *Biochim. Biophys. Acta* **428**:304–322, 1976.
68. Neville, M.C., Patterson, C.A., Rae, J.L., and Woessner, D.E. Nuclear magnetic resonance studies and water ordering in the crystalline lens. *Science* **184**:1072–1075, 1974.
69. Callaghan, P.T., Jolley, K.W., and Leleivre, J. Diffusion of water in the endosperm tissue of wheat grains as studied by pulsed field gradient nuclear magnetic resonance. *Biophys. J.* **28**:133–142, 1979.
70. Seitz, P.K., Clegg, J.S., and Hazlewood, C.F. Water proton magnetic resonance of metabolic and ametabolic *Artemia* embryos. *Biophys. J.* **17**:303a, 1977.
71. Sidorova, A.I., Kochnev, I.N., Moiseeva, L.V., and Khaloimov, A.I. Investigation of the state of tissue water by infrared spectroscopy. In: *Water in Biological Systems.* Kayushin, L.P., ed. [Special research report translated from the Russian by the Consultants Bureau.] Plenum Press, New York, 1969.

CHAPTER 2

# *BASIC PHYSICS FOR NMR AND NMR IMAGING*



## 2.1. INTRODUCTION

Nuclear magnetic resonance, a nuclear phenomenon related to the magnetic properties of nuclei, was first demonstrated by Bloch at Stanford and Purcell at Harvard. For this work, they received the Nobel Prize in Physics in 1952. NMR has been a major tool in chemistry and physics for the exploration of molecular structure. Even though it is a nuclear phenomenon, NMR does not involve nuclear disintegration and no ionizing radiation is emitted. The NMR properties of nuclei are studied using magnets and radio-frequency (rf) electrical circuits. Therefore, some of the basic laws of electricity and simple circuits will be reviewed in this chapter for those individuals who may not have had occasion to utilize them in some time.

## 2.2. ELECTRICITY AND MAGNETISM

**2.2.A.** *Charges and forces*

To express electrical quantities, a quantity of charge is needed as well as the fundamental quantitites of time, mass, and length. Such quantities are specified by the metric system of units (Systeme International d'Unités—SI). The basic unit of charge is the coulomb (C) and the smallest charge is that of the electron ($-1.6 \times 10^{-19}$ coulombs). The hydrogen nucleus, the proton, carries a charge of the same numerical value, but of the opposite sign.

If two opposite charges $q_1$ and $q_2$ are separated by a distance $r$ in space, the attractive force $F$ between them is given by

$$F = \frac{q_1 q_2}{4\pi\epsilon r^2} \quad ,$$

where $q_1$ and $q_2$ are the numerical values of the two charges and $\epsilon$ is the permittivity of the medium surrounding the charges (for a vacuum, $\epsilon_0 = 8.854 \times 10^{-12}$ farad/meter). Note the inverse square nature of the attractive force. If the charges are of the same sign, the force will be a repulsive one directed along the line separating the two charges. The force $F$ can also be written as

$$F = E_1 q_2 \quad ,$$

where $E_1$ is considered to be the electrical field intensity due to charge $q_1$ and is given by

$$E_1 = \frac{q_1}{4\pi\epsilon r^2} \quad .$$

Whether the electrical force is either toward the charge or away from it is designated as the central nature of the force. This and the inverse square behavior are similar to the behaviors of other natural forces, and many properties of electrical fields originate from these two relationships. The concept that the charge creates a field at a distance which affects other charges is important. The electrical field $E_1$ can also be considered to be due to an electrical potential $V_1$ given by

$$V_1 = \frac{q_1}{4\pi\epsilon r} \quad .$$

Note that the potential due to a single charge $q_1$ varies inversely with $r$. Thus, it should be clear that the concepts of potential and electrical field are convenient descriptive mechanisms to explain forces at a distance from the electrical charge. Certain specific arrangements of electrical charges are worth mentioning.

**2.2.A.*1*.** *The dipole.* Two charges $+q$ and $-q$ separated by a distance $s$ form an electrical dipole. The potential due to this dipole at a point at distance $r$ from the center of the dipole axis is given by

$$V_{\text{dipole}} = \frac{qs}{4\pi\epsilon_0 r^2} \cos\Theta \qquad \text{if } r^2 \gg s^2 \quad .$$

where $\Theta$ is the angle between the dipole axis and the direction at $r$. Note that this potential falls off as the square of $r$. The electrical field intensity of a dipole will thus fall off as the cube of the distance $r^3$.

**2.2.A.*2*.** *The linear quadrapole.* The linear quadrapole is an arrangement of four charges: $+q$ and $+q$, separated by a distance of $2s$ and two charges $-2q$, located on the center of the line joining the two positive charges. If $r$ and $\Theta$ have the same meaning as in the case of the dipole, the potential at the point at distance $r$ is given by

$$V_{\text{quadrapole}} = \frac{2q}{4\pi\epsilon_0 r^3} \frac{s^2}{r^3} \frac{(3\cos^2\Theta - 1)}{2} \quad .$$

Thus, the electrostatic potential due to a linear quadrapole varies inversely as the cube of the distance, and the field intensity $E$ will vary as the fourth power of the distance. Therefore, the fields of these charges cancel almost completely for $r \gg s$.

**2.2.B.** *Capacitance*

Because of the repulsion between similar charges, energy must be expended to charge a conductor. The charge which must be added per unit increase in potential is defined as the capacitance of the conductor: $C = Q/V$. The unit for capacitance is the coulomb/volt or farad (F); a microfarad is $10^{-6}$ farad. The capacitance of an isolated conductivity sphere of radius $R$ will be: $C = Q/V + 4\pi\epsilon_0 R$. Since $\epsilon_0 = 8.85 \times 10^{-12}$ farad/meter, the capacitance of a sphere of 1 meter would be $4 \times 3.14 \times 8.85 \times 10^{-12}$ farad, or approximately $100 \times 10^{-12}$ farad.

Let two parallel conducting plates of area $A$ be separated by a distance $d$. The charge which must be transferred from one to the other per unit potential difference is defined as the capacitance between two conductors. Capacitors consist of such pairs of conductors. The capacitance of such a capacitor is given by

$$C = \epsilon_0 A/d$$

where $A$ is in square meters and $d$ is in meters. If $A$ is 1 $cm^2$ and $d$ is 1 mm, $A/d$ will be 0.1 $m^{-1}$ and the capacitance will be approximately $10^{-12}$ farad.

Energy, $W$ of a charged capacitor of capacitance $C$ is given by

$$W = \frac{CV^2}{2} \quad .$$

When any number of capacitors are connected in series, the reciprocal of the equivalent capacitance $C_s$ equals the sum of the reciprocals of the individual capacitances:

$$\frac{1}{C_s} = \frac{1}{C_1} + \frac{1}{C_2} + \frac{1}{C_3} + \cdot \cdot \cdot + \frac{1}{C_n} \quad .$$

When any number of capacitors are connected in parallel, the equivalent capacitance $C_p$ equals the sum of the individual capacitances:

$$C_p = C_1 + C_2 + C_3 + \cdot \cdot \cdot + C_n \quad .$$

**2.2.C.** *Fields, currents, and resistance*

Electrical charge experiences a force under the influence of an electrical field that may cause the charge to move. In an electrical conductor, the electrons bump into other atoms as they move through it. Some conductors offer less resistance than others to such movement, and this property is called conductivity $\sigma$. The current density $J$ in a conductor is the rate of charge flow or current through a given cross-sectional area. It is given by

$$J = \sigma E = \frac{1}{\rho} E \quad ,$$

where $E$ is the electrical field and $\rho$ is the resistivity of the conductor. Consider a cylindrical conductor of cross-sectional area $A$ and length $l$, across which a potential $V$ is applied. Under the influence of this potential $V$, a current $I$ will flow through the conductor:

$$J = \sigma E = \sigma \frac{V}{l} \quad .$$

(The electrical field, also called the potential gradient, is given by $V/l$. The gradient, or slope, is the rate of change of that quality with respect to another, in this case, the distance.)

The electrical current $I$ is defined as $J \cdot A$ and is given by

$$I = J \cdot A = \frac{\sigma A V}{l} = \frac{V}{R} \quad ,$$

where $R = \rho l/A$, and $R$ is the resistance of the conductor. Note that resistance is directly proportional to the length of the conductor ($l$) but inversely proportional to the cross-sectional area ($A$) and depends upon the electrical property of resistivity ($\rho$). Therefore,

$$V = IR \quad ,$$

which is referred to as Ohm's law.

The resistance of a conductor changes with temperatures, most metallic conductors showing increased resistance with increased temperature. Certain metals, alloys, and compounds, below a certain transition temperature (usually a few degrees above absolute zero) characteristic of the material, exhibit a property whereby their electrical resistance disappears or nearly disappears. This property is called superconductivity, and such substances are used in the production of electromagnets having a very high magnetic field, the so-called superconducting magnets.

The resistance of a conductor is not the same for alternating current (AC) as for direct current (DC). In higher-frequency AC, the current is forced to flow mostly in the outer skin of the conductor, thus greatly reducing the effective cross-sectional area of the conductor and increasing the resistance. Thus, the AC resistance of a conductor is much greater than the DC resistance.

When a circuit has a number of resistances connected in series, the total resistance of the circuit is the sum of the individual resistances.

$$R_s = R_1 + R_2 + R_3 \ldots$$

When a circuit has a number of resistances in parallel, the total resistance is less than that of the lowest resistance in the group.

$$R_p = \frac{1}{\frac{1}{R_1} + \frac{1}{R_2} + \frac{1}{R_3} \cdot \cdot \cdot} \quad .$$

**2.2.D.** *Power and energy*

Power, the rate of doing work, is given by voltage multiplied by current. The unit of electrical power is the watt and is the work done by 1 ampere through 1 volt: $P = VI$, where $P$ is power in watts (W), $V$ is voltage in volts (V), and $I$ is current in amperes (A).

Electrical energy is the total work done by the electrical current and is given by the product of the power and the time during which it was used: $W = PT$, where $W$ is in watt-hours, $P$ is in watts, and $T$ is in hours. Note that watt-second is also a unit of energy.

**2.2.E.** *Inductance*

The electrical current flowing through a conductor produces a magnetic field. This changing magnetic field is capable of producing a flow of current through another conductor—this is called inductance. Inductance depends on the physical characteristics of the conductor. A coil has more inductance than a straight wire of the same material, and a coil with more turns has more inductance than a coil with fewer turns. The unit of inductance is the henry. The energy stored in the magnetic field of an inductor is given by

$$W = \frac{I^2 L}{2} \quad ,$$

where $W$ is energy in joules (J), $I$ is current in amps, and $L$ is inductance in henrys. When two or more inductances are connected in series, the total inductance is equal to the sum of the individual inductances. If the inductances are connected in parallel and the coils are separated sufficiently, the total inductance is given by

$$L_p = \frac{1}{\frac{1}{L_1} + \frac{1}{L_2} + \frac{1}{L_3} \cdots} \quad .$$

[NMR information can be received by a coil surrounding the sample of nuclei through the induction of a very small current in the coil when the spin system undergoes reorientation after a pulse of rf energy.]

**2.2.F.** *Alternating current*

When the voltage applied to an electrical circuit is such that it changes direction over time, it is called an alternating current (AC). Two reversals of direction make a cycle; the number of cycles occurring in one second is called the frequency of the current. The inverse of frequency is called the period of the current. AC is characterized by its frequency and waveform. The simplest form is that of a sine wave. Any complex periodic waveform can be considered to be a combination of a number of similar sine waveforms of different frequencies. The process of extracting these individual waveforms from a complex signal is called Fourier transform, which is often used in imaging techniques to extract data from a complex signal.

The phase of a current can be measured in units of time such as seconds. Phase is defined as the time interval between one event and the instant when a second, related event occurs. Since the current and voltages are repeating themselves after a cycle, the cycle or its parts can also be used as a unit of phase. A cycle is divided into 360 equal parts. Thus a phase difference of 180° means that the two currents are exactly out of phase with each other.

In AC circuits, capacitors and inductors have a property that exhibits behavior similar to that of a resistor in a DC circuit. This property is called reactance. The reactance of a capacitor $X_C$ is given by

$$X_C = \frac{1}{2\pi fC} \quad ,$$

where $X_C$ is capacitive reactance in ohms, $f$ is frequency in hertz (cycles/s), and $C$ is capacitance in farads. The reactance of an inductor is given by

$$X_L = 2\pi fL \quad ,$$

where $X_L$ is inductive reactance in ohms ($\Omega$), $f$ is frequency in hertz (Hz), and $L$ is inductance in henrys. Reactances in series and in parallel add up in a manner similar to resistances.

**2.2.G.** *Resonance in a series circuit*

Consider a resistance, a capacitor, and an inductor connected in series to a source of AC of variable frequency. At some frequency, the reactance of $C$ and $L$ will be equal and the voltage drop across the coil and capacitor will be equal and 180° out of phase. The current flow will be determined by the resistance $R$ only. At that frequency, the current through the circuit has its largest value and is said to be resonant, as given by

$$X_C = X_L \quad ,$$

$$\frac{1}{2\pi fC} = 2\pi fL \quad ,$$

$$f = \frac{1}{2\pi\sqrt{LC}}$$

The value of the reactance of either the inductor or the capacitor at the resonant frequency of a series resonant circuit is called the quality factor $Q$. Thus, $Q = X/r$, and the smaller the $Q$ value, the sharper the resonance.

These simple rules for electrical circuits are useful in daily laboratory practice; however, a basic understanding of electromagnetic theory is essential to follow most textbooks on the electromagnetic properties of materials and NMR techniques.

## 2.3. NUCLEAR MAGNETISM

### 2.3.A. *An exercise in units*

When one writes an equation showing a relationship between physical quantities, the units on either side of the equality must be the same. Consider a simple equation such as

$$\text{Force} = \text{Mass} \times \text{Acceleration}.$$

In SI units, mass is measured in kilograms (kg) and acceleration in meters per seconds squared ($m/s^2$). Therefore, the units of force are kg · m $s^{-2}$, which are called newtons (N).

Also, the force on a charge $q$ moving with a velocity $v$ in a magnetic induction $B$ has the magnitude:

$$q \cdot v \cdot B \sin \Theta$$

where $\Theta$ is the angle between $B$ and $v$. In vector notation, one would write this as $q[\overline{v} \times \overline{B}]$, and it would then be understood that this force is perpendicular to the plane containing $v$ and $B$. The SI units of charge and magnetic induction are, respectively, the coulomb and the tesla (T). One can therefore write the dimensional equation: newton = coulomb × meter/second × tesla, or cm $s^{-1}$ · T. Substituting the equivalent units for N from above, kg · m $s^{-2}$ = cm $s^{-1}$ T, or T = kg $cm^{-1}$ $s^{-1}$. Thus, in any relationship using SI units, one can replace the magnetic induction $B$ by mass/charge × time. Since current is the rate of flow of charge, we expect the dimension of current to be charge × $s^{-1}$, but replacing this value, as above, T = kg $s^{-2}$ $\text{amperes}^{-1}$.

### 2.3.B. *Two important formulae*

Consider one of the most often used formulae in NMR theory:

$$\mu = \gamma I \hbar = \gamma I \frac{h}{2\pi} \quad ,$$

where $\mu$ is the magnetic dipole moment of the nucleus, $I$ is the spin of the nucleus, $\gamma$ is the magnetogyric ratio, and $\hbar$ is Planck's constant ($6.625 \times 10^{-34}$ joule-second). This formula tells us that the magnetic dipole moment of a nucleus is a quantum concept. Nuclear angular momentum is also a quantitized value. Even though $I$ is called the nuclear spin or nuclear angular momentum, it is only the maximum value of the magnetic quantum number, $m_I$. The explanation of this concept would require a review of quantum mechanics, which is beyond the scope of this chapter.

### 2.3.C. *Units for nuclear magnetic moments*

When a dipole magnet is placed in a field the direction of which makes an angle $\Theta$ with the axis of the magnet, a torque acts on the magnet. The magnitude of the torque is given by

$$\text{Torque} = mB \sin \Theta = q_m lB \sin \Theta = Fl \sin \Theta.$$

The maximum value of the torque will be $Fl$ when sin $\theta = 1$, i.e., when $\Theta = 90°$, or when the magnet is perpendicular to the external magnetic field $B$. It is easy to see that the units of torque are the same as that of force × distance, or that of work. In SI units, this is the joule. Therefore, units of magnetic moment are the same as those of work/magnetic induction $B$, or joule/$Wb/m^2$, or joule/tesla.

In atomic spectra, the magnetic moments of atoms are measured in units of the Bohr magneton $\mu_B$, defined as the magnetic moment associated with an atomic electron in orbital motion with an angular momentum of $lh$. The magnetic dipole set up by a plane loop is $\mu_l = i \times s$, where $i$ is the current and $s$ is the area of the loop. For a revolving particle with velocity $v$ and charge $q$, in a circular orbit of radius $r$, this will be given by

$$\mu_l = \frac{e}{2\pi\gamma/v} \pi r^2 = \frac{evr}{2} \quad .$$

The orbital angular momentum of this particle is $mvr = \rho_l$, when $m$ is its mass. The ratio of the magnetic dipole moment to the angular momentum is called the magnetogyric ratio $\gamma$ and will be

$$\gamma = \frac{\mu_l}{\rho_l} = \frac{e}{2m} \quad .$$

Note that $\mu_l$ and $\rho_l$ are vectors and that this relationship holds in any direction. Actual measured values of nuclear magnetic moments are between $-3$ nuclear magnetons and $+10$ nuclear magnetons. The positive sign means that the magnetic moment of the nucleus is in the same direction as the nuclear spin, whereas the negative sign means that it is in the opposite direction. The magnetic moment of the proton is $+2.7027$ nuclear magnetons and the magnetic moment of the neutron is $-1.1913$, which indicates that both have complex charge distributions, even though the total charge on the proton is the same as that on an electron, and the neutron has zero charge.

Now consider the second important formula in NMR theory:

$$\omega = \gamma B \quad ,$$

where $\omega$ is angular velocity of precession in radians/second, $\gamma$ is magnetogyric ratio in radians/tesla/second, and $B$ is magnetic induction in tesla. Note that the units of $\gamma$ are obtained from this equation. Since $\gamma = \omega/B$, and angular velocity is measured in radians/second and the field in tesla, the units of $\gamma$ are radians/second/tesla. $\gamma$ is a nuclear constant. For protons, it is

$$\gamma_{\text{proton}} = 2.675 \times 10^8 \text{ radians s}^{-1}\text{ T}^{-1} \quad .$$

Thus, for protons, the resonant frequency, in a field of 0.1 T or 1 kG, is

$$\omega_{\text{proton}} = 2.675 \times 10^8 \times 0.1 \text{ radians s}^{-1} \quad ,$$

and the frequency for proton resonance $f$ is

$$f = \frac{2.675 \times 10^8 \times 0.1}{2 \times 3.14} \text{s}^{-1} \quad ,$$

$$f = 4.257 \text{ MHz} \quad .$$

The relationship between the resonance frequency and the magnetic induction field is linear. Thus, if the field is doubled to 0.2 T, the resonant frequency for protons will be 8.514 MHz. Also, $\gamma = \omega/B = 2\pi f/B$. Many tables give the value of the gyromagnetic ratio in terms of $\gamma/2\pi$ or $\gamma$. This being equal to $f/B$, the units are seconds/tesla ($\text{s}^{-1}/\text{T}^{-1}$). Thus, $\gamma$ for protons is quoted as 42.57 MHz/T. To find the resonant frequency in any other field, simply multiply 42.57 by the field in tesla (T) and the answer will be in megahertz (MHz). For other nuclei, i.e., phosphorus-31, whose $\gamma$ is 17.23 MHz/T, simply multiply $\gamma$ by the field in T to get MHz.

## 2.4. ELEMENTS OF RESONANCE

Resonance is a well-known and interesting phenomenon. The most noticeable aspect of resonance in a mechanical system is the extreme magnitude of the displacement and velocity that such a resonant system achieves under the influence of an external periodic agent that has the same frequency as the natural frequency of the resonating system. However, there is another important aspect of resonance. At resonance, the average rate of energy transfer per cycle from the external agent to the resonant system is maximal. It is this characteristic of resonance that is most useful in studying effects of resonance in microscopic particles, the position or velocity of which is difficult to measure. In NMR, energy from an external oscillating rf field is very effectively transferred to the nuclear particles. The magnetic component of the rf field interacts with the total macroscopic magnetic moment of the nuclear particles when the frequency of the rf field is exactly equal to the natural frequency of the nuclear system. In this case, the natural frequency happens to be the precessional frequency of the macroscopic magnetic moment in the external static large magnetic field $B$. This kind of precession is clearly understood in the classical sense. Any rotating body, when acted upon by a torque which tends to change the direction of its axis of rotation, precesses around another line that intersects the axis of rotation. In the case of the nuclear system, the total spin of the nuclear particles is equivalent to that of the rotating body, the torque is provided by the external large magnetic field $B$, and the resulting precession takes place around the direction of $B$. For each specific nucleus, this precessional angular frequency $\omega$ is exactly given by the equation $\omega = \gamma B$, where $\gamma$ is a characteristic nuclear constant called the gyromagnetic ratio. Thus when the rf field frequency $f$ is such that $2\pi f = \gamma B$, there will be resonance, and energy will be transferred from the rf field to the nuclear spin system, resulting in a tilting of the total magnetic moment away from the direction of $B$.

## 2.5. RELAXATION TIMES

Let us now turn to relaxation, which is also a simple phenomenon exhibited by many physical systems. By relaxation, we simply mean that an observable time elapses between the moment when a system in equilibrium is subjected to a momentary change in condition and the moment when the system is again in equilibrium. An electrical example of such a process would be the charging of a capacitor $C$ connected to a resistance $R$ by a constant voltage $V$. The capacitor does not receive its full charge $Q$ instantaneously but rather in an exponential fashion given by

$$Q = CV[1 - \exp(-t/RC)] \quad .$$

The time $RC$ is called the relaxation time of the capacitor resistance system. In NMR experiments, the equilibrium condition of the total nuclear magnetic moment $M$ is deliberately disturbed by applying a small rf field in a pulse to the system. The behavior of the magnetic moment over time is then studied.

There are two ways of looking at the NMR phenomena. The first is by the quantum mechanical description, which yields a proper understanding of energy levels of nuclear spin systems such as populations of water hydrogen protons, the populations of these energy levels, the probabilities of transitions between levels, the half-lives of the excited states, and the radiationless transitions by which the perturbed spin systems return to the equilibrium state. These radiationless transitions are called the relaxation processes.

There are two kinds of relaxation processes: the spin–lattice relaxation with a characteristic time constant, designated $T_1$, and the spin–spin relaxation process, called $T_2$. In spin–lattice relaxation, the spin system transfers its energy to its surroundings, referred to as the lattice, as a result of interactions with the local (resulting from neighboring nuclear magnets) fluctuating magnetic fields. Such fluctuating local fields can also contribute to the spin–spin relaxation process. However, local nonfluctuating magnetic fields can contribute only to the spin–spin relaxation process. Therefore the spin–spin relaxation rate is always equal to or less than $T_1$. In spin–spin relaxation, the spin system dephases itself without any change in energy as it interacts with local fluctuating and nonfluctuating magnetic fields. For this reason $T_2$ is sometimes referred to as the entropy effect. Thus, it is easy to see how the local environment of the nuclei would contribute to the relaxation process and determine the values of $T_1$ and $T_2$.

The second way of looking at NMR phenomena is the classical approach, in which NMR is viewed as the precession of the total nuclear macroscopic magnetization in the stationary, large, external magnetic field of flux density $B$, being perturbed by application of an rf field having a magnetic flux density of $B_1$ transverse to the direction of $B$. This approach was taken by Felix Bloch. He derived a set of equations known as the Bloch equations, which describe the behavior of the net magnetization $M$ as a function of time after perturbation. Bloch's $T_1$ is the same as the spin–lattice relaxation time defined by the quantum mechanical approach, and his $T_2$ is identical to the spin–spin relaxation time.

There are two other relaxation times which can be measured, $T_{1\rho}$ and $T_2^*$. The relaxation time $T_{1\rho}$ is called the spin–lattice relaxation in the rotating frame, and corresponds to the decay of the net magnetization aligned along $B_1$, which is perpendicular to $B$ and much smaller than it. Thus, the relaxation times $T_1$ and $T_{1\rho}$ are sensitive to molecular dynamic processes occurring at different frequencies. $T_1$ is most sensitive to motions corresponding to the Larmor frequency ($\omega = \gamma B$), and $T_{1\rho}$ is most sensitive to motions characterized by frequencies corresponding to $\omega_1 = \gamma B_1$. The magnetization decay rate constant in the transverse plane is not simply $1/T_2$ as previously described; it also has contributions from the fact that the large external field $B$ is not uniform. The measured decay rate is $1/T_2^*$, which can be used to derive a true $T_2$ if several other variables are known. Table 2.1 summarizes various characteristics of the different relaxation processes.

## 2.6. FUNDAMENTAL EQUATIONS OF ELECTROMAGNETISM

There are four electromagnetic field vectors, **E**, **B**, **D**, and **H**. The order of production of these vectors is as follows: electrical charges produce **D** and electrical current produces **H**; vectors **E** and **B** are produced, dependent on the medium, by **D** and **H**, respectively. The framework of electricity and magnetism can be completely built and explained by seven basic relationships between these vectors. These relationships are given in Table 2.2.

## 2.7. PHYSICS OF NMR IMAGING

In medical imaging, one measures and depicts, in an absolute or relative sense, a certain intrinsic property of the tissue to form a visual representation of the

**Table 2.1. Relaxation times: nomenclature and typical values**

| Symbol | Origin of name: Quantum mechanics | Origin of name: Classical | Formula | Values (ms) |
|---|---|---|---|---|
| $T_1$ | Spin–lattice relaxation | Longitudinal relaxation | $M(t) = M_0[1 - 2\exp(-t/T_1)]$ | 100–400 |
| $T_2$ | Spin–spin relaxation | Transverse relaxation | $M(t) = M_0 \exp(-t/T_2)$ | 10–50 |
| $T_{1\rho}$ | | $T_1$ in the rotating frame | $T_1 > T_{1\rho} > T_2$ | 50–150 |
| $T_2^*$ | | $T_2^*$, measured value | $1/T_2^* = 1/T + 1/T_2 + \gamma\Delta B/2$ | |

**Table 2.2. Fundamental equations of electromagnetism**

| | |
|---|---|
| Lorentz force | |
| (a) $F = q(\mathbf{E} + V \times \mathbf{B})$ | |
| (b) $F = Ids \times \mathbf{B}$ | |
| Maxwell's equations | |
| (a) $\nabla \times \mathbf{E} = -\dot{\mathbf{B}}$ | $\oint \mathbf{E} \cdot ds = -\dot{\phi}$ |
| (b) $\nabla \times \mathbf{H} = J + \dot{\mathbf{D}}$ | $\oint \mathbf{H} \cdot ds = I_{total}$ |
| (c) $\nabla \cdot \mathbf{D} = \rho$ | $\oint \mathbf{D} \cdot nds = q$ |
| (d) $\nabla \cdot \mathbf{B} = O$ | $\oint \mathbf{B} \cdot nds = O$ |
| Medium-dependent equations | |
| $\mathbf{D} = \epsilon \mathbf{E}$ and $\mathbf{B} = \mu \mathbf{H}$ | |

organs. For example, in X-ray imaging it is the differential attenuation of X-rays that results in an image. The differential attentuation comes about because of differences in electron density among the organs and cells. At the level of the interactions among X-ray photons and electrons and nuclei (at the level of Compton scattering, photoelectric effect, or pair production), the physics is complex and the details can only be worked out with the aid of quantum mechanical theory. However, the medical interpretation of an X-ray image does not require a detailed knowledge of physics, nor does the interpretation of ultrasound or nuclear isotope images. However, for NMR imaging, the situation is not the same. In NMR, an understanding of the underlying basic physics is necessary to interpret the image, especially since this is a new method for which there has been little basic research into the origins of image differences and for which there has been limited clinical experience. As methods become standardized, the operation of the machinery will become routine and more clinical data will become available. However, though the need for such basic understanding may be reduced, it will never be eliminated.

The basic NMR experiment consists of three steps, whether they are applied to a small biopsy or to the whole body. First, the sample containing spinning nuclei of the chosen type is placed in a static uniform magnetic field, $B_0$, such as an electromagnet or a superconducting magnet. An imaginary coordinate system is set up, such that the $z$ axis is aligned with the direction of the magnetic field. The magnetic moments of the nuclei interact with the magnetic field and begin precessing around the direction of $B_0$ with the angle $\Theta$ at the Larmor frequency $\omega_0$ given by $\omega_0 = \gamma B_0$, where $\gamma$ is the gyromagnetic ratio of the nuclei under study. The second step uses an rf wave of frequency $\upsilon_0 = \omega_0/2\pi$ to perturb the nuclear spin system from the equilibrium distribution it assumed in the static magnetic field. Energy can be absorbed by the nuclei under these resonance conditions. The duration of the rf pulse will determine the power, or total energy, absorbed by the spin system. The absorbed energy will result in an increase of the angle $\Theta$ between the direction of $B_0$ and the precessing magnetic moment. If the pulse lasts long enough, the net magnetization will be brought into the $xy$ plane; this is called a 90° pulse. A pulse twice as long will return the magnetization to the $z$ axis in the negative direction and is called a 180° pulse. A detector coil in the $xy$ plane can measure the net magnetization seen in the $xy$ plane only. The third step is to use the detector to monitor the appearance or disappearance of the net magnetization of the spin system as a function of time in the $xy$ plane. Using the Bloch equations, which describe this behavior, it is possible to determine several NMR parameters (proton density, $T_1$, $T_{1\rho}$, $T_2$) which may be displayed in an imaging mode. Each of these parameters contains different information about the behavior of the spin system. The mechanism and method of measurement of each of these parameters are covered in more detail in Chapters 3 and 4; however, they will be discussed briefly here as a background for the imaging techniques that may produce pictures based on a single parameter or a combination of them.

The spin–lattice or longitudinal relaxation time, $T_1$, of water protons in pure bulk water is about 3 s, whereas the $T_1$ of protons in most biological tissues falls in the range of 100–1,000 ms. One explanation of the apparently more efficient relaxation process in biological cells has been presented by Zimmermann and Brittin,[1] who postulated that water molecules in tissues are found in two environments, a "bound" or motionally hindered state near the surface of macromolecules and a "free" or bulk state in the remainder of the water. The bound molecules are assumed to have a much slower rotational and translational motion than the free water molecules. The protons of the hydrogens on these water molecules are constrained to experience the hindrance of the whole water molecule and therefore move about more slowly than those in bulk water. It is proposed that the two types of water molecules are in "fast exchange" with one another, allowing for an exchange of energy between the two populations and an average relaxation time shorter than that of bulk water. This explanation is very simplistic and by no means sufficient to explain all the behavioral differences in water observed in $T_1$ and $T_2$ data; however, it can serve a conceptual role in the understanding of the differences in intensity of images produced by $T_1$ data. One explanation for the enhanced $T_1$ values often seen in diseased tissues would be that those cells probably have more free water than other tissues.

In the $T_2$ relaxation process, energy is exchanged between two neighboring nuclei instead of between the nucleus and its surrounding fluctuating field. The transverse magnetization is due to the coherence of neighboring precessing nuclei. They lose coherence even in a uniform external field because the intrinsic localized magnetic field experienced by each nucleus is slightly different. This field is not affected by the speed of molecular rotational motion; however, it is smallest for small molecules and largest for large molecules. It is also more efficient than the fields that produce $T_1$ relaxation (i.e., $T_2 \ll T_1$) and is affected much less by the change in resonance frequency. On the other hand, when the resonance frequency changes, the speed of the molecular rotational motion that would be most efficient is different, i.e., a different number of molecules at this speed is available and therefore the $T_1$ relaxation time is different. Usually the higher the resonance frequency, the higher is $T_1$.

The amplitude of the net magnetization vector in the system is proportional to the number of nuclei in the volume of the sample under investigation. For protons, this may be related to proton density; however, the density of water in the sample is altered only by the inclusion of other molecules in the cells. Since the percent water content of tissues does not vary much, especially between soft tissues of the body, proton density does not provide much contrast, except in special applications, such as the imaging of bone. The relaxation times $T_1$ and $T_2$, however, vary due to other mechanisms, and they do not always change proportionally with water content. Therefore, there tends to be more difference in these parameters between tissues and they produce greater contrast in images.

### 2.7.A. *Imaging principles*

Once it was determined that NMR parameters contain differential information depending on water content, fat content, relaxation times, and physiological state, the idea of making a pictorial representation of these parameters was not far away. NMR has the advantage that these measurements are made in a nondestructive manner, using very low levels of energy.

A very simple analogy that can aid in understanding how NMR images are made can be constructed from, first, one-dimensional and then multidimensional information. If there is a large magnet with a wide space between poles, a large circular receiving coil may be placed in it. If the conditions of the resonance frequency are met, a sample placed in the coil and pulsed with rf will produce a net magnetization which can be detected. If the sample is a capillary tube of water much smaller than the diameter of the coil, the amplitude of the signal will change depending on how close the sample is to the center of the receiving coil. As the capillary tube is moved from the edge of the coil toward the center, the amplitude of the signal increases, and as the sample is moved back toward the edge, the amplitude decreases. In effect, the magnetic field properties of the receiving coil are mapped by the movement of the sample inside the coil. This is, in effect, a one-dimensional image. The imposition of a magnetic field gradient along one axis of the field will alter the magnetic field across the coil. Now movement of the sample from left to right across the coil will give information that, in effect, produces a two-dimensional image of the magnetic field across the coil. The addition of other field gradients will produce spatial information in three dimensions. The extension of such an analogy to whole body imaging requires that either the receiver "sweet spot" be moved through the body or that the body be placed on a table that can be moved in the $xy$ or $z$ plane. While these simple concepts were indeed used in early imaging attempts, the problem is actually much more complex in the clinical setting, with time becoming a significant factor that necessitates specialized techniques to produce images.

The main problems in producing an image from NMR data are the following:

(a) Spatial localization. The signal from water protons in the sample must be isolated in some manner so that its origin in three dimensions can be determined. For rapid processing, the spatial information must be encoded in the signal itself.

(b) Determination of relaxation times in defined volumes. Since the most valuable information resides not in simple proton density or in signal intensity, a method must be devised to collect relaxation time data simultaneously from multiple volumes located in space.

Conceptually, spatial localization can be achieved simply by restricting each measurement to a very small region of the object to be imaged and then moving to another region. Carrying this process sequentially throughout the object, a complete image can be obtained. This was the principle used in the first field focusing NMR (FONAR) instrument and is very similar to the rectilinear scanner used in nuclear isotope scans. In nuclear medicine scans, the focused collimator allows only a certain region to be examined, and then the scanner, in a rectilinear fashion, moves over the area of interest to get the "picture" of the distribution of the isotope in the organ. The major drawback to this method is lack of speed and spatial resolution. The more sophisticated Anger camera technique is able to simultaneously gather data from the entire volume of interest with multiple detectors, whose geometry can trace the coordinates of the origin of the event.

In current NMR imaging devices, the principal method used to localize events is the correlation of the frequency or phase of the rf signal with the position of the event. This can be accomplished by imposing a magnetic field gradient on the uniform, static magnetic field. If the gradient field is denoted by $B_{z'}$ and is a linear function along the $z$ axis, then the total magnetic field at a point $z'$ along the $z$ direction will be

$$B_{z'} = B_0 + B(z') \quad .$$

Hence, the Larmor frequency of the nuclei along the $z$ axis at a point $z'$ will be given by

$$\omega_{z'} = \gamma B_{z'} = \gamma B_0 + \gamma B(z')$$

or

$$(z') = \frac{\omega_{z'} - \omega_0}{\gamma B} \quad .$$

If one knows $\omega_{z'}$, $\omega_0$, $\gamma$, and $B$, then $z'$ or the position along the $z$ axis can be calculated. In a three-dimensional situation, the signal will be coming from an entire plane because the $z$ axis defines a plane. All the points in that plane resonate at the frequency $\omega_{z'}$. In practice, if the $z$ gradient is turned on during rf excitation and the pulse bandwidth is such that it generates signals of the type, $\omega_{z'} \pm \delta\omega_{z'}$, where $\delta\omega_{z'} = \gamma B \delta z'$, then the signals would be coming from a slice of thickness $2\delta z'$ centered at $z = z'$. Then, in the detection mode, if another gradient in the $y$ direction is turned on, one could detect signals coming from a single column volume with a fixed $y$ coordinate. By providing two gradients, a projection along a column volume can be obtained in a single period. If many such projections are obtained in the constant $z$ plane, then a distribution map of the plane can be calculated. Such methods are called projection reconstruction methods. Note that the two gradients simply form a convenient method of localizing the region from which the signal is coming. The nature of the signal depends on what type of rf pulse sequence is used.

Now consider excitation of the protons in the imaging slice by a frequency-selective pulse in the presence of a field gradient that is perpendicular to the imaging plane. This first gradient, say in the $z$ direction, is a slice-selection gradient. The second gradient, say in the $y$ direction, is applied after the rf excitation. Nuclei along the $y$ direction will precess at slightly different frequencies around this direction because of the magnitude of the field is $y$ position–dependent. This is called phase encoding. After a certain time, $\Delta t$, nuclei along the $y$ direction will have different phases (the angle through which they have rotated, given by the product of their frequency and time). After the $y$ gradient is turned off, the signals are read in the presence of an $x$ gradient. This creates different frequencies along the $x$ direction. The readout signal contains all the frequencies along the $x$ direction. The experiment is then repeated with different values of $y$ gradient as many times as the resolution requires. Fourier transform of the data collected in the $x$ and $y$ directions is required to get the values of signals from each pixel, or limiting volume. This is why the method of phase encoding is also referred to as the two-dimensional Fourier transform method.

If the slice-selectivity excitation rf pulse is wide-banded instead of narrow-banded, the entire volume of interest can be excited. To differentiate pixels along the $z$ direction, the $z$ gradient must be applied. The $z$ and $y$ gradients are applied for a time, $\Delta t$, followed by a readout $x$ gradient. Such data must undergo three Fourier transforms to unscramble the $x$, $y$, and $z$ coordinate information. Therefore, this method is called the three-dimensional Fourier transform method.

It should be noted that data obtained by any of the above methods give a functional value of NMR parameters measured for each pixel (voxel). This is stored in the computer as the pixel value and can be operated on by any number of image-processing algorithms. Some of these methods and the usual nomenclature are described in Section 2.7.C and Chapter 4.

**2.7.B.** *Hardware considerations*

For NMR imaging the following functions are necessary: (a) generation of a uniform magnetic field; (b) imposition and control of magnetic field gradients; (c) transmission and reception of rf signals; (d) conduction of the NMR experiment-pulse sequencer; (e) storage and processing of data; (f) display of data; and (g) sequential operations of all systems.

**2.7.B.*1*.** *Magnet.* Fields of 0.04–1.5 tesla have been used in imaging applications. Resistive magnets have been used at fields of up to 0.2 tesla. Permanent magnets may produce fields as high as 0.3 tesla, but above 0.3 tesla, superconducting magnets are usually used. The quality of current images produced at 0.2 tesla and above do not differ significantly or favor one type of magnet over another at the present time. The initial investment costs, maintenance and operating costs, and the size and weight of the magnet in a particular situation are the more important determining factors in the purchase of equipment. Consideration should also be given to the type of imaging applications desired. For nuclei other than protons, much higher fields, above 0.3 tesla are desirable, and superconducting magnets are the only choice. However, for proton imaging, the choice of a magnet is more open.

**2.7.B.*2*.** *Gradient coils.* The following are some of the factors affecting the design of field gradient coils: driving speed of the gradient currents; geometry of the magnet generating the static field; and material of which the magnet is constructed. The gradient coil power supply is of special concern, since large amounts of current are being delivered to inductive loads. In some imaging methods, the programmable sequence must control the current amplifiers of each set of gradient coils. Such flexibility is best achieved under direct microprocessor control.

**2.7.B.*3*.** *rf detector coils.* Many special probes (transmitting and receiving) have been designed for both head and body imaging. Signal-to-noise considerations are extremely important in such designs.

**2.7.B.*4*.** *NMR spectrometer.* Generation of rf pulses of sufficient power amplitude, their programmed sequence, and their phase detection are the functions of the NMR spectrometer.

**2.7.B.*5*.** *Data storage and processing.* The main functions are analog-to-digital conversion, signal averaging, and information storage (sometimes requiring very large capacity). Much of this technology, hardware and software, has been borrowed from radiological image processing. It is hoped that in the future hardware and software specialized for NMR applications will become available and will be more economical and efficient.

**2.7.B.*6*.** *Display of data.* Since distinctive tissue characterization and visualization of blood flow are strong points of NMR imaging, it is not surprising that color-coded display modes have been adapted in some NMR labs. However, the more traditional gray scale of radiology often finds favor with NMR workers. They gray scale will probably dominate until investigators are more familiar with the diversity offered by the multiple parameters that can be visualized with NMR. Since three-dimensional data may be obtained with NMR, whole new ways of displaying the data will be needed. The current methods of data collection are slow, cumbersome, and expensive. Storage of raw data may allow the sequential processing of several patients, but in some machines it may take all night to process the data. Sometimes the methods used for imaging can exceed the storage capacity of the available computer as well.

**2.7.B.*7*.** *Central controller.* The entire NMR imaging process is a sequence of operations that requires careful control. Since the time scale of many of the operations is on the level of microseconds, only a sophisticated computer can handle the imaging data. Especially troublesome is the difference in time between what is happening in the NMR spectrometer and what is being displayed on the screen. In some cases, the two events are minutes apart; at best, they are a few seconds apart.

The field of NMR imaging is still in its infancy. The rapid development of most components of imaging systems is continuing, and quantum jumps may be expected in the next few years. Most improvements will result in the upgrading of current components and methods, but certainly there will be some innovations that may make such systems obsolete. However, so much information can be found even in NMR images of low quality that existing systems will have value for a long time.

**2.7.C.** *Comparison of techniques*

Among NMR imaging methods, there are certain commonalities in pulse sequences. For this discussion of different types of imaging sequences, the following notations will be used: (a) Pulse widths capable of rotating the spin system by 90° or 180° will be referred to as 90° or 180° pulses. (b) The events in a pulse sequence will be summarized within parentheses, with each event separated by a dash, and if the sequence is repeated $n$ times, that number will be a subscript. (c) A time delay is denoted by $T$, with a subscript or superscript, such that $T_R$ is the pulse repetition time, $T_I$ is the interval affecting the magnetization or the interpulse interval, and $T'$ is the recovery time interval between the last pulse and the following sequence.

A few of the most commonly used imaging pulse sequences will be described, using the following notation: partial saturation or repeated free induction decay time $(90°–T_R)_n$; inversion recovery $(180°–T_I–90°–T')_n$; spin echo $(90°–T_I–180°–T')_n$; and multiple spin echo $[90°–(\tau–180°–\tau)_m–T']_n$, where $T_R = 2m\tau + T'$.

These notations describe succinctly the pulse sequences in detail. For example, the notation for the partial saturation sequence says that it is a sequence consisting of the application of 90° pulses separated by time intervals of $T_R$, repeated $n$ times. The inversion recovery notation reads as a sequence of a single 180° pulse followed by a 90° pulse after an interval $T_R$. The sequence is repeated at $n$ times to measure $T_1$.

Only the transverse magnetization residing in the $xy$ plane can be detected by the coils of most NMR machines. The intensity of the signal is related to the magnitude of the transverse signal through equations described by Bloch. These can be solved, given the initial conditions and the sequence used. Table 2.3 summarizes the results for signal intensity changes

**Table 2.3. Pulse sequences and what they measure**

| Sequence | Signal intensity proportional to: |
|---|---|
| Partial saturation | $p^*(1 - e^{-T_R/T_1})$ |
| Inversion recovery | $p^*(1 - 2e^{-T_I/T_1} + e^{-T_R/T_1})$ |
| Spin–echo | $p^*e^{-2T_I/T_2}(1 - 2e^{-(T_R - T_I)/T_1} + e^{-T_R/T_1})$ |
| | since $T_I \ll T_R$, |
| | $p^*e^{-2T_I/T_2}(1 - e^{-T_R/T_1})$ |

$p^*$, proton density.

over time for the commonly used NMR pulse sequences.

Consider the case of the partial saturation sequence. If the pulse repetition time $T_R$ is much longer than $T_1$, then the term $e^{-T_R/T_1}$ tends to be negligible, and the signal intensity will be proportional only to the proton density. In such a situation, tissue differentiation on the basis of $T_1$ is not possible. As the repetition time is decreased, the signals begin to depend on $T_1$ and a differential picture can be obtained.

In the case of inversion recovery, the image intensity is dependent on proton density, $p^*$; spin–lattice relaxation time, $T_1$; interpulse interval, $T_I$; and pulse repetition time, $T_R$. The dependence of the signal intensity on interpulse interval is peculiar. As the interpulse interval is altered, the contrast between two tissues with different $T_1$ values can first decrease and then increase.

In the spin–echo sequence the signal intensity is dependent on proton density, $p^*$; spin–lattice relaxation time, $T_1$; spin–spin relaxation time, $T_2$; interpulse interval, $T_I$; and pulse repetition time, $T_R$. In this case, tissue differentiation can be obtained because of $T_1$ and $T_2$ differences and by manipulating $T_I$ and $T_R$.

An indepth discussion of imaging techniques can be found in sources given in the Bibliography below.

## 2.8. REFERENCE

1. Zimmermann, J.R., and Brittin, W.E. Nuclear magnetic resonance studies in multiple phase systems. *J. Phys. Chem.* **61**:1328–1333, 1957.

## 2.9. BIBLIOGRAPHY

Further details on the physics of NMR imaging are available in the following articles and books.

Bottomley, P.A. NMR imaging techniques and applications: a review. *Rev. Sci. Instrum.* **53**:1319–1337, 1982.

Dixon, R.L., and Ekstrand, K.E. The physics of proton NMR. *Med. Phys.* **9**:807–818, 1982.

Fullerton, G.D. Basic concepts for magnetic resonance imaging. *Magn. Reson. Imaging* **1**:39–55, 1982.

Hoult, D.I. Medical imaging by NMR. in: *Magnetic resonance in Biology*, Vol. 1. Cohen, J.S., ed. John Wiley & Sons, New York, 1980.

Kaufman, L., Crooks, L.E. and Margulis, A.R. *Nuclear Magnetic Resonance Imaging in Medicine.* Igaku-Shoin, New York, 1981.

Lai, C.M., House, W.V., and Lauterbur, P.C. Nuclear magnetic resonance zeugmatography for medical imaging. *IEEE Trans. Biomed. Eng.* **78**:73–87, 1978.

Mansfield, P., and Morris, P.G. *NMR Imaging in Biomedicine.* Academic Press, New York, 1982.

Newton, T.H., and Potts, D.G. *Advanced Imaging Techniques.* Clavadel Press, 1983.

Partain, C.L., James, A.E., Rollo, F.D., and Price, R.R. *Nuclear Magnetic Resonance Imaging.* W.B. Saunders, Philadelphia, 1983.

Pykett, I.L. NMR imaging in medicine. *Scientific American* **246**:78–88, 1982.

Witcofski, R.L., Karstaedt, M., and Partain, C.L. *NMR Imaging.* The Bowman Gray School of Medicine at Wake Forest University, Winston-Salem, NC, 1981.

CHAPTER 3

# APPLICATIONS OF NMR IN BIOMEDICINE



## 3.1. INTRODUCTION

Nuclear magnetic resonance (NMR) spectroscopy, although discovered independently in 1946 by Bloch[1] and Purcell et al.,[2] remained only an analytical and research tool in chemistry and physics and, to a limited extent, in biology, until the beginning of the 1970s. With the development of Fourier Transform NMR techniques, a new area of research emerged, namely, the applications of NMR in medicine. There are two main advantages of NMR that may make it a routine clinical tool in the future. First, it is noninvasive and nondestructive and can therefore be utilized to acquire clinically useful information in intact living systems—humans, animals, or cultured cells—without causing any known damage to those systems. Second, in its use as a potential diagnostic tool, no medical hazards have yet been reported. Unlike x-rays, the radiation used in NMR is nonionizing, which makes its use in diagnosis attractive.

Much of the work done in earlier years, before the development of NMR imaging modalities, was carried out *in vitro* on tissue biopsy samples. With the development of high-field NMR spectrometers, it has been possible to obtain *in vivo* information on metabolism and on the metabolic status of normal and diseased tissues. During the last 7 years, exciting developments in the field of NMR imaging[3–5] have made it possible to obtain anatomical images of human beings.

Current efforts in the study of applications for NMR in biomedicine are focused on four main areas: (a) *in vitro* studies on human and animal tissues, mostly using $^{1}H$ NMR of water in tissues; (b) *in vivo* $^{31}P$ NMR of tissue metabolism; (c) flow instruments; and (d) NMR imaging.

The aim of this chapter is not to describe the details of such applications, especially in view of the fact that whole books have already been written on these topics.[6–9] Rather, this chapter will indicate, to those new to NMR, the sorts of research that investigators can undertake, depending on the facilities available to them.

## 3.2. *IN VITRO* STUDIES

### 3.2.A. *Proton NMR*

Considering the facts that water is the major constituent of living cells and that the proton is the most

NMR-sensitive nucleus in biological systems, it is not surprising to find that most of the extensive studies in NMR have been carried out on the water in tissues. Although some work had been done in the earlier years, notably by Odeblad,[10] on the properties of water in tissues and cells using low-resolution NMR techniques, much of the impetus for the current variety of *in vitro* investigations was provided by Damadian's study[11] on the water proton relaxation times in normal and malignant animal tissues. He suggested that the differences in the water proton relaxation times in normal and malignant tissues could be used as a diagnostic tool for cancer. Subsequently, several laboratories (see later sections and tables in this chapter) measured the water proton relaxation times in a large variety of normal and diseased tissues and unequivocally established that water proton spin–lattice relaxation times ($T_1$), spin–spin relaxation times ($T_2$), and spin–lattice relaxation times in the rotating frame ($T_{1\rho}$) are, in general, longer for malignant tissues and other diseased tissues than for their normal counterparts. Nevertheless, some overlap in the $T_1$ and $T_2$ values exists for normal and diseased tissues. It has also been established that even in nonmalignant tissues, longer relaxation times are sometimes observed.[12,13] A great deal of data exists for various tissues. However, not much data exist for $T_1$ and $T_2$ at low frequencies, particularly in human tissues. Also, not much information is available on the frequency dependence of $T_{1\rho}$ in animal or human tissues. Further, little work has been done on the frequency and temperature dependence of the relaxation times of protons in the water in human tissues. Such detailed *in vitro* studies might be helpful in deciding which frequency would give the best resolution[14,15] and contrast for each organ. In NMR imaging research, all *in vivo* proton NMR imaging is done at frequencies below 10 MHz, although attempts are being made to extend these studies with higher frequencies.

*In vitro* studies will be helpful in understanding the basic mechanisms responsible for the observed longer water proton relaxation times in diseased tissues. Ample evidence exists to show that the increased water content of the diseased tissue can partially be responsible for the longer relaxation times observed in pathological tissues (see later sections in this chapter for details). Nevertheless, it is becoming increasingly apparent that factors other than water content also influence the changes in the relaxation times, such as the presence of specific antigens,[16] the presence of specific viruses,[17] or changes in the cytoskeletal organization.[18] So, further *in vitro* studies should throw light on those factors that contribute to the relaxation time differences observed between normal tissues and their diseased counterparts. Much work needs to be done to this end.

Some investigators have sought a "systemic effect"; they have tried to detect changes in the relaxation times of cellular water in the organs not directly involved in the disease process, i.e., tumor, in animals.[19] Studies of water proton $T_1$ of the sera from humans with cancer[19] have indicated evidence for a "systemic effect" of elevated $T_1$ values in animals and humans with cancer. There are few available studies on the "systemic effect" in relation to other diseases.

Another area in which *in vitro* studies could possibly be extremely useful is in the development of contrast agents to improve contrast between tissues on the basis of differences in $T_1$, $T_2$, and proton density caused by these contrast agents. At present, most contrast agents that have been tried are stable, free radicals and paramagnetic relaxing agents, such as $Mn^{2+}$ or molecular oxygen.[20,21] The possibilities of this area of research remain to be explored, but *in vitro* studies would be useful to gain an understanding of the effects of the contrast agents before they are tested *in vivo,* using more expensive imaging equipment and possibly exposing patients to hazards.

### 3.2.B. *Other nuclei*

Other nuclei of interest that are amenable to NMR studies in living systems are sodium ($^{23}Na$), phosphorus ($^{31}P$), potassium ($^{39}K$), carbon ($^{13}C$), fluorine ($^{19}F$), and nitrogen ($^{15}N$). Of these, next to hydrogen, sodium and fluorine are the most NMR sensitive, based on their abundance in tissues and gyromagnetic ratios. Some *in vitro* studies have been conducted using $^{23}Na$ in tissues (see Table 9.22 in Chapter 9 for references). Potassium is an extremely difficult nucleus to study because of its low NMR sensitivity; it is not likely to be suitable for either *in vitro* research or *in vivo* imaging studies. However, *in vitro* relaxation times of $^{23}Na$ and $^{39}K$ in normal and malignant tissues have been measured,[22,23] and it was found that the malignant tissues exhibited longer relaxation times, as has been observed for cellular water protons. Similarly, *in vitro* spin–lattice relaxation times ($T_1$) for $^{31}P$ have been measured[24] in normal and malignant tissues. It was found that tumor tissues had almost twofold longer $T_1$ times, compared with normal tissue.[8] Furthermore, in the case of $^{31}P$ NMR, it is interesting to note that brain tissue exhibits a much shorter $T_1$ than does liver tissue, in contrast to the case of proton NMR of cellular water, in which liver tissue exhibits a much shorter $T_1$ than does brain tissue. The underlying mechanisms for these differences remain to be understood. High-resolution $^{31}P$ NMR is rapidly becoming a valuable tool in the study of *in vivo* metabolism, using topical magnetic resonance techniques. This will be discussed later in this chapter.

### 3.2.C. *Correlation between* in vitro *and* in vivo *studies*

It is apparent from the above discussion that many investigations can be conducted using equipment that is much cheaper than whole body NMR imaging equipment. A question can be raised as to whether such *in vitro* studies can be extrapolated to *in vivo* processes. A recent study by Ling and Foster,[25] on the changes in NMR relaxation times of cellular water associated with local inflammatory response to turpentine injection in the thigh muscle of rabbits, showed an interesting comparison between the *in vitro* and *in vivo* $T_1$ values. They found that the $T_1$ value measured *in vitro* around the injection site was elevated within 24 h after injection. A comparison of the $T_1$ values obtained for the rabbit tissue samples *in vitro* with the NMR images of dead rabbits showed the same zonal effect of changing $T_1$ values around the injection site. The authors suggested that *in vitro* $T_1$ measurements reflect changes that can be observed *in vivo* by NMR imaging techniques.

Obviously, this type of study requires facilities that enable *in vitro* and *in vivo* studies to be conducted with the same system. Nevertheless, such studies would aid in the understanding of the interrelationship between the NMR properties of tissues and physiological and pathological states.

## 3.3. *IN VIVO* STUDIES

### 3.3.A. *Proton and other nuclei*

Extensive *in vivo* NMR studies have been conducted on the $^{31}P$ nucleus of tissue metabolites. Some *in vivo* studies have been conducted on the cellular water in the tail of a live mouse and $^{23}Na$ and $^{19}F$ nuclei of blood constituents.

Weisman et al.[26,27] measured the water proton relaxation times ($T_1$ and $T_2$) of water in the tail of a live, normal mouse carrying a Cloudman S91 tumor, using a conventional pulsed NMR spectrometer with a slight modification of the probe. With this method, they found it possible to detect and follow the growth of cancer in a live animal by monitoring changes in $T_1$ and $T_2$.

NMR studies have been conducted on circulating blood[28] that reported the development of an exterior arteriovenous bypass system, with the NMR tube forming part of the circuit, and the observations of $^{31}P$, $^{23}Na$, and $^{19}F$ NMR signals from the circulating, whole blood of dogs. The $^{31}P$ signals were attributed to the 2,3-diphosphoglyceric acid in erythrocytes and the phospholipids on the circulating lipoproteins. The $^{19}F$ NMR signal was attributed to the anesthetic, methoxyflurane, in the bloodstream.

Bené et al.[29] developed methods to observe NMR signals of water *in vivo* using a very low field of 50–100 G for polarizing the proton spins and observing the free precession of the signals in the earth's magnetic field (~0.5 G). This enabled them to measure the $T_1$ and $T_2$ of water in blood and other physiological fluids, such as urine and amniotic fluid, *in vivo*. The technique, however, requires observation over large volumes of the sample because of the low sensitivity associated with extremely low fields. Nevertheless, it has the advantage that strong magnetic fields are avoided, and hence it can be useful diagnosing patients who cannot be subjected to strong magnetic fields, e.g., for examination of the amniotic fluid in pregnant women.

### 3.3.B. *$^{31}P$ NMR*

With the development of the high-field Fourier Transform (FT) high-resolution NMR spectrometers (which cost between $70,000 and $350,000), it has been possible to obtain $^{31}P$ NMR signals *in vivo*. The $^{31}P$ NMR signals are mainly from tissue metabolites, and hence have the obvious advantage of providing information about the metabolic status of the system under investigation. Also, the sensitivity of $^{31}P$ relative to nuclei other than the proton, based on its concentration in tissue and gyromagnetic ratio, is about the same as that of $^{23}Na$. Useful information can be obtained in a few minutes even using a low-frequency (36.4 MHz) instrument. Other high-frequency instruments permit better resolution and a more rapid (~30 s) observation of $^{31}P$ NMR signals from tissue metabolites. Basically, information is sought about the quantity of tissue metabolites and their nature in normal and diseased conditions. In other words, it is possible to study *in vivo* chemistry using $^{31}P$ NMR. The reader is referred to the extensive literature on *in vivo* $^{31}P$ NMR studies.[7,30–33]

The first application of *in vivo* $^{31}P$ NMR studies was in the determination of intracellular pH in red blood cells.[34] This was based on the fact that the chemical shift values for adenosine triphosphate (ATP), inorganic orthophosphate and sugar phosphate depend on the hydrogen ion concentrations at physiological pH. It was possible, for example, to study the interrelationship between intracellular pH and the function of the heart.[35–37] Another possible application is the study of the correlation between the energy status and pH of

**Table 3.1 NMR applications in biomedicine**

| Type of study | Cellular constituent investigated | NMR parameters frequently measured | Type of NMR equipment needed | Type of information obtained |
|---|---|---|---|---|
| • *In vitro* | | | | |
| $^1H$ NMR | Cellular water, lipids in tissues and cultured cells | $T_1$, $T_2$, $T_{1p}$, and D<br>Chemical shifts | Table-top pulsed spectrometer<br>High-resolution NMR | (a) Distinction between normal and diseased states<br>(b) Relationship between pathological changes and NMR properties of cellular water<br>(c) Possible contrast agents |
| $^{31}P$ NMR | Tissue metabolites in biopsy specimens | $T_1$, $T_2$, and possibly D<br>Chemical shifts | Conventional pulsed<br>High-resolution FT NMR | (a) Concentration of tissue metabolites *in vitro*<br>(b) Distinction between normal and diseased states<br>(c) Basic mechanisms of disease process, and their influence on NMR properties |
| $^{23}Na$ NMR | Intracellular and extracellular sodium | $T_1$, $T_2$ | Pulsed NMR, with facility to change field or frequency | (a) Distinction between normal and diseased states<br>(b) Concentration changes as a result of pathological processes<br>(c) Possible contrast agents |
| $^{39}K$ NMR | Intracellular and extracellular potassium | $T_1$, $T_2$ | Because of difficulty of obtaining signals in tissues with low-field, low-cost NMR instrument, need facility where high fields (~50 G) are available | (a) Distinction between normal and pathological conditions<br>(b) Concentration changes as a result of disease processes<br>(c) Basic mechanisms |
| • *In vivo* | | | | |
| $^1H$ NMR | Cellular water, lipids | $T_1$, $T_2$ | *In vivo* measurements on small animals can be done using low-field pulsed NMR instruments[28] | (a) *In vivo* information, such as effect of growth rate of tumor on NMR parameters<br>(b) *In vivo* detection of neoplasms<br>(c) *In vivo* flow measurements |
| $^1H$ NMR, in earth's field | Cellular water, lipids | $T_1$, $T_2$ | Very low field of 50–100 G (0.005–0.01 T); can be homebuilt[28] | (a) *In vivo* and *in vitro* detection of tumors<br>(b) *In vivo* and *in vitro* investigation of physiological fluids under normal and pathological conditions |

| | | | | |
|---|---|---|---|---|
| $^{31}P$ NMR | Tissue metabolites | Chemical shifts and intensities | Need high-field FT NMR instruments, operating at 72–100 MHz or higher | (a) Changes in the tissue metabolite concentrations<br>(b) Changes in intracellular pH<br>(c) *In vivo* chemistry of normal and diseased tissues<br>(d) Discovery of specific metabolites not detectable by conventional techniques<br>(e) *In vivo* study of the effects of drugs on the functions of different organs<br>(f) Viability test on organs for transplantation in sterile conditions<br>(g) Size of damaged regions in disease states and assessment of their metabolic states<br>(h) Specific metabolic disorders (e.g., enzyme deficiencies) such as McArdle's syndrome[40] and Duchenne muscular dystrophy[41] |
| Flow measurement | Water, lipids, $^{23}Na$, $^{31}P$ | Intensity of the NMR signal | Modification of either simple pulsed NMR instruments or FT high-resolution instruments needed, or imaging equipment | (a) Flow velocity<br>(b) Flow distribution as a function of velocity<br>(c) Different types of flow in normal and pathological states |
| • NMR imaging<br>Proton NMR imaging | Water, lipids in organs | *In vivo* proton density; $T_1$ and $T_2$ mapping | Need NMR imaging equipment | (a) Anatomical details comparable to X-ray and CT scans<br>(b) Specially suited for disorders such as hydrocephalus in children, pulmonary edema, etc.<br>(c) *In vivo* imaging of blood flow<br>(d) *In vivo* detection of cancer, benign lesions, etc.<br>(e) Particularly suited for soft-tissue contrast |
| Other nuclei ($^{31}P$ and $^{23}Na$) | Tissue metabolites and ions | $^{31}P$ or $^{23}Na$ density; possibly $T_1$ and $T_2$ in future | Equipment suitable for $^{31}P$ or $^{23}Na$ imaging, i.e., capable of working at higher fields | (a) *In vivo* mapping of the tissue metabolite distribution, etc. (still in infancy at this time) |

D, diffusion coefficient; $T_1$, spin–lattice relaxation time; $T_2$, spin–spin relaxation time.

the heart and functional changes during treatment designed to protect the ischemic myocardium following coronary infarct. Other possibilities include the estimation of the size and location of ischemic zones and the testing of organ transplant viability.

*In vivo* $^{31}P$ NMR has become a very valuable technique in the study of diseased tissue. Essentially, the method consists of comparing the $^{31}P$ NMR spectra of normal with diseased tissue, e.g., in muscle. For example, Glonek et al.[30] reported a significant lowering of the concentrations of all the phosphates in the intact muscle of a patient with nemaline rod myopathy, compared with normal values found by *in vitro* studies. It is also possible to follow the metabolic state of muscle as a function of time and to detect unsuspected phosphates that are involved in a particular disease process.

The investigations reported above were carried out using conventional FT high-resolution $^{31}P$ NMR instruments, with slight modifications. A variation of the technique is to use surface coils[38] (used to observe NMR signals) to get information about the metabolic state in a localized region of the organ, thus obtaining high-resolution NMR signals noninvasively. Another way to obtain this information is to use a special type of field homogeneity coil to shape the field so that the static field is homogeneous over a certain volume but is inhomogeneous elsewhere.[39] This makes it possible to obtain high-resolution NMR signals from a specified region; the technique is called topical magnetic resonance (TMR) (see ref. 31 for a review). TMR is being used to explore a variety of medical problems, such as cardiac physiology, cerebral infarction, metabolic diseases, etc.

An elegant example of the application of TMR was the confirmation of the diagnosis of McArdle's syndrome (which is the lack of phosphorylase activity in skeletal muscle).[40] The diagnosis of this inborn metabolic disease is confirmed if an ischemic exercise of the forearm fails to generate lactic acid. In other words, in the patient with McArdle's syndrome, there is no decrease in intracellular pH associated with ischemic exercise, as would be expected in a healthy person. Other interesting applications of TMR have included the study of energy metabolism in a patient with phosphofructokinase deficiency and a study of alterations in ATP and phosphocreatine contents of resting muscle in a patient with muscular dystrophy.[41]

### 3.4. FLOW MEASUREMENTS

Flow of the blood and of other body fluids can be measured *in vivo* noninvasively by monitoring the proton NMR.

The possibility of using NMR for blood flow measurements was suggested by Singer.[42,43] A detailed account of the current thinking in this area can be found in a book written by Singer in 1981.[44] Measurements of blood flow in femoral arteries were carried out by Battocletti et al.[45] It has been possible to measure the velocity of blood flow accurately, as well as flow distribution as a function of velocity (the relative numbers of molecules flowing in particular velocity intervals). These experiments can be conducted with relatively inexpensive, commercially available instruments, with slight modifications of the NMR probe.

Studies of blood flow are helpful in understanding the flow patterns in normal and diseased conditions. Recent studies by Grant and Back[46] on the effect of flow on NMR images showed that different modes of flow can be distinguished by NMR imaging. The study of flow patterns by NMR imaging was termed by these authors, "NMR rheotomography." They suggested that NMR rheotomography may prove useful for "the non-invasive diagnosis of structurally-originating cardiovascular defects."[46]

### 3.5. INSTRUMENTATION

*In vitro* studies on water or other mobile protons in tissues, such as in lipids, can be carried out with table-top commercial machines which are currently available. These machines cost from $30,000 to $40,000. A list of these machines and their manufacturers is given in the Appendix to this book. Other nuclei, such as $^{23}Na$ and $^{31}P$, can be investigated with the same instruments if the magnetic field can be increased using the magnet provided.

Designs are available in the literature for building pulsed NMR instruments, which can be used for *in vitro* studies.[47–56]

*In vivo* $^{31}P$ NMR studies are usually conducted either with commercially available pulsed FT NMR spectrometers or with newly developed techniques such as TMR.

Equipment used for NMR imaging is by far the most expensive; currently available imaging machines cost more then $1 million.

### 3.6. NMR IMAGING

One of the most recent applications of NMR in biomedicine is NMR imaging. The concept of NMR imaging was proposed and developed by various groups; however, the work of Damadian,[11] Lauterbur,[4] and Mansfield and Grannell[5] is most significant. The general technique consists of isolating a small volume of the subject under study and performing a standard NMR experiment to obtain NMR parameters such as nuclide densities, relaxation times, etc., from that volume. This can be carried out sequentially through the entire volume or, alternatively, methods have been devised such that the information from the entire volume is obtained simultaneously. Ingenious methods have been devised to accomplish the details of the spatial localization and NMR parameter measurements.[57-60] The key to spatial localization is the fact that a field gradient is superimposed on a stable magnetic field. Therefore, the resonance frequency, which is directly proportional to the total magnetic field, will be different when the field is different. Hence, spatial coordination can be linked to the frequency measurements.

There are many commercial instruments on the market now for doing research and development work. Various field strengths, from 0.04 T to 1.5 T, are being used. Permanent and resistive magnets produce a field of up to 0.3 T. Fields of 0.3 T and above are being produced by superconducting magnets. Table 3.2 lists the current manufacturers and types of magnets in use for imaging instruments.

NMR imaging machines have been installed in many clinical settings, notably, Hammersmith Hospital, London, England; Aberdeen Royal Infirmary, Aberdeen, Scotland; University of Nottingham, Nottingham, England; Radiologic Imaging Laboratory, University of California at San Francisco, San Francisco, California, U.S.A.; Baylor College of Medicine, Houston, Texas, U.S.A.; Cleveland Clinic, Cleveland, Ohio, U.S.A.; Massachusetts General Hospital, Boston, Massachusetts, U.S.A.; and The Mayo Clinic, Rochester, Minnesota, U.S.A. A few machines are also used in private clinics. Results of the work being done at these and other places have been published.

**Table 3.2 NMR imaging equipment**

| Manufacturer | Permanent magnet system | Resistive magnet system | Super-conductive magnet system |
|---|---|---|---|
| General Electric | | x | x |
| Technicare | | x | x |
| Picker International | | x | x |
| Siemens | | | x |
| Philips | | x | x |
| Diasonics | | | x |
| Elscint | | | x |
| FONAR | x | x | |
| OMR | x | | |
| M & D (Scotland) | | x | |
| Toshiba (Japan) | | x | |
| Bruker (F.R.G.) | | x | x |

The following statements summarize the clinical work published so far:

1. Many pathological conditions, such as infarction, infection, demyelination, edema, malignancy, etc., produce long $T_1$ values, compared with normal tissues.
2. White matter lesions in demyelinating diseases can be seen clearly.
3. Multiple sclerosis lesions are better demonstrated with NMR imaging than with computerized tomography (CT).
4. Gray–white matter contrast is greater with NMR than with CT.
5. The lack of bone artifacts in NMR scanning is advantageous.
6. Because NMR provides excellent soft-tissue differentiation without the use of contrast agents and does not use ionizing radiation, it is a promising modality for many sensitive organs.
7. NMR has been found to be useful in the evaluation of bony and soft-tissue tumors.
8. Distinct anatomical structures in normal kidney can be seen clearly using NMR techniques.
9. NMR imaging has the ability to detect cerebral and spinal cord diseases.
10. NMR imaging is more sensitive than ultrasound or nuclear scanning techniques in the diagnosis of cirrhosis.

We would like to point out important areas in which more work needs to be done:

A. The current work does not clearly indicate what particular field is optimal for the imaging of a certain organ.
B. Long-range stability and economic superiority of a particular magnet type—permanent, resistive, superconducting—remain to be established.
C. The technique of choice for any specific disease on a particular system has not been universally established.
D. The time needed to make a single scan on certain machines is still too long to be clinically viable.
E. Cost-effectiveness of the NMR imaging technique in relation to other modalities needs to be ascertained.
F. Methods of routinely calibrating a machine in the clinical setting, and comparing them with other machines, are still very crude and unsatisfactory.

G. Installation problems, such as accommodating the size and weight of the machines and providing magnetic shielding to reduce peripheral fields to less than 1 G, are formidable for most machines.

The clinical advantages and disadvantages can be summarized as follows:

**3.6.A.** *Advantages*

1. Because of the potential to detect biochemical changes, early detection of tissue damage is possible.
2. Bone does not produce artifacts.
3. The effect of patient motion is less damaging than in CT.
4. Soft-tissue contrast is very good, even better than in CT.
5. Three-dimensional imaging is possible.
6. Different directional image planes through the organ can be obtained without moving mechanical parts of the machine and without moving the patient.
7. There are no mechanical moving parts in the machine; this should result in longer, stable working periods and fewer service requirements.
8. NMR does not use ionizing radiation.

**3.6.B.** *Disadvantages*

1. The imaging techniques are not yet disease specific
2. Extensive education for interpretation of NMR scans is necessary. The images do not resemble the more familiar X-ray films.
3. The time needed for interpretation of data is longer. Extensive professional involvement will be needed initially.
4. Patients with pacemakers, metal implants, and even claustrophobia (on some machines) will have to be excluded from study.
5. The lack of cost-effectiveness for certain machines may prohibit smaller, nonresearch-oriented institutions from using them.
6. Calcification is not imaged.
7. Site preparation, maintenance, and service problems can be formidable.

## 3.7 REFERENCES


1. Bloch, F. *Physiol. Rev.***70:**460, 1946.
2. Purcell, E.M., Torrey, H.C., and Pound, R.V. *Physiol. Rev.* **69:**37–38, 1946.
3. Damadian, R. U.S. patent 3,789,832. Filed 17 March 1972.
4. Lauterbur, P.C. *Nature* **243:**190, 1973.
5. Mansfield, P., and Grannell, P.K. *J. Phys. [C]* **6:**L422, 1973.
6. Kaufmann, L., Crooks, L.E., and Margulis, A.R., eds. *Nuclear Magnetic Resonance Imaging in Medicine.* Igaku-Shoin, New York, 1981.
7. Gadian, D.G. *Nuclear Magnetic Resonance and Its Applications to Living Systems.* Clarendon Press, Oxford, England, 1982.
8. Damadian, R., ed. *NMR in Medicine, NMR 19: Basic Principles and Progress.* (Diehl, P., Fluck, E., and Kosfeld, R., series eds.), Springer-Verlag, Berlin, 1981.
9. Mansfield, P., and Morris, P.G. *NMR Imaging in Biomedicine. Advances in Magnetic Resonance, Supplement* 2. Waugh, J.S., ed. Academic Press, New York, 1982.
10. Odeblad, E., and Lindstrom, G. *Acta Radiol.* **43:**469, 1955.
11. Damadian, R. *Science* **171:**1151, 1971.
12. Gwan Go., K., and Edzes, H. *Arch. Neurol.* **32:**462, 1975.
13. Misra, L.K., Kasturi, R.R., Kundu, S.K., Harati, Y., Hazlewood, C.F., Luthra, M.G., Yamanashi, W.S., Munzaal, R.P., and Amtey, S.R. *Magn. Reson. Imaging* **1:**75, 1982.
14. Coles, B.A. *J.N.C.I.* **57:**389, 1976.
15. Diegel, J.G., and Pintar, M.M. *J.N.C.I.* **55:**725, 1975.
16. Nadkarni, J.S., Nadkarni, J.J., Ranade, S.S., Chaughule, R.S., Kasturi, S.R., and Advani, S.H. *Ind. J. Cancer* **13:**76, 1976.
17. Wagh, U.V., Kasturi, S.R., Chaughule, R.S., Shah Smita, And Ranade, S.S. *Physiol. Chem. Phys.* **9:**167, 1977.
18. Beall, P.T., Brinkley, B.R., Chang, D.C., and Hazlewood, F.C. *Cancer Res.* **42:**4124, 1982.
19. Beall, P.T., Medina, D., Chang, D.C., Seitz, P.K., and Hazlewood, C.F. *J.N.C.I.* **59:**1431, 1977.
20. Brady, T.J., Goldman, M.R., Pykett, I.L., Buonanno, F.S., Kistler, G.M. *Radiology* **144:**343, 1982.
21. Lauterbur, P.C., Dias, M.H.M., and Rudin, A.M. *In: Frontiers of Biological Energetics* Dutton, P.L., Leigh, J.S., and Scarpa, A., eds. Academic Press, New York, 1978.
22. Goldsmith, M., and Damadian, R. *Physiol. Chem. Phys.* **7:**263, 1975.
23. Damadian, R., and Cope, F.W. *Physiol. Chem. Phys.* **6:**309, 1974.
24. Zaner, K.S., and Damadian, R. *Science* **189:**729, 1975.
25. Ling, R.C., and M.A. Foster *Phys. Med. Biol.* **27:**853, 1982.
26. Weisman, I.D., et al. *Science* **178:**1288, 1972.
27. Weisman, I.D., Bennett, L.H., Maxwell, L.R., Sr., and Henson, D.E. *NMR in Medicine.* Damadian, R., ed. *NMR 19: Basic Principles and Progress.* Springer-Verlag, Berlin, 1981.
28. Burt, T.C., Eisman, A., Schofield, J.C., and Wyrwicz, A.W. *J. Magn. Reson.* **46:**176, 1982.
29. Bené, G.J., Borcard, B. Hiltbrand, E., and Magnin, P. *In: NMR in Medicine.* Damadian, R., ed. *NMR 19: Basic Principles and Progress.* Springer-Verlag, Berlin, 1981.
30. Glonek, C.T., Burt, C.T., and Barany, M. *In: NMR in Medicine.* Damadian, R., ed. *NMR 19: Basic Principles and Progress.* Springer-Verlag, Berlin, 1981.
31. Shaw, D. In: *Nuclear Magnetic Resonance Imaging in*

*Medicine* Kaufmann, L., Crooks, L.E., and Margulis, A.R., eds. Iguk-Shoin, New York, 1981.

32. Gadian, D.G., Radda, G.K., Richards, R.E., and Seeley, P.J. In: *Biological Applications of Magnetic Resonance*. Shulman, R.G., ed. Academic Press, New York, 1979.
33. O'Neill, I.K., and Richards, C.P. *Annu. Rep. NMR Spectros.* **10A:**133, 1980.
34. Moon, R.B., and Richards, J.H. *J. Biol. Chem.* **248:**7276, 1973.
35. Hollis, D.P., Nunnally, R.L., Taylor, G.J., Weisfeldt, M.L., and Jacobus, W.E. *J. Magn. Reson.* **29:**319, 1978.
36. Hollis, D.P. (1979) *Bull Magn. Reson.* **1:**27, 1979.
37. Garlick, P.B., Radda, G.K., Seeley, P.V. and Chance, B. *Biochem. Biophys. Res. Commun.* **75:**1086, 1977.
38. Ackerman, J.J.H., Grove, T.H., and Wong, G.G., et al. *Nature* **237:**736, 1980.
39. Gordon, R.E., Hanley, P.E., Shaw, D., et al, *Nature* **287:**736, 1980.
40. Ross, B.D., Radda, G.K., Gadian, D.G., Rocker, G., Esiri, M., and Falconer-Smith, J. *New Engl. J. Med.* **304:**1338, 1981.
41. Edwards, R.H.T., Dawson, J. M., Wilkie, D.R., Gordon, R.E., and Shaw, D. *Lancet* **1:**725, 1982.
42. Singer, J.R. *Science* **130:**1652, 1959.
43. Morse, S., and Singer, J.R. *Science* **170:**440, 1970.
44. Singer, J.R. In: *Nuclear Magnetic Resonance Imaging in Medicine*. Kaufmann, L., Crooks, L.E., and Margulis, A.R., eds. Igaku-Shoin, New York, 1981.
45. Battocletti, J.H., Halbach, R.E., and Salles-Cunha, S.X., et al. *IEEE Trans. Biomed. Eng.* **67:**1359, 1979.
46. Grant, J.P., and Back, C. *Med. Phys.* **9:**188, 1982.
47. Clark, W.G. *Rev. Sci. Instrum.* **35:**316, 1964.
48. Lowe, I.J., and Tarr, C.E. *J. Phys. [E]* **1:**320, 1968.
49. Redfield, A.G. In: *Introductory Essays*. Pintar, M.M., ed. Springer-Varlag, Berlin, 1976.
50. Ellett, J.D., Jr., Gibby, M.G., Haeberlen, U., Huber, L.M., Mehring, M., Pines, A., and Waugh, J.S. *Advances in Magnetic Resonance 5*. Waugh, J.S., ed. Academic Press, New York, 1971.
51. Engle, J.L. *J. Magn. Res.* **37:**547, 1980.
52. Clark, W.G., and McNeil, J.A. *Rev. Sci. Instrum.* **44:**844, 1973.
53. Stokes, H.T. *Rev. Sci. Instrum.* **49:**1011, 1978.
54. Hoult, D.I. (1978) *Prog. NMR Spectros.* **12:**41, 1978.
55. Karlicek, R.F., Jr., and Lowe, I.J. *J. Magn. Reson.* **32:**199, 1978.
56. Gordon, R.E., Strange, J.H., and Webber, J.B.W. *J. Phys. [E]* **11:**1051, 1978.
57. Damadian, R., Minkoff, L., and Goldsmith, M. *Physiol. Chem. Phys.* **9:**97, 1977.
58. Hutchison, J.M.S., Edelstein, W.A., and Johnson, G. *J. Phys. [E]* **13:**947, 1980.
59. Holland, G.M., Hawkes, R.C., and Moore, W.S. *J. Comput. Assist. Tomogr.* **4:**429, 1980.
60. Hinshaw, W.S. *J. Appl. Physiol.* **47:**3709, 1976.

CHAPTER 4

# *MEASUREMENT OF RELAXATION TIMES*



## 4.1. INTRODUCTION

Before we discuss the techniques for the measurement of relaxation times used in nuclear magnetic resonance spectrometry, we will introduce the basic terms necessary for understanding these methods.

### 4.1.A. *Radio-frequency pulse*

In pulsed NMR experiments, the radio-frequency (rf) radiation is applied in the form of a pulse, i.e., the rf radiation is on for a short time and then turned off. The rf pulse is characterized by an amplitude, as shown in Fig. 4.1b, and a width, $t_w$. $B_1$ is a measure of the power delivered to the sample coil, and the width of the pulse, $t_w$, determines the angle through which the magnetization vector is tipped away from the equilibrium direction.

### 4.1.B. *Free induction decay signal*

A free induction decay (FID) signal is the signal induced in the NMR coil due to the precession of nuclear spins after the rf pulse in pulsed NMR experiments is turned off. It can be explained in more detailed terms as follows:

Consider a frame of reference rotating at the fre-

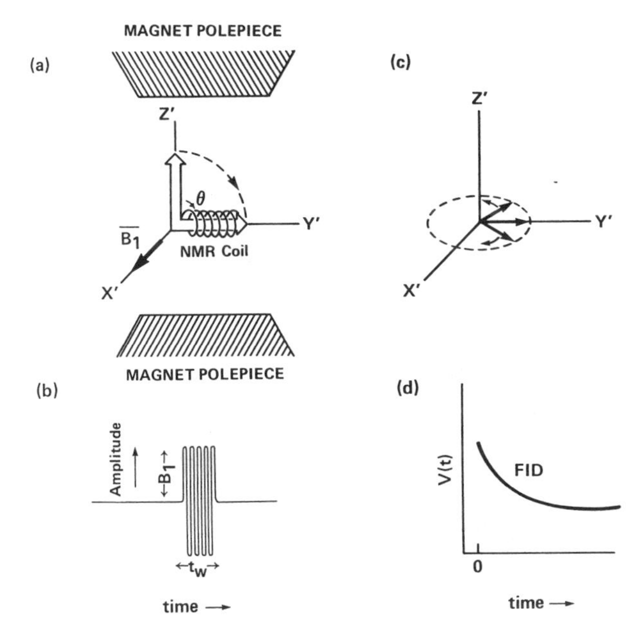


Fig. 4.1. **a:** Three-dimensional coordinate system in the rotating frame. The static magnetic field is along the direction of the $z'$-axis. The rf field, $B_1$, is applied along the direction perpendicular to the static field direction, i.e., in the $x'$-axis. The NMR coil is in the $y'$-axis direction. **b:** Typical rf pulse one observes on an oscilloscope. **c:** Free precession of the nuclear spins in the $x'y'$ plane in the rotating frame of reference. **d:** Typical free induction decay (FID) signal.

quency of rf radiation; this frame of reference is called the rotating frame. The axes of the frame are as represented in Fig. 4.1a.

The rf pulse is applied along the $x'$-axis while the magnetization is along the direction of the static field $B_0$ ($z'$-axis) in equilibrium. Typically, the rf pulse applied is a 90° pulse which tips the magnetization by 90° away from the equilibrium directions, as shown in Fig. 4.1a. The magnetic moment vectors (e.g., proton spins of cellular water), which are now in the $x'y'$ plane, start dephasing, as shown in Fig. 4.1c. The NMR coil is generally along the $x'$- or $y'$-axis and a voltage is induced in this coil due to the precession of the magnetic moments. This signal is called the FID signal (Fig. 4.1d) because it is induced after the rf pulse is turned off. The FID signal decays to zero, with a time constant usually referred to as $T_2$.

### 4.1.C. *90° and 180° pulses*

If an rf pulse of strength $B_1$ is applied for a time $t_w$ (i.e., width $t_w$) to the spin system, then the magnetization vector **M** is tipped by an angle $\theta$ from the equilibrium direction, i.e., the direction of the static field (see Fig. 4.1a and b):

$$\theta = \gamma B_1 t_w \quad , \tag{4.1}$$

where $\gamma$ is the magnetogyric ratio of the nucleus, $B_1$ is the strength of the rf field, and $t_w$ is the width of the rf pulse. If $\theta = 90°$, then the pulse is referred to as a 90° pulse, and if $\theta = 180°$, then it is referred to as a 180° pulse. Similarly, one can flip the magnetization by 270°, 360°, etc., by increasing $t_w$.

### 4.1.D. *Adjustment of 90° and 180° pulses*

The 180° pulse is characterized by a zero signal in the coil located in the $x'y'$ plane. Hence, start from $t_w =$ 0 μs and begin increasing $t_w$ until the FID signal becomes a flat zero signal—the pulse of this width ($t_w$) is referred to as a 180° pulse.

There are two ways of adjusting the 90° pulse. First, start from $t_w = 0$ μs and increase $t_w$ until you see the first maximum in the FID signal—this corresponds to a 90° pulse. Alternatively, you can adjust $t_w$ to correspond to a 180° pulse and set the pulse width to be half of this value, to obtain a 90° pulse.

### 4.1.E. *Spin–echo signal*

Suppose a 90° pulse is applied to the spin system and after a time interval $\tau$ a 180° pulse is applied. Then the spin isochromats* that are spinning faster and those that are spinning slower catch up with one another to rephase and form a net magnetization. Thus, the magnetization gets refocused along the $-y'$-axis at time $2\tau$. The signal due to this magnetization refocusing is called the spin–echo signal. The spin–echo signal can be considered to be two FID signals back-to-back. Details regarding the way the spin–echo forms can be found in ref. 1.

## 4.2. MEASUREMENT OF SPIN–LATTICE RELAXATION TIME ($T_1$)

The spin–lattice relaxation time ($T_1$) is, in general, defined as an exponential recovery to the equilibrium value of the magnetization ($M_\infty$) from the zero value of the magnetization. The spin–lattice relaxation time $T_1$ is thus the time constant required for the magnetization to return to its thermal equilibrium value. This is

*The term "spin isochromat" is defined in the glossary of physical terms for NMR in Chapter 10. For the benefit of the reader, we will give the definition here as well: When a sample is kept in the NMR coil and a 90° pulse is applied, different portions of the spin system in the sample precess at slightly different frequencies due to the inhomogeneity of the applied static magnetic field. Those portions of the spin system that experience the same homogeneous applied magnetic field are called spin isochromats because all the spins in this portion precess at the same Larmor frequency.

defined mathematically as

$$M_z = M_\infty\left[1 - \exp\left(-\frac{\tau}{T_1}\right)\right] \quad , \qquad (4.2)$$

where $M_z(\tau)$ is the magnetization built up along the $z'$ direction (i.e., the direction of the static field $B_0$) at any instant of time, and $M_\infty$ is the equilibrium value of the magnetization (i.e., $\tau = \infty$).

In practice, in the case of cellular water for example, the spin–lattice relaxation is nonexponential in tissues, whole cells, and organs, and can be resolved into at least two components (see Chapter 7). However, in most of the work done in normal and diseased tissues in laboratories around the world, only average relaxation times have been reported. To resolve the recovery of equilibrium into two or more components, a large number of data points must be collected ($\tau$ values) for a long period of time.

While there are several methods for the measurement of $T_1$, only three commonly used methods will be described here. For more details, the reader is referred to the references suggested at the end of this chapter.

**4.2.A.** *Inversion recovery sequence*

The inversion recovery sequence (180°–$\tau$–90° pulse sequence) is most commonly used in *in vitro* studies on animal and human tissues, as well as *in vivo* NMR imaging.

In this sequence, the first pulse of rf radiation (180° pulse) prepares the spin system by tipping the magnetization 180° away from the thermal equilibrium direction (the direction of the static field $B_0$, i.e., $z'$-axis). The second rf pulse, applied after an interval of time $\tau$, is used to monitor the magnetization that has built up along the $z'$-axis after this time interval. This sequence is repeated at time periods greater than $5T_1$. The height of the FID signal is measured by sampling the FID after the 90° pulse in the sequence. The sequence is shown schematically in Fig. 4.2. The corresponding picture for the build-up of the magnetization as a function of $\tau$ is given in Fig. 4.3.

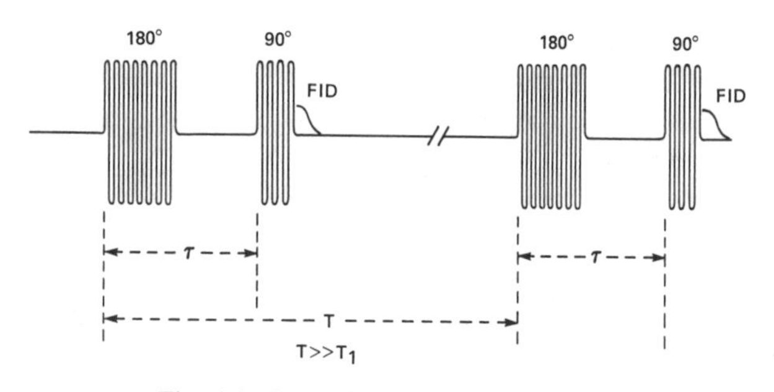


Fig. 4.2. Inversion recovery sequence.

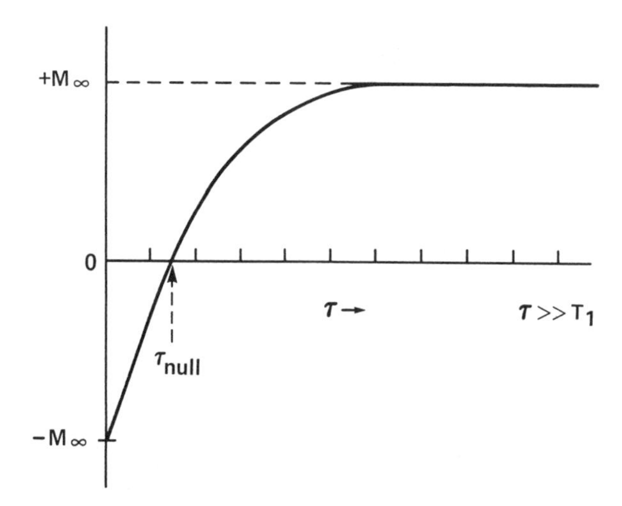


Fig. 4.3. Build-up of magnetization as a function of $\tau$ in inversion recovery sequence.

The magnetization is $-M_\infty$ after the 180° pulse because the 180° pulse inverts the magnetization; it starts to recover to the thermal equilibrium value of $+M_\infty$, as shown in Fig. 4.3, for $\tau \gg T_1$. This is represented mathematically as

$$M_z(\tau) = M_\infty\left[1 - 2\exp\left(-\frac{\tau}{T_1}\right)\right] \quad . \qquad (4.3)$$

**4.2.A.*1*.** *Analysis of the data.* A simple way to calculate $T_1$ from the measured $M_z(\tau)$ values in this method is to observe the $\tau$ value for which the signal after the 90° pulse becomes zero. This is referred to as $\tau_{null}$ (Fig. 4.3). $T_1$ can be obtained from

$$\tau_{null} = 0.693\ T_1 \quad . \qquad (4.4)$$

While this is a very quick way of determining $T_1$, it is not usually recommended if an accurate $T_1$ is desired. Nevertheless, a determination of $T_1$ from $\tau_{null}$ helps the operator to determine the correct period to be used between $M_z$ excursions.

The data, $M_z(\tau)$ determined for different $\tau$ values, can be plotted on a semilogarithmic plot of $[M_\infty - M_z(\tau)]$ or $[1 - M_z(\tau)/M_\infty]/2$ versus $\tau$.

That is, Eq. 4.3 can be written in either of the following ways:

$$\left[1 - \frac{M_z(\tau)}{M_\infty}\right] = 2\exp\left(-\frac{\tau}{T_1}\right) \qquad (4.5)$$

or

$$M_\infty - M_z(\tau) = 2M_\infty \exp\left(-\frac{\tau}{T_1}\right) \quad . \qquad (4.6)$$

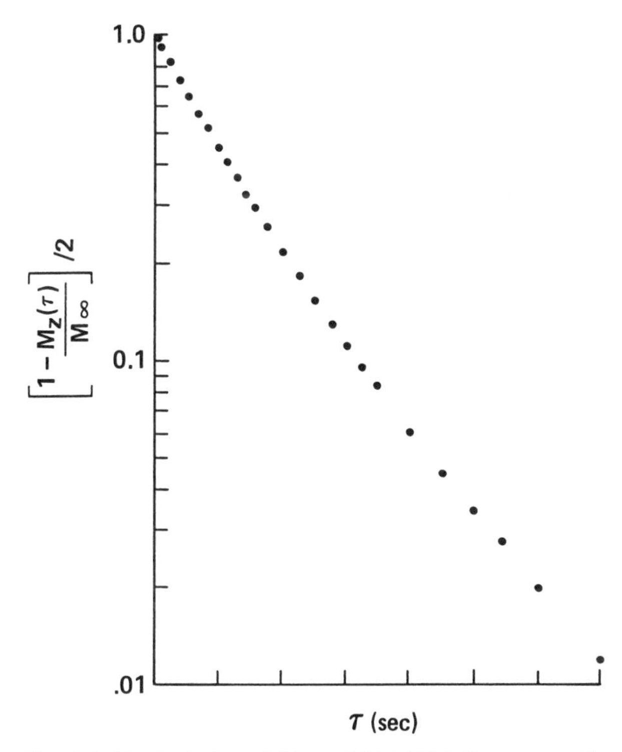


Fig. 4.4. Typical plot of $[1 - M_z(\tau)/M_\infty]/2$ versus $\tau$ for determination of $T_1$ according to the 180°–$\tau$–90° method for water in tissues and cells. Notice the slightly nonexponential nature of the plot. For pure water, such a plot would be a single straight line. The plot shown is for an arbitrary biological system.

Usually, a fit of the data is carried out on a computer or a calculator with statistical programs, and $T_1$ is obtained from the slope of the straight line of the above equation. A typical plot according to Eq. 4.5 is shown in Fig. 4.4 for an arbitrary biological system.

Alternatively, $T_1$ can be determined by fitting the data to an exponential curve on a computer. Modern NMR machines, whether home-built or commercial, have either a minicomputer- or microprocessor-controlled system so that $T_1$ values may be calculated and printed out online.

**4.2.B.** *Saturation recovery sequence*

The saturation recovery sequence (90°–$\tau$–90° pulse sequence) differs from the previous methods in that the first pulse in the pair tips the magnetization by 90° instead of 180°. The rest of the process is the same as that for the inversion recovery sequence. In this method, also, the second pulse is used to measure the height of the FID signal after waiting for a time $\tau$ after the application of the first 90° pulse. The time interval between the sequences, or the period $P$, is chosen such that $P \gg T_1$ (on the order of $5T_1$). The sequence is shown in Fig. 4.5.

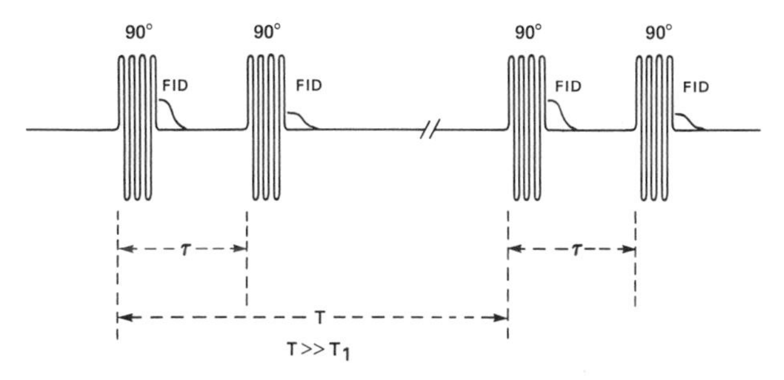


Fig. 4.5. Saturation recovery sequence.

The corresponding expression for the recovery of the magnetization to the equilibrium value $M_\infty$ is

$$M_z(\tau) = M_\infty\left[1 - \exp\left(-\frac{\tau}{T_1}\right)\right] \quad . \tag{4.7}$$

This can be represented mathematically as shown in Fig. 4.6.

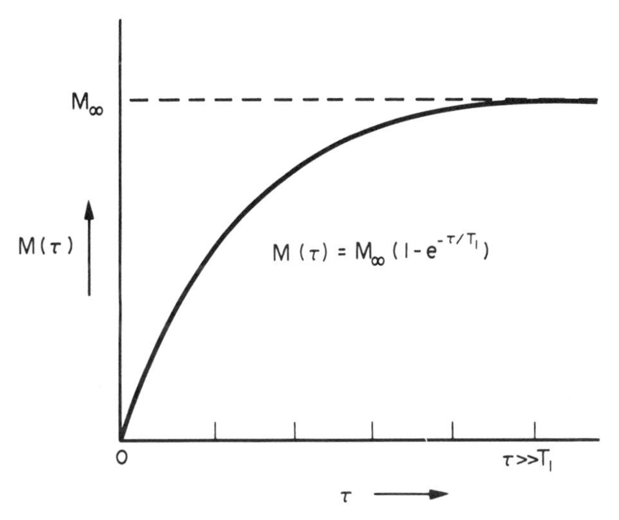


Fig. 4.6. Magnetization build-up to equilibrium value in 90°–$\tau$–90° pulse method (saturation recovery sequence).

**4.2.B.*1*.** *Analysis of the data.* The $T_1$ value can be determined by plotting the data on a semilogarithmic plot of $[1 - M_z(\tau)/M_\infty]$ or $[M_\infty - M_z(\tau)]$ versus $\tau$. That is, Eq. 4.7 can be rewritten as

$$\left[1 - \frac{M_z(\tau)}{M_\infty}\right] = \exp\left(-\frac{\tau}{T_1}\right) \quad . \tag{4.8}$$

A typical plot of this on a semilogarithmic graph looks like Fig. 4.4, except the ordinate is $[1 - M_z(\tau)/M_\infty]$ instead of $[1 - M_z(\tau)/M_\infty]/2$. It can also be plotted as

$$[M_\infty - M_z(\tau)] = M_\infty \exp\left(-\frac{\tau}{T_1}\right) \quad . \tag{4.9}$$

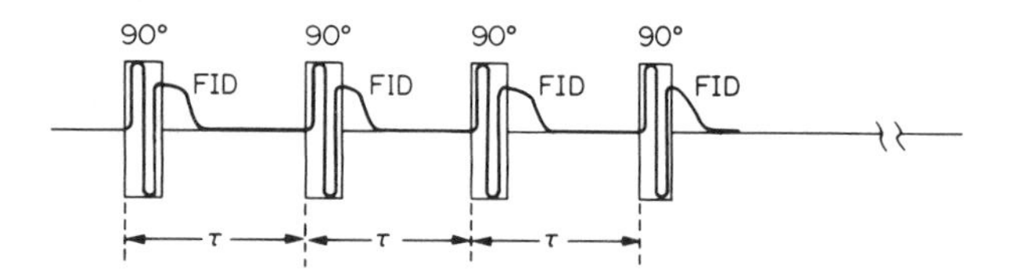


Fig. 4.7. Steady-state sequence.

Such a fit can be carried out on a computer or calculator with a least-squares fit program. Alternatively, $T_1$ can also be determined by fitting Eq. 4.7 to an exponential function using a computer.

**4.2.C.** *Steady-state sequence or progressive saturation recovery*

This method can be considered as a special case of the saturation recovery sequence described above. It is equivalent to the case in which the time interval between the pulse pairs in the 90°–$\tau$–90° method is made zero. Then the sequence becomes an infinite train of 90° pulses spaced at intervals of $\tau$. The signal after each 90° pulse reaches a steady value dependent upon $\tau$ and $T_1$. This sequence is represented in Fig. 4.7. The approach of the signal size (magnetization) to the equilibrium value is the same as that in the saturation recovery sequence.

The signal size after each 90° pulse builds up to the thermal equilibrium value $M_\infty$, according to

$$M_z(\tau) = M_\infty \left[1 - \exp\left(-\frac{\tau}{T_1}\right)\right] \quad . \qquad (4.10)$$

The $T_1$ value can be obtained from this equation in the same way as in the saturation recovery sequence method.

**4.2.D.** *Comparison of the three methods*

With these three methods, the reader may wonder which one to use for the determination of $T_1$. Each method has advantages and disadvantages, but the inversion recovery sequence (180°–$\tau$–90°) method is the one recommended for the most accurate work. The three methods are compared in Table 4.1.

**4.2.E.** *Tips to avoid systematic errors in the measurement of $T_1$*

(1) Position the sample in the most homogeneous part of the rf and static magnetic fields. Adjust the 180° pulse accurately. If there is difficulty in getting a flat zero FID signal from the 180° pulse, remove some sample and readjust the position of the sample in the coil so that it is in the most homogeneous part of the rf ($B_1$) and static ($B_0$) magnetic fields.

(2) $M_\infty$ should be measured very accurately by determining it as many times as possible, preferably at the beginning, middle, and end of the measurement. It would be ideal if $M_\infty$ were determined each time after $M_z(\tau)$ is determined.

(3) Use $\tau$ values extending up to three to four times the $T_1$ to get an accurate measurement of $T_1$.

(4) If a diode detector is used, make sure that the detector is calibrated.

**Table 4.1. Comparison of methods for measurement of spin–lattice relaxation time ($T_1$)**

| Method | Advantages | Disadvantages |
|---|---|---|
| Inversion recovery sequence (180°–$\tau$–90°) | 1. Variation of the magnetization is from $-M$ to $+M$ and thus one can get more data points, which lead to more accurate measurements<br>2. Commonly used in imaging experiments | 1. Takes time because of the waiting period of $5T_1$ between pulse pairs |
| Saturation recovery sequence (90°–$\tau$–90°) | 1. Takes less time, since the time interval between the pulse pairs need not be $5T_1$<br>2. For broad lines, yields much more accurate line shapes and intensity if $B_1$ (strength of rf field) is small | 1. Limited to cases where $T_2 \ll T_1$ |
| Steady-state sequence (progressive saturation recovery method) | 1. Very quick compared with the above two methods<br>2. Suitable for long $T_1$ and for long time averaging of repetitive signals to improve signal-to-noise ratio<br>3. Use for imaging and *in vivo* applications | 1. Limited to cases where $T_2 \ll T_1$<br>2. Because of the relatively higher repetition rate, the demands of the rf transmitter are high for short $T_1$ |

(5) While adjusting the 180° pulse width, make sure that the phase of the rf is set correctly before adjusting $t_w$. Otherwise, an erroneous 180° pulse setting can result.

### 4.3. MEASUREMENT OF THE SPIN-SPIN RELAXATION TIME ($T_2$)

As in the case of the spin–lattice relaxation time, there are basically three methods for measuring the spin–spin relaxation time ($T_2$). The later methods are improvements over the earlier one.

The decay constant of $T_2$ of the FID includes contributions from natural spin–spin relaxation ($T_2$) and the magnetic field inhomogeneity ($\Delta B_0$):

$$\frac{1}{T_2^*} = \frac{1}{T_2} + \frac{\gamma \Delta B_0}{2} \quad . \tag{4.11}$$

In liquids and biological systems such as tissues, cells, and organelles, the magnetic field inhomogeneity contributes to $T_2$, and hence the natural spin–spin relaxation time $T_2$ cannot be obtained from the FID. A method was developed by Hahn for the measurement of $T_2$ in liquids and liquid-like systems. It is called the Hahn spin–echo method and is described below.

#### 4.3.A. *Hahn spin–echo method*

In the Hahn spin–echo method, a 90°–$\tau$–180° pulse sequence is used. At time $2\tau$, a spin–echo signal is observed. To understand how this spin–echo signal is formed, the reader is referred to refs. 1–5. The pulse sequence used is shown in Fig. 4.8. The echo will look negative if phase-sensitive detection is used. If diode detection is used, then the echo will be positive.

The height of the echo $M$ ($2\tau$) is measured for different values of $2\tau$ and decays as a function of $2\tau$ according to the equation:

$$M(2\tau) = M_\infty\left[\exp\left(-\frac{2\tau}{T_2}\right)\right] \quad , \tag{4.12}$$

where $M_\infty$ is the value of the FID for $\tau = 0$.

$T_2$ can be determined from the slope of a semilog plot of $M(2\tau)/M_\infty$ versus $2\tau$. A typical plot of $M(2\tau)/M_\infty$ versus $\tau$ is shown in Fig. 4.9 for a biological system.

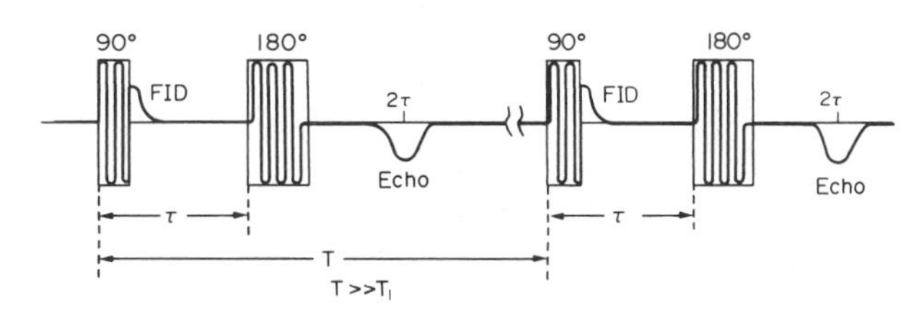


Fig. 4.8. The Hahn spin–echo pulse sequence.

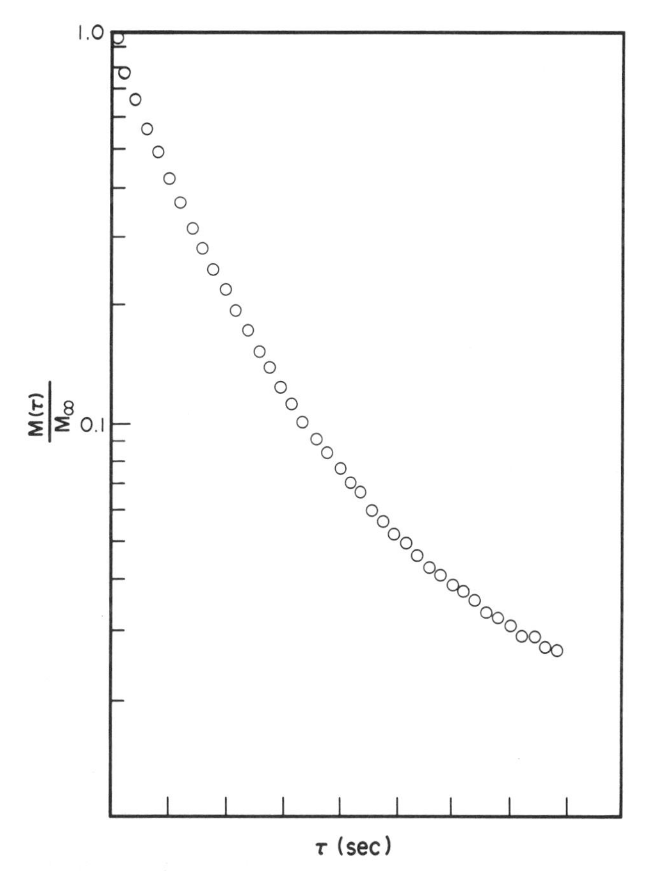


Fig. 4.9. Typical plot of $M(\tau)/M_\infty$ versus $\tau$ for the Hahn spin–echo or Carr-Purcell or Carr-Purcell-Meiboom-Gill sequence for a biological system. Notice the nonexponential nature of the plot; for pure water, such a plot would be a single straight line.

The time interval between one 90°–$\tau$–180° sequence and the other should be at least $5\,T_1$.

The technique described above does not yield true $T_2$ if there is molecular diffusion, as is the case in liquids and biological systems. Hence, Carr and Purcell modified the Hahn spin–echo method to reduce the contributions from molecular diffusion; this modification is referred to as the Carr-Purcell sequence.

#### 4.3.B. *Carr-Purcell sequence*

In this method, the rf pulses are applied in the sequence of 90°–$\tau$–180°–$\tau$–180°–$\tau$–180°–$\tau$. . . . All the rf pulses are applied along the $x'$-axis. There will be echoes at $2\tau$, $4\tau$, $6\tau$, etc., and these echoes will be

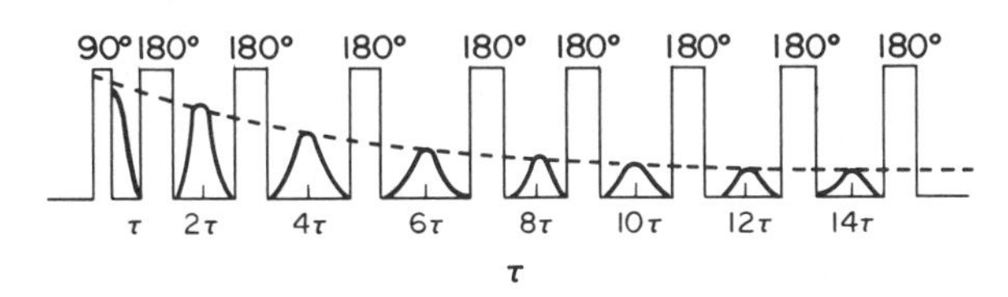


Fig. 4.10. Carr-Purcell sequence.

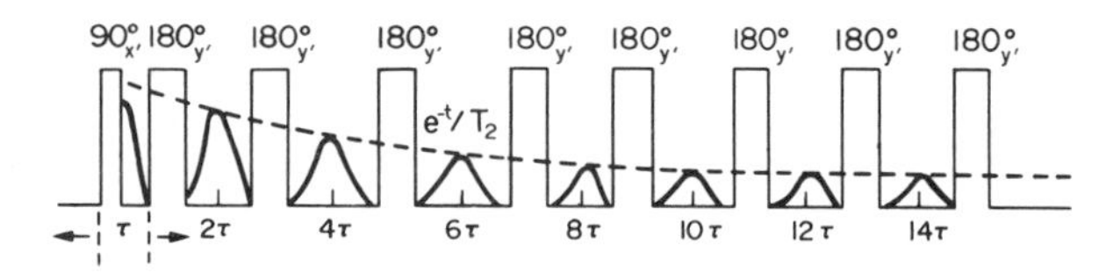


Fig. 4.11. Carr-Purcell-Meiboom-Gill sequence.

alternatively positive and negative if phase-sensitive detection is used. In diode detection, all echoes will have a positive sign. A typical Carr-Purcell (CP) sequence in the diode detection mode is shown in Fig. 4.10.

In the CP sequence, the echo height $M(\tau)$ is given by

$$M(\tau) = M_\infty\left[\exp\left(-\frac{t}{T_2} - \frac{1}{3}\gamma^2 G^2 D\tau^2 t\right)\right] \quad , \quad (4.12)$$

where $\gamma$ is the gyromagnetic ratio, D is the self-diffusion coefficient of the nucleus in the magnetic field gradient, and $G$ is the field gradient due to the inhomogeneity in the field. By making $\tau$ very short, it is possible to eliminate the contribution of the second term in the exponent, which arises due to diffusion. Then, $T_2$ can be obtained in the same way as described above, i.e., by measuring the height of the successive echoes and plotting echo heights versus time.

### 4.3.C. *Carr-Purcell-Meiboom-Gill method*

The Carr-Purcell-Meiboom-Gill (CPMG) sequence is essentially the same as the CP sequence, except that a phase shift of 90° is introduced in all the 180° pulses. This implies that if the 90° pulse is applied along the $x'$-axis, then all the 180° pulses are applied along the $y'$-axis. When the 180° pulses are applied with this phase shift, all the echoes form along the $y'$-axis (along $B_1$). The result is that all even-numbered echoes have correct amplitude, while the odd-numbered echoes have slightly reduced amplitudes (but not cumulatively). All the echoes will have the same sign, even in phase-sensitive detection mode. A typical CPMG sequence (Fig. 4.11) will look the same as that in Fig. 4.10 (CP sequence), even in the phase-sensitive detection mode. A detailed explanation of how this method compensates for slight missettings of the 180° pulse lengths can be found in ref. 1. The $T_2$ value is determined by measuring the height of the echoes as a function of $t$ and is obtained in exactly the same way as in the CP method.

### 4.3.D. *Comparison of the three methods*

The three methods of measuring the spin–spin relaxation time ($T_2$) are compared in Table 4.2.

### 4.3.E. *Tips to avoid systematic errors in the measurement of $T_2$*

(1) For an accurate measurement of $T_2$, the 180° pulses should be adjusted very accurately.

(2) While adjusting the 180° pulse, make sure that the phase of the rf is set correctly before adjusting the $t_w$.

(3) Use a spacing of at least 5 times $T_1$ between CP sequences (or CPMG sequences).

(4) Keep the sample in the most homogeneous part of the static field and also at the center of the rf coil (NMR coil), so that the rf field ($B_1$) is homogeneous.

## 4.4. APPENDIX

Quite often someone new to the field of NMR biomedical applications will hear terms such as "90° phase shift introduced into the 90° pulse or 180° pulse," or "an FID having a gaussian shape or a

**Table 4.2. Comparison of methods for measurement of spin–spin relaxation time ($T_2$)**

| Method | Advantages | Disadvantages |
|---|---|---|
| Hahn spin–echo | 1. Can be done with relatively simple spectrometers | 1. Takes a long time<br>2. Does not give true $T_2$ if there is molecular diffusion |
| Carr-Purcell sequence | 1. Relatively much faster, as a train of echoes is observed at one time<br>2. The effect of diffusion on the decay can be eliminated by making $\tau$ very short | 1. The 180° pulses should be set very accurately, since otherwise the cumulative effect of slight missettings can cause large errors in $T_2$ |
| Carr-Purcell-Meiboom-Gill sequence | 1. As fast as the Carr-Purcell sequence<br>2. Eliminates the effects due to diffusion | 1. The field/frequency ratio should be very stable<br>2. Needs extra electronics for introducing the phase shift |

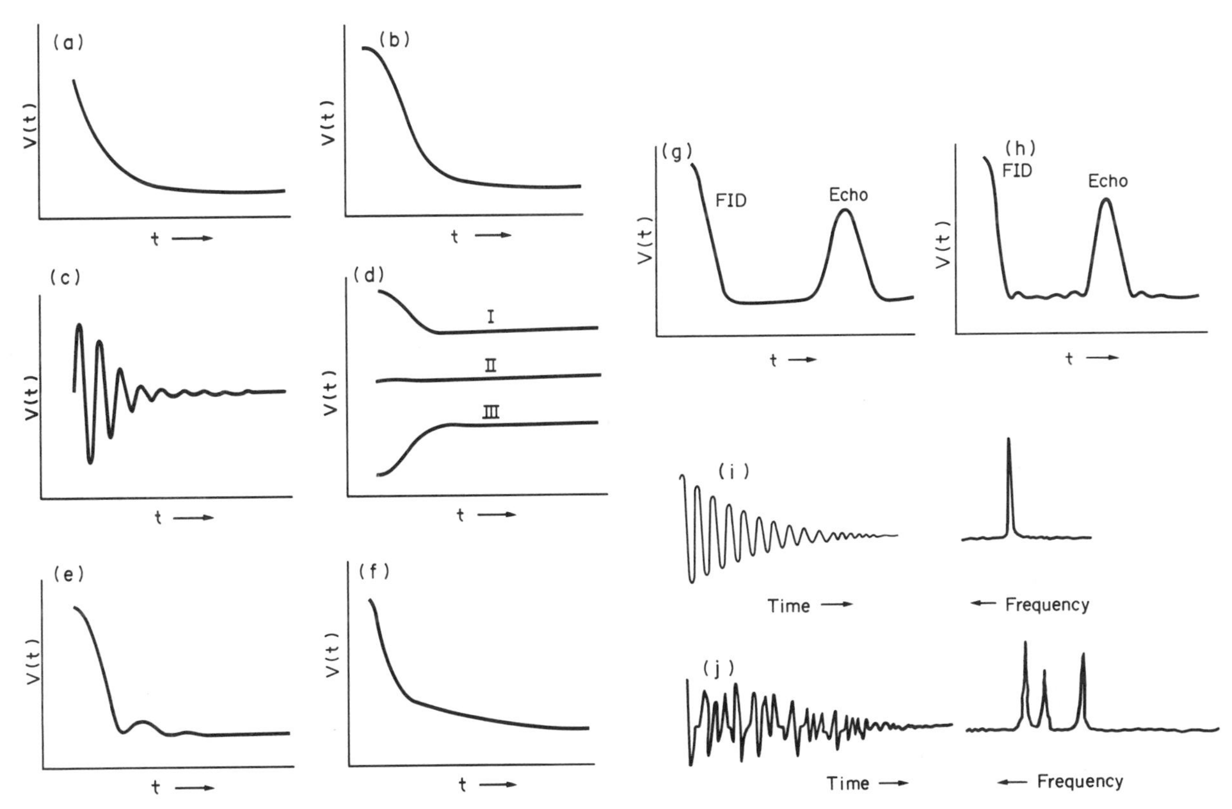


Fig. 4.12. **a:** Free induction decay (FID) signal when the decay is exponential, i.e., in the case of short $T_2$ values. **b:** FID when the FID has a gaussian shape. In the case of long $T_2$ (e.g., liquids), the FID shape is determined by the gaussian distribution of the inhomogeneities in the static magnetic field, resulting in a gaussian shape. **c:** FID signal when phase-sensitive detection is used and the field is set slightly off resonance. In diode detection, such beats will not be observed even if the field is off resonance. **d:** (I) FID signal when phase-sensitive detection is used and the phase difference between the reference and the signal is set to zero; (II) FID signal when phase-sensitive detection is used and the phase of the signal has a 90° phase shift with reference to the reference rf (i.e., a phase shift of 90° is introduced into the rf being applied to the spin system); (III) FID signal in phase-sensitive detection mode when the phase of the signal is different from that of the reference by 180°. **e:** FID signal when a field gradient is applied. **f:** FID signal when there are two different spin–spin relaxation times ($T_2$). **g:** FID and spin–echo signal when there is a gaussian field distribution, such as for $T_2$ values in liquids. **h:** FID and spin–echo signal when a field gradient is applied. **i:** FID consisting of one frequency component and its Fourier transform into the frequency domain. **j:** FID consisting of three frequency components and its Fourier transform into the frequency domain.

lorentzian shape." As an aid to the beginner, in this Appendix are shown the typical FID shapes under different conditions. In Fig. 4.12, the vertical scales represent the voltage $V(t)$, while the horizontal axis represents time $t$.

## 4.5. REFERENCES

1. Farrar, T.C., and Becker, E.D. *Pulse and Fourier Transform NMR*. Academic Press, New York, 1971.
2. Fukushima, E., and Roeder, S.B.W. *Experimental Pulse NMR: A Nuts and Bolts Approach*. Addison-Wesley, Reading, MA, 1981.
3. Rushworth, F.S., and Tunstall, D.P. *Nuclear Magnetic Resonance*. Gordon and Breach Science Publishers, Inc., New York, 1973.
4. Shaw, D. *Fourier Transform NMR Spectroscopy*. Elsevier, Amsterdam, 1976.
5. Mullen, K., and Pregosin, P.S. *Fourier Transform NMR Techniques: A Practical Approach*. Academic Press, London, 1976.

CHAPTER 5

# *PRACTICAL METHODS FOR BIOLOGICAL NMR SAMPLE HANDLING*

## 5.1. INTRODUCTION

The quality of information obtained from nuclear magnetic resonance (NMR) experiments on excised tissues and fluids depends on a number of physical and biological factors. While numerous investigators have conducted their own internal control experiments for many of these factors, only a few reports on these practical matters have been published. Chang et al.[1] followed changes in spin–lattice relaxation time ($T_1$) and spin–spin relaxation time ($T_2$) in rat leg muscle from within a few moments of excision through rigor mortis. This firm tissue showed little alteration in $T_1$ and $T_2$ for 3–5 h at room temperature. Tanaka et al.[2] reported different rates of change in $T_1$ values for normal and cancerous uterine tissues. While $T_1$ values changed little over the first few hours, tumor tissue $T_1$ values decreased and normal tissue $T_1$ values increased after 5–15 h of refrigeration. Valensin et al.[3] found no change in $T_1$ for human cultured cells for 36 h. Relaxation times for muscle[4–6] and kidney[4,7] change after freezing and thawing because of ice crystal disruption of macromolecular organization. Freezing hysteresis[8] precludes storing cells or tissues for NMR studies in a frozen state.

We will report here our experience with animal tissues, human biopsy material, cultured cells, and blood components. We will describe, for a number of different types of samples, variation due to sample preparation, methods of storage, length of storage, and temperature. Protocols which have successfully resulted in reproducible relaxation times under various types of conditions are provided as examples.

NMR spectroscopy is rapidly becoming a useful and practical tool in medical diagnosis.[9,10] Water proton signal intensity and relaxation times are used to measure water content in biological samples and to quantitatively determine the biophysical properties of that water. The spin–lattice relaxation time $T_1$, the spin–spin relaxation time $T_2$, and the self-diffusion coefficient $D$ of water protons can yield additional information about the relative mobility of water molecules in a sample. These three parameters can distinguish normal tissues from pathological conditions in excised biopsy material, and can now be used *in vivo* in whole body NMR imaging machines.[9]

The first measurements on biological tissues, such as red cells, cervical mucus, and vaginal cells, were done by Odeblad in the 1950s and 1960s. Since 1971, numerous investigators[2,11–15] have utilized NMR to distinguish cancerous tissue from the normal tissue of origin. Such work has led to the development of NMR whole body imaging for the localization of tumors and the distinction of various body dysfunctions.[9,16] During the decade of the 1970s, certain problems were recognized as NMR was used by more and more investigators. First, the machinery needed to be properly tuned and operational for biological samples in terms of frequency, reasonable sample size, and temperature of operation such that the excised samples (as well as whole bodies) would not deteriorate, become denatured, or degrade during the measurements. In addition, many practical methods of sample acquisition and handling were devised that improved the reproducibility of NMR numbers. The inherent variation in animal and human groups makes the interpretation of NMR data extremely difficult unless sample handling protocols are as carefully designed and implemented as any other part of the experimental design. In our laboratory, we have spent considerable time on the standardization of sample protocols, and we have collected data on animal and human tissues, cell suspensions, and blood sera. Some practical hints are given in this chapter for the standardization of protocols to obtain the most reproducible biological NMR results for diagnostic and research purposes.

## 5.2 GENERAL CONSIDERATIONS

### 5.2.A. *Magnetic field considerations*

The term "nuclear magnetic resonance" reminds one that all NMR machines include a source of a strong magnetic field. This may be a permanent magnet, an electromagnet, or a super-conducting magnet. It is a good idea for the biologist to keep this in mind, if for no other reason than to save the cost of several new watches. (Digital watches are usually not susceptible to magnetic effects, but credit cards with magnetic strips can be wiped clean if passed near the magnet.) The homogeneity of the magnetic field is an important factor in the production of NMR data and is discussed elsewhere. However, the theoretical perfection of the magnetic field itself may also be a problem because it is affected by the presence of masses of magnetic materials in the area of the laboratory (such as drainage pipes in walls, reinforcing steel in concrete, and other instruments). Specific methods for correcting field problems have been published.

The magnet may also impede sample preparation by grabbing scissors, tweezers, and wire-frame eyeglasses from technicians. Especially with the larger whole body magnets, extraneous ferromagnetic metal objects must be removed from the laboratory. Stories are told of large magnets grabbing floor-waxing machines and vacuum cleaners, as well as nails and pens from pockets. Any magnetic material in the sample will also affect NMR parameters detrimentally. One source of such material would be rusty razor blades or scissors used to mince tissue. Only very clean stainless steel scalpel blades and clean, dry stainless steel tweezers and scissors should be used to prepare NMR samples. For safety and to prevent contamination, the best materials for containing NMR samples are clean, dry glass or plastic. These precautions will make life with a magnetic field much easier.

On rare occasions the presence of magnetic or paramagnetic material, or free radicals in the sample itself may reduce $T_1$ and $T_2$ values below those expected. For example, both fungus and plants are often grown in solutions containing ferric chloride as a nutrient salt. The magnetic ferric ion acts as a relaxation center and true measurements cannot be achieved with such materials. However, when investigated, biological animal tissues and blood, even hemolyzed serum, contain insufficient amounts of *free* paramagnetic materials to affect NMR relaxation times[17–19] (see Table 5.9). Apparently, most heavy metal ions are processed in the body and are bound to proteins such as hemoglobin, apoferritin, or ceruloplasmin where they do not act as strong relaxation centers. It has been suggested, however, that malignant melanoma tissue may contain high enough levels of the free radical form of melanin to suppress $T_1$ and $T_2$ values in such tumors. Not enough is known about the effects of vitamin and mineral supplementation, antibiotics, and unusual biochemical products of diseases of metabolism to discount the possible contribution of free radicals and paramagnetic materials to NMR values. In fact, some investigators are seeking to develop paramagnetic NMR imaging contrast agents to improve resolution of images by reducing extracellular relaxation times.

**5.2.B.** *Sample hydration and fat content*

Any action taken during sample preparation which may change the water content of the samples could potentially alter relaxation time values and is to be avoided. Evaporation into the air can be prevented to some degree by transporting samples in tightly sealed containers of approximately the volume of the sample itself. For example, 800 ml of sera should be sealed in a screw-capped 1-ml Pro Vial instead of a 10-cc Vacutainer tube.

In the following sections, certain problems with loss of water from samples will be pointed out. Obviously, tissues should not be rinsed with saline or blotted with surgical towels or paper. To check the variation of water content in tissues or cells, portions of the wet samples may be placed in dry glass tare vials with ground glass stoppers (scintillation vials washed six times, without plastic caps but sealed with aluminum foil squares, may also be used) and carefully weighed. Open tares are dried at 105°C in an oven and transferred to a desiccator to cool while still open. After cooling, tares are closed and quickly weighed. Samples are dried and reweighed every 24 h until they come to a constant weight. The number of days of drying depends on the sample size and type (i.e., 0.1 g of skeletal muscle in a 15-ml tare usually comes to constant weight in 3 days). Ether-extractable fat may be determined by filling the tare, after drying and weighing, with reagent-grade petroleum ether, capping it, and extracting for 24 h under a fume hood. The ether is decanted and refilled three times; then the tare is redried to a constant weight. Care should be taken in this procedure because the ether is highly flammable. Fat determinations are recommended because mobile protons on lipids may make a contribution to NMR proton relaxation times. Human biopsy material is often difficult to work with because of high fat content, and interpretation of human data will probably require knowledge of the fat content.

**5.2.C.** *Control experiments*

NMR data collected on a group of samples or individuals will display certain variations (between tissues or between individual patients). Therefore, as much data as possible should be collected. Each investigator will want to check the reproducibility of results obtained with his or her machinery by repeatedly measuring a single sample. However, the reproducibility with a sample of doped water is not always the same as for a piece of fatty tissue. Machine variation should be measured for each type of sample and several determinations of each sample averaged, if possible, for greater accuracy.

It is recommended that a standard set of control experiments be run for each and every type of tissue or sample being studied. These should include: (a) reproducibility of water content during the steps of sample acquisition and handling, (b) reproducibility of measurements on the same sample and small pieces of the same sample, (c) time stability at various holding temperatures and at machine temperature, and (d) reproducibility with storage (for example after freezing and thawing of sera). When these conditions have been determined for both normal and pathological states, one may proceed to collect the data for the study. However, natural biological variation will far exceed machine error, and careful attention to the biological variables will be necessary to keep total measurement error within ±10%. All the machine control and calibration will be worthless if, for example, muscles from a set of frogs are measured, some of which are male and some female, or half of which have not been fed in 3 weeks. Both mechanical and biological control experiments and errors should be quoted in published reports of studies.

## 5.3. SELECTION AND HANDLING OF TISSUES

### 5.3.A. *Animal tissues*

**5.3.A.*1*.** *Animal models.* Inbred animal models of disease offer numerous advantages over diverse human populations when an investigator wishes to prove basic principles with a technique such as NMR. The majority of NMR tissue experiments have been conducted in frog, rat, and mouse models. One reason for this is illustrated in our study of tumor biopsy samples from mouse mammary adenocarcinoma. These were compared with surgical biopsy material of human mammary adenocarcinoma (Table 5.1). The $T_1$ range for human samples is much greater than that for the animal models, possibly due to the population variation among humans, the diversity of tissue types, the infiltrating nature of cancer, the selection of tissue from the excised sample of the larger human breast, and uncontrolled variables of age, heredity, and general health. It is apparent that the mouse samples have a much smaller range than the human samples.

NMR scientists have collected a reservoir of $T_1$, $T_2$, and $D$ data on many types of animal tissues in normal and diseased states. A few examples are given in Table 5.2. It should be kept in mind that these results were collected at different resonance frequencies and various temperatures, and with different sampling protocols. In several cases, values for the same animal tissue vary among laboratories for undetermined reasons.

**Table 5.1. Animal models versus human populations**

| Mammary tissue | $T_1$ range* (ms) Mouse mammary cancer | Human breast biopsies |
|---|---|---|
| Normal breast | 207–343 | 273–730 |
| Preneoplastic or hyperplastic nodules | 213–482 | 217–906 |
| Carcinoma | 409–902 | 276–876 |

*Measured at 30 MHz and 25°C (from ref. 12 and Hazlewood et al., 1976, unpublished data).

**5.3.A.*2*.** *Preparation.* Questions may arise as to the method of sacrifice used for animals whose tissues are to be taken for NMR study. Both cervical dislocation and intraperitoneal injection of Nembutal in rats, mice, and chickens have produced similar NMR numbers. Stress and dehydration of the animal should certainly be avoided. Ether anesthesia is not recommended since it may produce great stress during death and sometimes causes pulmonary edema. Rapid and careful sample preparation seems to be the most important parameter.

**Table 5.2. NMR data for $T_1$ values from various biological systems**

| Tissue (resonance frequency) | $T_1$ (ms) Normal | Neoplastic | Reference |
|---|---|---|---|
| Human (24 MHz) | | | |
| Lung | 788 | 1,110 | Damadian et al., 1973[32] |
| Breast | 367 | 1,080 | Damadian et al., 1974[33] |
| Skin | 616 | 1,047 | |
| Stomach | 765 | 1,238 | |
| Intestine | 641 | 1,122 | |
| Muscle | 1,023 | 1,413 | |
| Bladder | 891 | 1,245 | |
| Bone | 554 | 1,027 | |
| Liver | 570 | 823 | |
| Spleen | 701 | 1,113 | |
| Ovary | 989 | 1,282 | |
| Thyroid | 882 | 1,072 | |
| Mouse (100 MHz) | | | Iijima et al., 1973[34] |
| Muscle | 850 | 880 | |
| Liver | 420 | 480 | |
| Brain | 840 | 930 | |
| Human (30 MHz) | | | Medina et al., 1975[15] |
| Breast | 682 | 874 | |
| Human (25 MHz) | | | Kasturi et al., 1976[23] |
| Breast | 415 | 819 | |
| Stomach | 841 | 997 | |
| Cervix | 825 | 1,089 | |
| Skin | 360 | 1,037 | |

Tissues should be removed rapidly from carcasses with clean, dry instruments and placed on Teflon cutting boards or watch glasses. If several tissues are taken, some may be covered with squares of nonabsorbent Parafilm until ready to cut. A piece of each organ, about 2 g in weight, should be taken to prevent the edges from drying while the tissue is being stored.

*Resist the temptation*
*to place the tissues on ice*
*or in the refrigerator.*

The natural inclination is to cool and freeze everything to prevent decomposition of vital biochemical species. Yet, NMR relaxation times are more than a simple function of water content.[20–25] It is not only the amount of dry solid, but the type of molecule and its conformation that determine the extent of water–macromolecule interactions. We have previously shown that cooling below 15°C may depolymerize cytoskeletal elements and change values for the self-diffusion coefficient for water in cells.[26]

**5.3.A.*3*.** *Storage parameters.* Only after testing the effects of cold temperatures over several hours should cooling be added to a protocol. *Storage overnight in the refrigerator is not highly recommended for many kinds of mammalian samples.* While numerous investigators do refrigerate tissues and get similar values of $T_1$ over many hours with firm tissues, such storage of soft tissues, cells, or sera is not recommended. Even skeletal muscle, which has very stable NMR values for hours after death, does not give the same values after storage in the refrigerator overnight (Tables 5.3 and 5.4). Other samples change at various rates over short (Table 5.4) and long (Table 5.3) storage. The $T_1$ and $T_2$ parameters may be affected by drying of the samples or by biochemical degradation. If samples must be stored, however, refrigeration (after placing them in the NMR tube) may be better than freezing or room temperature.

*Firm tissues* may be excised and placed on Teflon blocks for cutting. In general, small 1 to 2-mm$^3$ cubes are cut by firm, quick strokes of the scalpel. (Mincing with scissors forces a great deal of fluid out of the tissue.) Cubes of tissue are loaded into an NMR tube and gently pushed to the bottom with a long Wilmad Glass NMR pipette (only Wilmad pipettes will reach the bottom of Wilmad NMR tubes). Glass rods do not allow air to escape as the tissue is forced into the tube. Some fluid may be extruded on to the sides of the tube, but with firm tissues this is probably extracellular fluid. We have obtained very reproducible numbers on skeletal muscle using this technique. A sample protocol for the preparation of firm tissues, such as rat gastrocnemius muscle, is given in Example A, below.

*Soft tissues* tend to have high fluid content (>80%), and this fluid will extrude or ooze from the tissue during preparation. Fluid may also be lost as tissues are forced down the sides of long narrow NMR tubes. In handling soft tissues, the best procedure is to purchase at a grocery store some clear, plastic straws (4 mm in diameter or of a size to fit inside glass NMR tubes); transparent, colored ones also may be used. The sharp edges of the straw can be used to cut into soft tissue and secure a plug of tissue in the end of the

**Table 5.3. Effect of long-term refrigeration and storage on $T_1$ parameters in tissue**

| | $T_1$ values (ms) of tissues stored at 5°C* | | | | | | | | |
|---|---|---|---|---|---|---|---|---|---|
| | | | | | Carcinoma of the breast (cell lines) | | | | |
| Storage time | Arm muscle | Normal breast | Nipple | Lymph node | 2221 | 2156 | 2024 | 1113 | 1190 |
| Fresh | 748 | 730 | 546 | 636 | 650 | 490 | 375 | 621 | 695 |
| 8 h | | | | | | | 360 | | |
| 10 h | | | | | | | | | |
| 14 h | | | | | | | | | |
| 19 h | | | | | | 505 | 375 | | 682 |
| 22 h | | | | | | | | | |
| 24 h | 710 | | | | | 502 | | | 697 |
| 48 h | 688 | 673 | 448 | 548 | 420 | 470 | | | 696 |
| 72 h | | | | | | | | | |
| 120 h | | | | | | 453 | | 543 | 716 |
| $T_1$ (%) | −8 | −7.8 | −17.9 | −13.8 | −30 | −7.5 | 0 | −12.5 | +18.9 |

*Measured at 30 MHz and 25°C [Beall et al. (1981, unpublished data)]; tissues were stored in NMR tubes at 5°C between measurements.

*Note:* Different tissues' $T_1$ values change at different rates under refrigeration. When dealing with any one particular type of tissue, appropriate control experiments should be run to determine the stability of that tissue in storage.

**Table 5.4. Effect of storing tissue biopsy samples at 25°C**

| | $T_1$ (ms) | | | |
|---|---|---|---|---|
| Time | Human breast cancer* | Rat leg muscle* | Uterus (normal)† | Uterus (tumor)† |
| 0 | 375 | | | |
| 30 min | 360 | 608 | | |
| 45 min | 360 | | | |
| 1 h | 360 | 610 | | |
| 1.25 h | 360 | | | |
| 1.5 h | 360 | 620 | | |
| 1.75 h | 360 | | | |
| 2 h | 360 | 640 | | |
| 3 h | 365 | 680 | | |
| 4 h | 370 | 670 | | |
| 5 h | 380 | 660 | 200 | |
| 6 h | | 650 | 210 | |
| 9 hr | | 650 | 220 | 540 |
| 18 h | | 650 | 270 | 450 |
| 28 h | | 650 | 270 | |

*Measured at 30 MHz and 25°C. Data from refs. 1, 20, and Beall et al. (1981, unpublished data).

†Measured at 10 MHz and 25°C. Data from ref. 2.

*Note:* Some tissue samples are very stable in storage at 25°C or 5°C for several hours. Others, which undergo a physiological change in function such as rigor mortis, may change rapidly after death. In general, however, NMR parameters are stable in excised tissue, if not dehydrated, for several hours.

straw. A sharp, clean scalpel can be used to cut the plug off and the straw can then be forced down into a soft sheet of dental wax or a cut-to-size Teflon plug can be inserted. The soft tissue is thus loaded from the proper end and loses little fluid. The straw is inserted inside a glass NMR tube and the sample adjusted to fit in the proper position in the coil. After measurements, the sample can be forced into a tare vial by blowing a strong puff of air into the opposite end of the straw. (Straws, Parafilm, and caps should be tested for any effect on proton signal.) Example B, below, demonstrates how soft human breast biopsy material can be processed.

Animal experiments have taught us many things about handling tissue for NMR. Tissues can be divided into those designated "soft" (brain, liver, spleen, kidney, intestine) and those designated "firm" (skeletal muscle, cardiac muscle, skin). (See Examples A and B, below.)

### 5.3.A.4. *Example A: Protocol for firm tissue*

1. *Equipment:* Clean, dry surgical scissors; two pairs of forceps, one toothed, one narrow with long tips; clean, new disposable scalpel; 4 in × 4 in square Teflon cutting board; Wilmad long glass pipettes (12 in); Wilmad 505 thin-walled glass NMR tubes; Parafilm.
2. *Animal:* Sacrifice animal by injection of ~30 mg/kg sodium pentobarbital (Nembutal) in abdomen (IP); allow ~15 min for anesthesia to take effect. Place animal abdomen-down on counter, grasp tail in right hand, place thumb and forefinger firmly behind skull, jerk tail to dislocate cervical vertebrae. (Dispose of carcass in required manner.)
3. *Tissue:* Remove ~2 g of selected, *firm* tissue from organ. In this example, the tissue will be the gastrocnemius muscle from the lower leg of a rat.
   a. Use scissors to clip skin around ankle.
   b. Grasp skin in toothed forceps and pull back above the knee.
   c. Locate ligament of insertion of muscle, place blunt forceps under it and pull toward knee, freeing muscle from other tissue.
   d. With forceps (not fingers) grasp muscle and cut connections. Muscle may still pulsate or contract. Place on Teflon board or cover with Parafilm.
4. *Preparation for NMR:* Clean tissue by dissecting away all extraneous material; pick a visually uniform section of tissue and slice away debris until ~15-$mm^3$ cube is left. With firm downward strokes of scalpel blade, cut into cubes ~1–2 $mm^3$ in size. (Mincing with scissors causes a release of fluid.) Within a few seconds, transfer some pieces to the top of a 505 NMR tube a few at a time (and some to a vial for water and fat content measurements). Push to bottom of NMR tube with a Wilmad pipette. Pack as close as possible to a 5–10mm depth, for most machines. (Larger samples from most small animals will give a heterogeneous mixture of tissue types.) Seal with Parafilm (not with plastic caps) and store at room temperature (20–30°C). Measure as soon as possible, but for reproducibility, establish a standard delay time. It is best to prepare each sample as needed; however, when several animals are prepared together, time the steps of preparation and measure all samples at the same number of minutes after excision from the animals.
5. *Recurrent Problems:*
   a. Experience has shown that since NMR tubes with tissue cannot be centrifuged, packing can cause nonsymmetrical free induction decay (FID) values and echoes. Pack samples tightly to the same preset height, avoid air bubbles, and keep fat to a minimum.
   b. The machine should be retuned for each sample (this is absolutely required if samples are

different tissues). Rotating the sample tube will allow adjustment of the 90° (FID) pulse to be as symmetrical as possible. Excess sample height can make it impossible to get the 180° pulse flat (remove some sample with pipette or wooden swab stick).

6. *After Experiment:* Remove sample pieces by flushing tube with 10% buffered formalin from a Wilmad pipette. Tissues will float to top; place them in vial of formalin for pathological analysis. Tubes must be soaked thoroughly with laboratory detergent solution, then rinsed from the bottom with water; then run distilled water through the tubes and the Wilmad pipettes. (Commercial NMR tube washers are available.) Tubes must be oven-dried and stored in a desiccator to prevent condensation inside the tubes in humid air. (Long Wilmad pipettes may also be washed and reused.)

**5.3.B.** *Human biopsy material*

**5.3.B.*1*.** *Surgical suite problems.* Human tissue biopsies are secured from surgery or the morgue. The first steps of sample preparation are therefore out of the NMR investigators' hands. No matter how interested and enthusiastic clinical colleagues may be in NMR research, their priorities must place the patient's welfare first. In the confusion of the surgical suite, many individuals may handle a sample before it gets out the door to the research technician. The surgeon always has the responsibility to visually examine the tissue before it is removed from the room. Typically, such a scenario would involve the removal of a piece of tissue with forceps from the wound or incision, which may be irrigated with saline. The tissue is placed on a sterile, cotton surgical towel, the blood washed away with saline, and the tissue fingered and rolled on the absorbent surface. If the patient requires some immediate attention or the surgical procedure requires continuous effort, the biopsy material may sit on the towel for a few minutes. Then the surgeon may turn and cut off some tissue for pathology and some for the research project. Even if the tissue for research is delivered in the clean, dry, sealed container provided, the only way to really assure proper handling is for the NMR investigator to scrub up and attend surgery.

**5.3.B.*2*.** *Pathological confirmation.* Human biopsy material often is not a uniformly textured piece of tissue. It will arrive with attached blood vessels, connective membrane, fluid-filled cavities, blood clots, and globs of fat. The research technician will need instruction as to what portion of the tissue is of interest; it should then be dissected away from the contaminating elements. We found this to be especially difficult with human mammary adenocarcinoma. This type of cancer often infiltrates surrounding tissue and is mixed with fat, cartilage, and necrotic tissue. The investigator must make decisions about tissue selection. In fact, it is probably unacceptable to divide selected biopsy material in half, and then send half to pathology and hold half for NMR measurements. *The exact piece of tissue used for NMR study should be sent for pathological analysis.*

Completed NMR sample tubes should be filled with buffered formalin to wash tissue up from the bottom. These tissue pieces are transferred to scintillation vials filled with formalin and sent for confirming pathological analysis. In our experience, the ratio of cancer cells to normal cells and fat may vary dramatically among even a few microtome slices. In human breast biopsies of small nodules, the content of cancer cells in the material was, at best, 15–60% (Table 5.5). Fat content also varied widely among tissues and contributed to lower apparent $T_1$ and $T_2$ values for such tissues (Table 5.5). While breast is certainly a difficult tissue to study as excised samples or in whole body imaging, there are many other tissues which will create as many problems in handling and data interpretation. Certainly, brain, kidney, and all glandular tissue are in this category.

**5.3.B.*3*.** *Example B: Protocol for human breast biopsy*

1. *Equipment:* Clean, dry container of the approximate volume of sample, with tight lid (clearly labeled "DO NOT REFRIGERATE"); clean, dry scissors; forceps; disposable scalpel; 4 in × 4 in Teflon cutting board; NMR tubes; long Wilmad pipette; some transparent 4-mm diameter straws; sheet of dental wax or Teflon plug for straw.
2. *Surgical Suite Procedure:* Ask the surgeon to do *nothing* to the sample (stress that no absorbent towels, saline, or rubbing be used). Ask the surgical team the approximate size of the sample they will get. Supply a 2-ml (Pro Vial), screw-capped vial for small samples, an unused orange-capped Vacutainer test tube for 3–10 ml samples, or a plastic snap-capped dish for larger ones. Preselection of pathological tissues by the surgeon can relieve the technician of the decision as to which portion of a large tissue sample to measure. Usually, a piece 1–2 $cm^3$ in size will be adequate for several NMR samples. (Large NMR biopsy samples cause confusion due to the

**Table 5.5 Effect of tissue fat content on NMR values in human breast biopsies, as determined by histological serial sections**

| Sample | % Fat | % Carcinoma | % Other | $T_1$* | $T_2$* | Diagnosis |
|---|---|---|---|---|---|---|
| 2234 | 0 | 50 | 50 | 973 | 125 | Cancer |
| 2233 | 0 | 20 | 80 | 817 | 100 | Cancer |
| 2267 | 0 | 15 | 85 | 670 | 58 | Cancer |
| 1295 | 0 | 0 | 100 | 878 | 64 | Fibroadenoma |
| 1281 | 0 | 10% other | 90 | 850 | 63 | Fibroadenoma |
| 2221 A | 5 | 25 | 70 | 680 | 93 | Cancer |
| 2209 A | 5 | 60 | 35 | 536 | 134 | Cancer |
| 2222 | 5 | 20 | 75 | 421 | 67 | Cancer |
| 2209 B | 5 | 50 | 45 | 316 | 64 | Cancer |
| 2211 A | 5 | 80 | 15 | 246 | 34 | Cancer |
| 1260 | 10 | 0 | 90 | 557 | — | Fibrosis |
| 1290 | 10 | 0 | 90 | 590 | 34 | Fibroadenoma |
| 2221 B | 10 | 30 | 60 | 614 | 72 | Cancer |
| 1270 A | 10 | 30 | 60 | 604 | 49 | Cancer |
| 1254 | 10 | 45 | 45 | 695 | 52 | Fibroadenoma |
| 1257 | 25 | 5 | 70 | 648 | 42 | Fibroadenoma |
| 1270 B | 60 | 5 | 35 | 500 | 80 | Cancer |

*Measured at 30 MHz and 25°C. Data from Hazlewood et al. (1979, unpublished data).

*Note:* The composition of tissue has a dramatic effect on NMR values. $T_1$ values depend on the percent malignant tissue and the percent fat in the sample. The higher the percentage of fat in the tissue, the lower is the apparent $T_1$ in many cases, even when substantial numbers of tumor cells are present. For this reason, every effort should be made to dissect away fat tissue from the tumor, and fat determinations should be done to aid in data interpretation.

heterogeneity of the sample.) You or your technician should be waiting to receive the sample (If it goes to the pathology department, some well-meaning person will place it in the refrigerator.) Clearly label all containers by patient number; when you pick up the sample, check that this has been done. (If samples must be held from early surgery until your work hours, it may be best to allow them to be refrigerated in a marked box in the pathology department refrigerator. Each investigator should check the effect of refrigeration on their type of study tissue.)

3. *Preparation for NMR:* Prepare firm tissues as given in Protocol A, above. For soft tissues such as brain, liver, kidney, and breast, examine the biopsy sample and dissect away fat, blood vessels, etc., until a fairly uniform piece is left on the cutting board. Take a portion for water and fat content determinations. Then position a transparent straw near the tissue and load by pushing one ~5-$mm^3$ piece into the open end. An alternative is to use the sharp edges of the straw to cut a plug of tissue and to slice the tissue off with the scalpel. Seal straw by pushing it into soft sheet of dental wax (available from medical supply stores) or pushing a tight-fitting Teflon plug into straw. (Test for effects of the wax or straw in your machine.) Place loaded straw into a 505 Wilmad NMR tube, or a larger one if necessary. After experiment, unload tissue into a vial of formalin by blowing a sharp puff of air into the straw.
4. *Problems*:
   a. The greatest problem with biopsy material is knowing what portion of the tissue is pathological and being able to dissect this out for NMR study. Training (by a pathologist) for sample selection is usually necessary.
   b. Fat is a large problem, especially in breast biopsy material, and can cause apparent reductions in $T_1$ and $T_2$. Try to dissect away as much fat as possible, and do fat determinations on tissues to aid in data analysis.
   c. Pathological analyses should be done on the same piece of tissue used for NMR. Also, several sections should be examined to determine the proportions of normal and pathological tissue.

**5.3.C.** *Cells and cell suspensions*

NMR measurements are also being made on cells and cell suspensions in efforts to improve the diagnostic ability of NMR and to understand the molecular basis of relaxation time differences.

**5.3.C.*1*.** *Blood cells.* For humans, the most commonly studied cells are red blood cells and lymphocytes. Interactions between water and the hemoglobin matrix of red blood cells have been used to distinguish normal erythrocytes from the sickle cells in sickle cell anemia.

Examination of the whole blood of cancer patients has also yielded some variations from normal, probably due to a systemic effect of cancers on the sera.[27–29] Lymphocytes from leukemia patients display measurable differences from normal white cells on NMR examination.[30]

Whole blood for NMR measurements should be collected in purple-capped Vacutainer tubes. The presence of anticoagulant may have a slight osmotic effect on cell water content; however, we are familiar with this type of treated cell. The most difficult problems with blood cells are the separation of cell types without their exposure to osmotic shrinking or swelling, and the avoidance of placing blood samples on ice. Cooling red blood cells causes swelling, and white blood cells may contain cold labile cytoskeletal elements. Separation of white blood cells on density gradients may result in osmotic shrinkage. Because it cannot be assumed that the amount of shrinking and swelling is exactly the same for normal and pathological cells, even the precaution of handling all cells in the same manner may not produce distinguishing results. The simplest way to examine blood cells is to look at whole blood at room temperature. If measurements are made on blood that is at 37°C, there is the possibility of altering NMR parameters due to the activity of enzymes.

Some groups have added $MnCl_2$ to the extracellular volume to reduce relaxation times to very small values outside the cells. In this way, the intracellular water is more NMR-visible. Since red cells are only slowly permeable to $MnCl_2$, this method works rather well; however, it will not work for permeable cell types, such as most cancers.

For white blood cells, collection of the buffy coats and recentrifugation through isotonic serum in tall, narrow tubes or capillary tubes may produce good samples. By using narrow capillary tubes, most red cells will be left at the bottom and the white cell column can be broken away from the red cells.

In looking at cancers and tumors with NMR, it would be useful to know how much of the relaxation time differences are attributable to intracellular organizational differences and how much to hydration differences, vascularization, necrotic tissue, and fat content. Preparation of cultured cells and single cell suspensions is outlined in Example C, below.

**5.3.C.2.** *Primary cultures.* The most appropriate comparison, however, is between normal and cancerous cells, isolated and grown from unpassaged primary epithelial cells. When all connective tissue, fat, and extraneous cells were removed, primary cell cultures of normal, preneoplastic, and neoplastic mouse mammary cells still displayed significant NMR relaxation time differences (Table 5.6). These differences are not due to differences in cellular hydration. High voltage electron micrographs suggest subtle differences in the organization of the fibrous proteins of the cytoskeleton in the three cell types. If this is true, a subtle difference in cellular macromolecular organization could account for detectable differences in water–macromolecule interactions that are reflected in relaxation times. These types of cells can be held for several hours (Table 5.7).

**5.3.C.3.** *Cultured cells.* In a series of human breast cancer cell lines, we have demonstrated a wide range of NMR relaxation times. In these cells $T_1$ differences correlated well with the percentage of fully polymerized microtubule complexes in the cells (Table 5.8). $T_1$ values did not correlate with hydration (Table 5.8). Intracellular macromolecular organization is critical to relaxation times. Cooling of these cells below 15°C caused elevation in $T_1$ and $T_2$, even when cells were brought back up to 25°C for measurements. Depolymerized microtubules could not reform in anoxic cell pellets. Cultured cells are delicate and require special handling, like that outlined in Example C, below, to achieve reproducible results. Cell viability testing by the trypan blue exclusion method can be used to

**Table 5.6. NMR data from primary mouse mammary cell cultures**

| Tissue | Whole tissue* | | Primary cell cultures† | | |
|---|---|---|---|---|---|
| | $T_1$ (ms) | $T_2$ (ms) | $T_1$ (ms) | $T_2$ (ms) | % $H_2O$* |
| Normal, pregnant | 380 ± 41 | 39 ± 2 | 916 ± 24 | 158 ± 6 | 90.8 ± 0.4 |
| Hyperplastic alveolar nodules (D2-HAN) | 451 ± 21 | 53 ± 1 | 1,029 ± 24 ($p < 0.005$)‡ | 187 ± 7 ($p < 0.01$) | 90.0 ± 0.5 ($p > 0.7$) |
| Mammary adenocarcinoma (D2) | 920 ± 47 | 91 ± 8 | 1,155 ± 42 ($p < 0.001$) | 206 ± 8 ($p < 0.001$) | 91.4 ± 0.2 ($p > 0.03$) |

*Relaxation times and % $H_2O$ content are means ± SE.
†$n$ = 15 samples of each type, for a total of 45 samples.
‡$p$ calculated between test group and normal group by Student $t$ test.
Data from refs. 12 and 21.

**Table 5.7. Effect of storing HeLa cell pellets on the NMR parameters**

| Random cell pellet measurement | Time after harvest | $T_1$ (ms) of sample stored at 25°C* | $T_1$ (ms) of sample held at 5°C* |
|---|---|---|---|
| 1 | 30 min | 635 | 657 |
| 2 | 1 h | 637 | 657 |
| 3 | 2 h | 640 | 657 |
| 4 | 3 h | 655 | 657 |
| 5 | 4 h | 630 | 630 |
| 6 | 2 days | 510 | 500 |

*Measured at 30 MHz and 25°C.

*Note:* Pellets of cultured cells may be stored for several hours at room temperature. If stored in the refrigerator, values may be reproducible but inaccurate, since the cytoskeleton depolymerizes. Pellets should not be held for more than 4 h before measurement. (Valensin et al. claim to store human cancer cells, HEp-2, for up to 36 h without observing changes in $T_1$.[3])

estimate the ratio of live to dead cells after NMR measurements.

### 5.3.C.4. *Example C: Protocol for cultured HeLa cells*

1. *Equipment:* Hard Teflon spatula or edge (not a rubber policeman); 10% trypan blue solution; slides; coverslips; microscope; NMR tubes and centrifuge adapters; Wilmad NMR thick-walled 503 (3 mm inner diameter) tubes; cotton swabs on long wooden sticks; Wilmad pipettes. (To make centrifuge adapters for NMR tubes: Fill 12-ml plastic conical-bottomed centrifuge tubes with liquid epoxy resin. Grease NMR 503 tubes with silicon grease and suspend them in the center of the resin. When resin is hardened, it can be pulled away from the sample tube, and, after cleaning, the adapter can hold NMR tubes for centrifugation.)
2. *Cell Preparation:* For cells grown in ascitic fluid from animals or in suspension cultures, place a thick slurry of cells directly in NMR tubes in adapters and centrifuge at room temperature. For cells grown as monolayers in dishes, do not use the usual trypsin enzyme release because it changes membrane permeability to ions and water and causes cells to swell. We have successfully used a hard Teflon edge to scrape cells from dishes while bathing them in culture medium. Cells may be returned to a temperature of 37°C for ~30 min in the incubator to heal any membrane damage. Monitor viability of cells by the trypan blue exclusion test: mix a drop of dye with a drop of cells; count blue, dead cells against cream-colored, live cells under the microscope.
3. *Preparation for NMR:* After harvesting, soft-pellet cells at ~500 *g* for 20 min. Transfer the soft slurry by pipette to NMR tubes and centrifuge in adapters at 1,000–2,000 *g* for 1 h. Each cell type may pack differently and control experiments

**Table 5.8. Correlation between $T_1$ and microtubule complexes in human breast cancer cell lines***

| Cell line (MDA-MB) | No. of samples | Approximate doubling time in culture (days) | NMR relaxation times (ms) | | | Microtubule complex‡ | |
|---|---|---|---|---|---|---|---|
| | | | $T_1$† | $T_2$† | % $H_2O$† | % Type I | % Type II |
| 231 | 9 | 1 | 934 ± 78 | 123 ± 31 | 86.7 ± 1.3 | 84 | 16 |
| 157 | 4 | 1–1.5 | 907 ± 10 | 135 ± 4 | 87.4 ± 0.1 | 7 | 93 |
| 361 | 2 | 1.5 | 849 ± 25 | | 88.4 ± 0.2 | | |
| 134 | 3 | 1.5–2 | 717 ± 64 | 145 ± 39 | 88.9 ± 0.4 | 7 | 93 |
| 453 | 4 | 1.5–2 | 770 ± 15 | 113 ± 18 | 87.4 ± 2.1 | 2 | 98 |
| 330 | 6 | 1.5–3 | 752 ± 39 | | 88.5 ± 1.4 | 55 | 45 |
| 435 | 4 | 6 | 607 ± 9 | 112 ± 13 | 88.5 ± 0.9 | 34 | 66 |
| 331 | 3 | 5–7 | 549 ± 136 | 75 ± 20 | 88.5 ± 0.1 | 95 | 5 |
| 431 | 1 | 12–14 | 521 | 126 | | 94 | 6 |
| 436 | 3 | 16–18 | 499 ± 49 | 100 ± 20 | 84.3 ± 0.4 | 80 | 20 |
| 231 (slow) | 12 | 2 | 622 ± 70 | 104 ± 18 | | | |
| 157 (slow) | 4 | 2.5 | 669 ± 52 | 102 ± 18 | 85.2 ± 1.1 | | |

*Data from ref. 22.

‡200 cells visually scored. Type I, cells with a full cytoplasmic microtubule complex; type II, cells with apparently fewer distinct microtubules, including both intermediate and diffuse patterns of immunofluorescence.

†Means ± SD.

*Note:* In a series of human breast cancer cells, $T_1$ values correlated with cytoplasmic organization. $T_1$ values did not correlate with cellular hydration in all cases. Cells with diffuse, unorganized tubulin protein had high $T_1$ values, suggesting freedom of motion for water in the system. Cells with highly organized fibrous networks of microtubules had low $T_1$ values, suggesting restricted freedom of motion of water.

should be done to insure reproducibility of extracellular volumes, which will run ~5–10%. (By centrifuging at different speeds a plateau at which $T_1$ values no longer change with increased speed will be reached.) After centrifugation remove the supernatant by pipette and allow the tube to sit for 5 min. Use a cotton swab to remove liquid from the walls and the top of the cell pellet. (A band of liquid above the cells will make interpretation of NMR results difficult.) Seal cell pellets with Parafilm and measure them as soon as possible.

4. *Problems:*
   a. For each cell type a plateau $T_1$ must be established for centrifugation speed and time.
   b. The centrifuge should have a controlled temperature that will not exceed 25°C.
   c. Extracellular volume can be determined for cell pellets by the inclusion of [$^{14}C$]polyethylene glycol as an extracellular marker; however, some cells, e.g., lymphocytes, may pinocytose the marker.
   d. Water and fat determinations may be made on cell pastes by preweighing NMR tubes or by sucking the paste up into a Wilmad pipette with a rubber tube (place a filter between your mouth and the cells). The paste can be transferred to glass vials for replicate water and lipid determinations. (There is too little lipid in most cultured cells to measure accurately.)

Ascitic cancer cells, leukemia cells, and established cancer cell culture lines are "clean" cell populations that have been compared with normal tissues and isolated normal cells. Table 5.9 lists some of these results.

**5.3.D.** *Blood and body fluids*

In the previous section we discussed the handling of blood cells which are to be centrifuged or separated from plasma. Interestingly, a great deal of information on the health of the patient may also be obtained from $T_1$ and $T_2$ values for plasma and serum. The preparation protocol for human serum is outlined in Example D, below.

**5.3.D.*1*.** *Plasma and serum.* Blood plasma contains all the blood proteins and coagulation has been theoretically prevented by the addition of anticoagulants. However, insufficient mixing of the anticoagulant in the plasma can lead to some clotting and the loss of variable amounts of blood protein during centrifugation. Experience has shown that the protein content of clotted sera, however, is very reproducible. Blood taken in orange-topped Vacutainer tubes and clotted at room temperature (do not heat or cool) produces a constant serum product. It can be drawn off with a disposable Pasteur pipette and aliquotted into 1 ml screw-topped Pro Vials. It should be frozen at −70°C immediately. To thaw, immerse in cool tap water and shake. To measure, wipe off the vial, resuspend the serum with a pipette, fill a long pipette to a depth of ~3 in, place the pipette tip in the bottom of the NMR tube to expel air and fill it without bubbles. Sera can be thawed and refrozen once and still give good NMR values. However, after more than two thawings, increased $T_1$ values correlate with visible protein precipitation (Table 5.10). Because sera are homogeneous samples, they give very reproducible $T_1$ and $T_2$ values. The only sera that are difficult are ones containing high lipid amounts, which are usually taken just after a meal. Lipid can reduce apparent $T_1$ and $T_2$ values, so fasting sera samples should be used in NMR experiments. Data on blood, plasma, and sera, and the systemic effect of cancers on sera have been collected (Table 5.11). An elevation in $T_1$ and $T_2$ in the blood of cancer victims may have value as a cancer screening method.[31]

**5.3.D.*2*.** *Other body fluids.* On occasion, body fluids other than plasma and serum, such as bile, urine, necrotic fluid, ascitic fluid, mucus, lymph, or cerebrospinal fluid, may be utilized for NMR measurements. In these cases, ideally the fluid should be filtered through a nonabsorbent plastic or glass filter to remove particulate matter. Filtered fluids will give more reproducible values. No experience with temperature effects or storage methods has been reported for these types of samples.

**5.3.D.*3*.** *Example D: Protocol for human serum*

1. *Equipment:* Blood collection kit; orange-topped Vacutainer tubes (or purple-topped for plasma); screw-capped Pro Vials; clean, dry Pasteur pipettes; Wilmad NMR tubes and long pipettes; Parafilm.
2. *Blood Sampling:* Since the lipid and protein content of blood serum may vary through the day, serum should be drawn from a patient who has fasted since the midnight before (no food, no water, and especially no coffee, which acts as a diuretic). If this is not possible, use blood from patients who have not eaten for at least 3 h. Be sure to record which is the case. Blood may be drawn from the vein at the elbow, but an 18–20

**Table 5.9. Selected NMR data on cultured cell systems**

| Cell line | Frequency (MHz) | $T_1$ (ms) | $T_2$ (ms) | % $H_2O$ | Reference |
|---|---|---|---|---|---|
| HeLa cells | 30 | 667 | | 86.0 | Beall et al., 1976[20] |
| | 25 | 990 | | | Ranade et al., 1975[13] |
| | | | | | Wagh et al., 1977[35] |
| Human breast cancer | 30 | | | | Beall et al., 1982[22] |
| 231 | | 934 ± 78 | 123 ± 31 | 86.5 ± 1.3 | |
| 453 | | 770 ± 15 | 113 ± 18 | 87.4 ± 2.1 | |
| 436 | | 499 ± 49 | 100 ± 20 | 84.3 ± 0.4 | |
| Human leukocytes | 24 | | | | Ekstrand et al., 1977[30] |
| Normal | | 715 ± 23 | — | — | |
| Leukemic | | 855 ± 42 | — | — | |
| Human colon | 30 | | | | Beall et al., 1982[31] |
| Normal adult | Primary | 1,214 ± 46 | 207 ± 7 | 92.3 ± 0.4 | |
| Normal fetal | Primary | 1,058 ± 13 | 221 ± 3 | 90.6 ± 0.4 | |
| Cancer LS 180 | | 643 ± 60 | 119 ± 12 | 85.5 ± 2.4 | |
| Cancer LS 174 T | | 744 ± 16 | 121 ± 3 | 86.6 ± 0.5 | |
| | | | Paramagnetic metal concentration | | |
| | | | Fe (μg/g cells) | Cu (μg/g cells) | |
| NRK (rat) | | 910 | 52 | 7.2 | Ader and Cohen, 1979[18] |
| NMRK (rat) | | 970 | 35 | 6.2 | |
| BALB 3T3 (mouse) | | 1,440 | 154 | 11.2 | |
| BALB R4 (mouse) | | 1,490 | 68 | 6.3 | |

*Note:* $T_1$ and $T_2$ values vary over a wide range for pellets of cultured cells. A portion of this difference is due to insufficient centrifugation and high extracellular volumes of water. Other contributing factors are nuclear-to-cytoplasmic mass ratios, intracellular organelles, and relative surface areas of membranes and cytoskeletons.

**Table 5.10. Effect of storing human sera at 25° and 5°C on $T_1$ values**

| Time | $T_1$ measured at 25°C (ms) | | | $T_1$ measured at 5°C (ms) | | |
|---|---|---|---|---|---|---|
| | Sample A | Sample B | Sample C | Sample A | Sample B | Sample C |
| Fresh | 1,609 | 1,553 | 1,609 | 1,675 | 1,553 | 1,609 |
| 1 h | 1,681 | | | | | |
| 1.5 h | 1,680 | | | | | |
| 2 h | 1,690 | | | | | |
| 3 h | 1,660 | | | | | |
| 8 h | | | 1,621 | | | |
| 24 h | 1,610 | 1,542 | 1,628 | 1,592 | 1,542 | 1,628 |
| | 1,625 | | | | 1,544 | |
| 48 h | 1,607 | 1,545 | | 1,647 | 1,590 | |
| Frozen and then thawed | | | | | | |
| 12 days frozen | | | 1,637 | | | |
| 2 months frozen | | | 1,644 | | | |

**Table 5.11. Systemic effects of elevated serum $T_1$ values from mice and humans with mammary cancer**

| | $n$ | Mean $T_1$ of sera ($\pm$ SD) (ms) | Serum protein (g/100 ml) |
|---|---|---|---|
| Mouse mammary cancer* | | | |
| Normal virgin | 24 | 1,554 ± 19 | 4.95 |
| Normal virgin | 21 | 1,503 ± 9 | 5.4 |
| Protein malnourished | 5 | 1,565 ± 6 | 3.8 |
| Ductal hyperplasia CD-1 | 4 | 1,575 ± 22 | — |
| HAN C4 | | 1,564 ± 13 | 5.6 |
| HAN C3 | | 1,463 ± 12 | 5.7 |
| Ductal papilloma (benign) | 4 | 1,719 ± 22 | |
| Mammary carcinoma C4 | 17 | 1,801 ± 46 | 5.6 |
| Mammary carcinoma C3 | 19 | 1,625 ± 23 | 5.8 |
| Human mammary cancer† | | | |
| Normal females (young, fresh sera) | 10 | 1,535 ± 59 | — |
| Normal males (frozen) | 30 | 1,503 ± 146 | 7.8 ± 1.0 |
| Normal females (frozen) | 30 | 1,534 ± 178 | 7.8 ± 1.5 |
| Biopsy females (small tumors, preoperative) | 10 | 1,612 ± 45 | — |
| Patients in remission (chemotherapy and radiation) | 20 | 1,460 ± 91 | 7.7 ± 0.9 |
| Patients with cancer (but not on drugs and not in remission) | 22 | 1,618 ± 112 | 7.1 ± 1.3 |
| Patients with metastases (at time serum was taken) | 5 | 1,734 ± 71 | 6.5 ± 0.6 |

*Data from ref. 19.
†Data from ref. 28.

gauge needle should be used to reduce hemolysis. If a standard glass or plastic syringe is used, the needle must be removed before transferring blood to a clotting tube. (Serum separator tubes should not be used, since we have noted a small change in $T_1$ values when these are used). After a standard clotting time of 30 min at room temperature, loosen the clot from the sides of the tube with a glass rod. Centrifuge the sample at 1,000 *g* for 30 min at 25°C. Avoid high temperatures during centrifugation.

3. *Preparation for NMR:* Draw sera off, pool it, and mix the sample using a clean Pasteur pipette. It should be immediately aliquotted into 2 ml Pro-Vials and frozen to −70°C in a Revco freezer or in liquid nitrogen. Keep one vial for fresh determinations. Fill a long Wilmad pipette to about 3 in above the tip with serum and introduce it into the bottom of the NMR tube before the serum is expelled, to prevent evaporation as the sera run down the sides of the tube. Seal samples with Parafilm; they can be kept at room temperature for 8 h without changes occurring.
4. *Problems:*
   a. Sera with high lipid content will be cloudy and may give reduced $T_1$ and $T_2$ values. Also, as the samples sit, lipid may separate like cream to the top. These sera should be mixed by flicking your forefinger against the end of the tube before measuring.
   b. Theoretically, iron from the degradation of hemoglobin from lysed red cells could affect $T_1$ values. However, the addition of large amounts of lysed cells to sera has not changed $T_1$ values in our experience. If sera are red, this should be noted.
   c. Sera kept at temperatures higher than 34°C for any length of time will have active proteases and lipases which may alter the macromolecular composition of the sample. However, too cool a temperature may cause solidification of lipid droplets in the sera.

## 5.4. ANALYSIS OF NMR DATA FROM BIOLOGICAL SAMPLES

Early work in the field of biological NMR has come under criticism for the lack of statistically significant sample sizes and over-interpretation of data. Survey studies including only one or two samples of normal tissue to compare with one or two samples of cancerous tissue were later improved by in-depth studies. However, to be accepted by the scientific community as proof that NMR parameters can distinguish various tissues from one another or from cancerous tissues, more care must be taken with experimental design.

Cumulative errors in NMR measurements include those caused by variation in instrumentation reproducibility and stability, calculation errors, and biological errors inherent in a population, and those caused by sample handling. (A detailed example utilizing a set of actual data is given in Chapter 7.)

### 5.4.A. *Instrumentation parameters*

Instrumentation error can be minimized if the magnetic field stability and measurement temperature are controlled. The reproducibility of an instrument measuring 10 samples of doped water, 10 times each, can be calculated. This error should be less than ±5%.

However, in working with biological samples the slope of the ln $(h/h_0)$ (the proton signal/original signal) versus time (recovery of the net magnetization of the system over time) rarely demonstrates true exponential behavior and will not be a straight line. Therefore, criteria must be set up for each type of tissue to fit the data for calculations of $T_1$ and $T_2$. The type of fitting program and the time span used in fitting should be stated. The variance of fit of the data to a straight line will add error in the calculation of the relaxation time. Sera, being dilute solutions of proteins, give exponential data. Tissue culture cell pellets and very wet tissues usually have small variance of fit. However, firm tissues and heterogeneous biological samples may give nonexponential relaxation data. For muscle, at least three slowly exchanging fractions of water have been delineated. Liver, a somewhat soft tissue, also gives a curved fit.

Investigators' prejudice toward a particular model of the behavior of water inside cells and tissue will determine how they curve-fit their data. Commonly, $T_1$ and $T_2$ values quoted in articles will be "initial" values. A range of time, excluding the first 10–50 ms of magnetization decay (because of short- or quick-relaxing lipid protons) and including approximately one-half of the relaxation time (300–500 ms for $T_1$ and 30–50 ms for $T_2$), is chosen for all samples of one type. Data are fit to the line of the best least-squares fit. Some investigators may choose to use the same time range for all samples, which results in different variances according to the tissue studied. Others may alter the time range to fit the line within a tolerable variance for each tissue type. Some instrumentation has sufficiently low signal-to-noise ratios and sufficiently flat 180° pulses to continue data collection to 3–5 times the $T_1$ or $T_2$ value. In this case, the curve can be drawn by eye or by computer program to yield multiple populations of protons, referred to as $T_1$-fast and $T_1$-slow, or $T_1$-short and $T_1$-long. Any reference to such fractions must include the fitting range or the cutoff variance for each fraction for each type of tissue.

### 5.4.B. *Biological parameters*

While the calculation of any one relaxation time for a single sample contains instrument reproducibility and fitting errors, biological error and variation greatly increase the problems of interpretation of NMR data. Errors in sample handling and preparation have been described earlier in this chapter. They include any factors or procedures that alter the water or electrolyte concentration of the sample and factors that change the three-dimensional structure of the macromolecular lattice of the cell. Primarily, these are blotting or washing the tissue, time delay, and temperature changes. Also, variables such as age, weight, sex, hormone status, and general health of the patient must be considered.

### 5.4.C. *Comparison of normal and pathological samples*

In addition to variation in measurements on a single sample are the problems of comparing samples of different types. Assuming all precautions have been taken to prepare and measure the samples in the same manner, some comparisons among them can be made on the basis of the mean, standard deviation, and standard error of the mean. Usually Student's *t* test is used to compare two groups of equal size and *p* values are quoted in the reports. However, certain problems with published NMR data in this respect have been pointed out:

a. To be of value, the sample of both groups should be of sufficient size; usually, 10–20 samples per group would be a minimum.
b. Two very unequal groups, such as 4 normal livers and 10 hepatomas, should not be used to make claims about the significance of results.
c. Comparison between very small groups of three or four samples each is not wise.
d. Fortuitous standard deviations that are much lower than the reproducibility of the instrumentation either should not be considered significant or else control experiments for that specific case should be done.

Data which have a broad range due to biological variation should not, however, be discarded if a sample of sufficient size is available to utilize more sophisticated statistical methods.

Certainly, thought should be given to the methods of data analysis at the beginning of the experimental design process, so that sample size and control experiments are considered. Similar considerations of water and fat content determination and other external factors must be made early on in the experimental protocol.

If the experiment is well designed, the samples handled carefully, the proper controls done, and the data analyzed scientifically, the quality of published biological NMR data will be improved. Increased confidence in the findings and conclusions from this area of research will benefit many branches of biological and medical science.

## 5.5 AXIOMS FOR SAMPLE HANDLING

1. Know the NMR machine and understand its limitations.
2. Design the experiment to include control experiments for the calculation of error.
3. Use large enough samples so results will clearly be significant.
4. Handle and store all samples appropriately.
5. Do not *assume* any condition of temperature, storage, or reproducibility for a particular sample type.
6. State the method of data calculation and analysis in the report for publication.

## 5.6. REFERENCES


1. Chang, D.C., Hazlewood, C.F., and Woessner, D.C. The spin–lattice relaxation times of water associated with early post mortem changes in skeletal muscle. *Biochim. Biophys. Acta* **437**:253–258, 1976.
2. Tanaka, K., Yamada, Y., Shimizu, T., Sano, F., and Abe, Z. Fundamental investigation (in vitro) for a non-invasive method of tumor detection by nuclear magnetic resonance. *Biotelemetry* **1**:337–350, 1974.
3. Valensin, G., Gaggelli, E., Tiezzi, E., Valensin, P.E., and Bianchi Bandelli, M.L. The water proton spin–lattice relaxation times in virus infected cells. *Biophys. Chem.* **10**:143–146, 1979.
4. Rustgi, S.N., Peemoeller, H., Thompson, R.T., Kydon, D.W., and Pintar, M.M. A study of molecular dynamics and freezing phase transition in tissues by proton spin relaxation. *Biophys. J.* **22**:439–452, 1978.
5. Belton, P.S., Packer, K.J., and Sellwood, T.C. Pulsed NMR studies of water in striated muscle: II. Spin–lattice relaxation times and dynamics of the non-freezing fraction of water. *Biochim. Biophys. Acta* **304**:45–52, 1973.
6. Peemoeller, H., Pintar, M.M., and Kydon, D.W. Nuclear magnetic resonance analysis of water in natural and deuterated mouse muscle above and below freezing. *Biophys. J.* **29**:427–435, 1980.
7. Morariu, V.V., Kiricuta, I.C., and Hazlewood, C.F. Nuclear magnetic resonance investigation of the freezing of water in rat kidney tissues. *Physiol. Chem. Phys.* **10**:517–524, 1978.
8. Beall, P.T. NMR relaxation times of water protons in cultured cells under freezing and osmotic stress conditions. In: *Proceedings of the Conference on the Biophysics of Water*. Franks, F., ed. John Wiley & Sons, New York, 1982, pp. 323–326.
9. Mansfield, P., and Morris, P.G. *NMR Imaging in Biomedicine*. Academic Press, New York, 1982.
10. Damadian, R., NMR scanning. In: *NMR in Medicine*. Damadian, R., ed. Springer Verlag, New York, 1981, pp. 1–16.
11. Damadian, R. Tumor detection by nuclear magnetic resonance. *Science* **171**:1151–1153, 1971.
12. Hazlewood, C.F., Chang, D.C., Medina, D., Cleveland, G., and Nichols, B.L. Distinction between the preneoplastic and neoplastic state of murine mammary glands. *Proc. Natl. Acad. Sci. U.S.A.* **69**:1478–1480, 1972.
13. Ranade, S.S., Chaughule, S., Kasturi, S.R., Nad Karni, J.S., Talwalkar, G.V., Wash, U.V., Korgaonkar, K.S., and Vijayaraghavan, R. Pulsed nuclear magnetic resonance studies on human and malignant tissues and cells in vitro. *Indian J. Biochem. Biophys.* **12**: 229–234, 1975.
14. Hollis, D.P., Economou, J.S., Parks, L.C., Eggleston, J.C., Saryan, L.A., and Czeisler, J.L. Nuclear magnetic resonance studies of several experimental and human malignant tumors. *Cancer Res.* **33**:2156–2160, 1973.
15. Medina, D., Hazlewood, C.F., Cleveland, G., Chang, D.C., Spjut, H.J., and Moyers, R. Nuclear magnetic resonance studies on human breast dysplasias and neoplasms. *J.N.C.I.* **54**:813–818, 1975.
16. Damadian, R., Goldsmith, M., and Minkoff, L. NMR in cancer. XVI. Fonar image of the live human body. *Physiol. Chem. Phys.* **9**:97–100, 1977.
17. Block, R.E. Factors affecting proton magnetic resonance line widths of water in several rat tissues. *FEBS Lett.* **34**:109–112, 1973.
18. Ader, R., and Cohen, J.S. Relaxation time measurements of bulk water in packed mammalian density-dependent and -independent cells grown in culture. *J. Magn. Reson.* **34**:349–356, 1979.
19. Beall, P.T., Medina, D., Chang, D.C., Seitz, P.K., and Hazlewood, C.F. A systemic effect of benign and malignant mammary cancer on the spin–lattice relaxation time, $T_1$, of water protons in mouse serum. *J.N.C.I.* **59**:1431–1433, 1977.
20. Beall, P.T., Hazlewood, C.F., and Rao, P.N. Nuclear magnetic resonance patterns of intracellular water as a function of HeLa cell cycle. *Science* **192**:904–907, 1976.
21. Beall, P.T., Asch, B.B., Chang, D.C., Medina, D., and Hazlewood, C.F. Distinction of normal, preneoplastic, and neoplastic mouse mammary primary cell cultures by water nuclear magnetic resonance relaxation times. *J.N.C.I.* **64**:335–338, 1980.
22. Beall, P.T., Brinkley, B.R., Chang, D.C., and Hazlewood, C.F. Microtubule complexes correlated with growth rate and water proton relaxation times in human breast cancer cells. *Cancer Res.* **42**:4124–4130, 1982.
23. Kasturi, S.R., Ranade, S.S., and Shah, S. Tissue hydration of malignant and uninvolved human tissues and its relevance to proton spin–lattice relaxation mechanisms. *Proc. Ind. Acad. Sci.* [*B*] **84**:60–74, 1976.
24. Chaughule, R.S., Kasturi, S.R., and Ranade, S.S. Proton spin–lattice relaxation times and water content in human tissues. *Proc. Nucl. Phys. Solid State Phys. Symp.* **17c**, 1974.
25. Ranade, S.S., Shah, S., Korgaonkar, K.S., Kasturi, S.R., Chaughule, R.S., and Vijayaraghavan, R. Absence of correlation between spin–lattice relaxation times and water content in human tumor tissues. *Physiol. Chem. Phys.* **8**:131–135, 1976.
26. Beall, P.T., Hazlewood, C.F., and Chang, D.C. Microtubule organization and the self-diffusion coefficient of water in baby hamster kidney cells as a function of temperature. *J. Cell Biol.* **95**:334a, 1982.

27. Koivula, A., Suominen, K., Timonen, T., and Kivinitty, K. The spin–lattice relaxation time in the blood of healthy subjects and patients with malignant blood disease. *Phys. Med. Biol.* **27**:937–947, 1982.
28. Beall, P.T., Medina, D., and Hazlewood, C.F. The systemic effect of elevated tissue and serum relaxation times for water in animals and humans with cancer. In: *NMR Basic Principles and Progress*. Diehl, P., ed. Springer Verlag, Berlin, 1981, pp. 39–57.
29. McLachlin, L.A. Cancer induced increases in human plasma proton NMR relaxation rates. *Phys. Med. Biol.* **25**:309–315, 1980.
30. Ekstrand, K.E., Dixon, R.L., Raben, M., and Ferree, C.F. Proton NMR relaxation in the peripheral blood of cancer patients. *Phys. Med. Biol.* **22**:925–931, 1977.
31. Beall, P.T., Hazlewood, C.F., and Rutzky, L.P. NMR water relaxation times of human colon cancer cell lines and clones. *Cancer Biochem. Biophys.* **6**:7–12, 1982.
32. Damadian, R., Zaner, K., and Hor, D. Human tumors by NMR. *Physiol. Chem. Phys.* **5**:381–402, 1973.
33. Damadian, R., Zaner, K., and Hor, D. Human tumors detected by nuclear magnetic resonance. *Proc. Natl. Acad. Sci. U.S.A.* **71**:1471–1473, 1974.
34. Iijima, N., Saitoo, S., Yoshida, Y., Fujii, N., Koike, T., Osanzi, K., and Hirose, K. Spin–echo nuclear magnetic resonance in cancerous tissue. *Physiol. Chem. Phys.* **5**:431–435, 1973.
35. Wagh, U.V., Kasturi, S.R., Chaughule, R.S., Smita, S.S., and Ranade, S.S. Studies on proton spin–lattice relaxation time ($T_1$) in experimental cell cultures. *Physiol. Chem. Phys.* **9**:167–174, 1977.

CHAPTER 6

# DEPENDENCE OF RELAXATION TIMES ON PHYSICAL PARAMETERS



## 6.1. INTRODUCTION

Since Damadian[1] published his report on the water proton relaxation times in various tissues and their malignant counterparts, many laboratories have carried out NMR investigations on a variety of biological systems *in vitro*. The investigations were carried out to explore the diagnostic value of the NMR parameters and to understand the details of the relaxation mechanism in particular systems. For example, blood and muscle have been studied extensively. Much of the vast amount of data that exists in the literature has been summarized in Tables 9.2–9.25 in Chapter 9 of this book. As is evident from the tables, different groups have measured proton relaxation times at different resonance frequencies and at slightly different temperatures. Relaxation times and other NMR parameters are strongly dependent on temperature and frequency. There are many other factors that can affect the observed NMR parameters. For example, the $T_1$ of water protons in blood depends not only on frequency and temperature but also on pH, hematocrit, and hemoglobin content. Similarly, in solid tissues, the water content is often correlated with the relaxation times of water protons. Thus a dehydrated muscle will exhibit a shorter water proton $T_1$ than will a fully hydrated muscle at the same temperature and frequency. Caution should be exercised in comparing the relaxation times for the same type of tissue when they were measured by different groups. Some of the biological factors that can influence the measured water relaxation times are discussed in Chapter 5 of this book.

In this chapter, examples of existing data which show the effect of physical parameters, such as temperature and frequency, on the measured water proton relaxation times are collected and presented.

## 6.2. DEPENDENCE OF $T_1$ AND $T_2$ ON PHYSICAL PARAMETERS

### 6.2.A. *Dependence on frequency and temperature*

Both the spin–lattice relaxation time ($T_1$) and the spin–spin relaxation time ($T_2$) of water protons in biological tissues are frequency dependent. For liquid water, however, both $T_1$ and $T_2$ are frequency independent. For example, the $T_1$ of pure water is ~3.2 s and $T_2$ is approximately 1.7 s at any frequency and the same temperature (24–25°C). The $T_1$ of water in tissues, however, is reduced by a factor of 2.5–23 and $T_2$ by a factor of 5.8–221 compared with pure water[2];

they are frequency dependent. NMR can also be used to measure self-diffusion coefficients. The self-diffusion coefficient ($D_{H_2O}$) measured for water in tissues is approximately one-half that observed for bulk water.[2] Various models (see refs. 2–5 for reviews of existing models) have been devised to explain the observed shorter relaxation times and diffusion coefficient for water protons in tissues. Similarly, the frequency dependence of the tissue water proton relaxation process can be explained in more than one way.

The relaxation of water protons in tissues is not easy to understand. However, the following simple, qualitative considerations will help elucidate the kind of temperature and frequency dependence observed for water proton relaxation times in tissues.

In a simple model, the relaxation rate of water protons can be considered to arise from interactions between two neighboring spins (protons, in case of water) at a distance of $r$, and is given by

$$\frac{1}{T_1} = \frac{3\gamma^4\hbar^2}{10r^6}\left(\frac{\tau_C}{1 + \omega^2\tau_C^2} + \frac{4\tau_C}{1 + 4\omega^2\tau_C^2}\right) \quad , \qquad (6.1)$$

where $\hbar = h/2\pi$ and $h$ is Planck's constant; $\gamma$ is the gyromagnetic ratio of protons; and, $\omega = 2\pi\nu$, where $\nu$ is the Larmor frequency.[6]

$\tau_C$ is a characteristic time describing the motion of water molecules and is referred to as the correlation time associated with this interaction. For example, in this case, the rotational motion of the water molecules gives rise to the interaction.

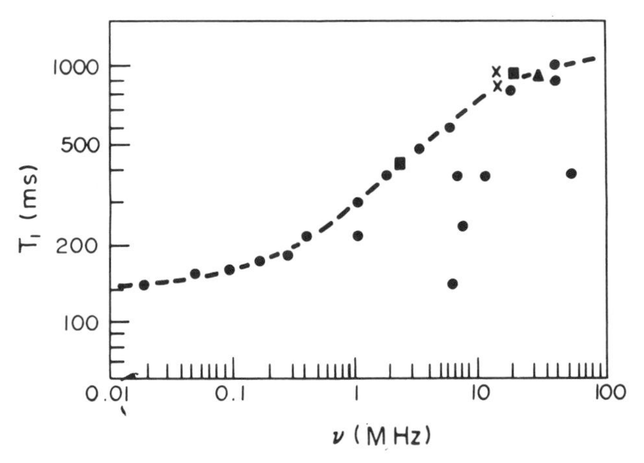


Fig. 6.1. Water proton spin–lattice relaxation time ($T_1$) in blood as a function of frequency. Data from various types of blood—venous, *in vivo,* fresh, and stored—from various sources—human, dog, chicken, cow, and mouse—are shown. The temperatures are not exactly the same. The graph is intended to give an idea of the frequency dependence rather than the exact values at each frequency. (Adapted from Brooks et al., 1975,[7] with permission.) [Some data taken from Koivula et al., 1982[39](x); Ratkovic et al., 1974[43] (■); and Ling et al., 1980[23] (▲).]

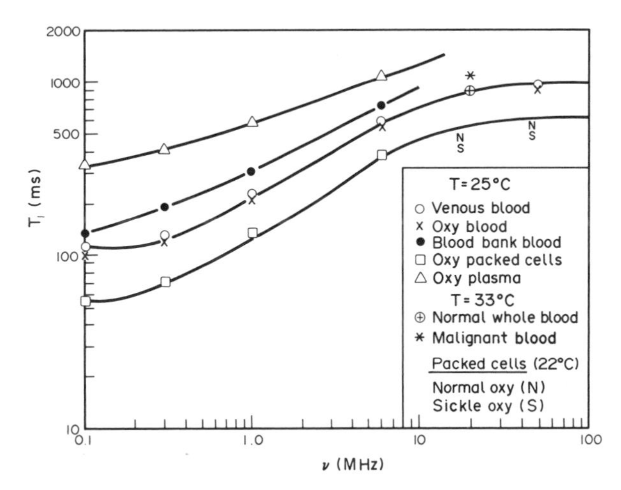


Fig. 6.2. Frequency dependence of proton spin–lattice relaxation time ($T_1$) for various types of blood, packed cells, and plasma. All samples are from normal donors, unless otherwise indicated. Points marked (N) and (S) are from the data of Zipp et al., 1976,[8] and points marked by (⊕) and (∗) are from the data of Koivula et al., 1982.[39] All other points are from the work of Brooks et al., 1975.[7]

For $\omega^2\tau_C^2 \ll 1$, $T_1$ is independent of frequency. For bulk water, $\tau_C \approx 10^{-12}$ s, and hence $T_1$ is independent of frequency. For $\omega^2\tau_C^2 \geq 1$, $T_1$ begins to increase and is proportional to $\omega^2$ (or $\nu^2$). This is the situation for many tissues, although deviations are observed. The detailed models for the relaxation of water in tissues can be found in several review articles.[2,4,5]

The temperature dependence of the relaxation times can also be predicted using simple terms, although it must be noted once again that the temperature dependence of water relaxation times is not fully described by the model. $T_1$ and $T_2$ become temperature dependent because of the temperature dependence of $\tau_C$ in Eq. 6.1 for the relaxation rate. The temperature dependence of $\tau_C$ is assumed to be given by the Arrhenius relation:

$$\tau_C = \tau_C^0 \exp(E_a/kT) \quad , \qquad (6.2)$$

where $E_a$ is the activation energy for the particular molecular motion and $\tau_C^0$ is the preexponential factor which reflects the jump frequency of the molecular motion, $k$ is the Boltzmann constant, and $T$ is the temperature.

For $\omega\tau_C \ll 1$:

$$\frac{1}{T_1} \propto \tau_C \propto \tau_C^0 \exp(E_a/kT) \quad . \qquad (6.3)$$

**Table 6.1. Some useful parameters for NMR studies of blood**

| Description of blood | pH | Hematocrit* (%) | Hemoglobin* (g%) |
|---|---|---|---|
| Venous blood | 7.26§ | 45.8<br>42.0†<br>(34–49)‡ | 16.3<br>14.3†<br>(11.8–16.2)‡ |
| Oxy blood | 7.73§ | 44.2 | 15.7 |
| Stored blood (36 days old) | 6.54§ | 39.2 | 13.5 |
| Oxy cells | 7.88§ | 77.9 | 26.6 |
| Oxy plasma | 8.01§ | — | — |
| Deoxy plasma | 7.84§ | — | — |
| Umbilical artery (whole blood) | 7.21<br>(7.05–7.38)‡ | — | — |
| Umbilical vein (whole blood) | 7.32<br>(7.23–7.42)‡ | — | — |
| Malignant venous blood | — | 34*† | 10.9*† |
| Arterial whole blood | | | |
| Infant (1–4 weeks old) | 7.377<br>(7.315–7.439)‡ | — | — |
| Infant (4–16 months old) | 7.432<br>(7.366–7.498)‡ | — | — |
| Adult | 7.424<br>7.392 | — | — |
| Capillary whole blood | | | |
| Men | 7.390<br>(7.36–7.42)‡ | — | — |
| Women | 7.398<br>(7.366–7.430)‡ | — | — |
| Arterial plasma, adult | 7.39<br>(7.35–7.43)‡ | — | — |
| Venous plasma, adult | 7.398<br>(7.378–7.418)‡ | — | — |
| Erythrocytes, adult | 7.209<br>7.19 | — | — |

*Values are means for the range of values that exists for these parameters.
†Data from Koivula et al., 1982.[39]
‡Values in parentheses are the 95% range around the mean.
§Temperature was 25°C (data from Brooks et al., 1975[7]). All other pH values were measured at 38°C (data from Geigy Scientific Tables, p. 560, Diem and Leutner, 1973[42]).

It can be seen from this equation that $T_1$ decreases as the temperature is decreased to this limit. On the other hand, for $\omega\tau_C \gg 1$:

$$\frac{1}{T_1} \propto \frac{1}{\tau_C} \propto \frac{1}{\tau_C^0} \exp\left(-E_a/kT\right) \quad . \qquad (6.4)$$

This implies that $T_1$ increases as temperature is decreased in this limit. However, for most of the tissues in the frequency and temperature (0–37°C) ranges normally used in biomedical NMR applications, $T_1$ decreases with a decrease in temperature (Eq. 6.3).

**Table 6.2. Blood viscosity**

Dynamic viscosity (*in vivo*) for whole blood, 2.30–2.75 centipoise

Kinematic viscosity (*in vitro* at 37.5°C) for serum, 1.15 (centistoke) (95% range of values, 1.08–1.22)*

Osmolality, Serum, mean 289 (mOsmol/kg $H_2O$) (range, 281–297)

Data from ref. 42.
*For a plot of hematocrit versus absolute viscosity of blood, see p. 557 in ref. 42.

Except for blood, no systematic frequency and temperature dependence studies of relaxation times have been made on human tissues. Data for muscle and blood in animals are available over a wide range of frequencies, and for a limited number of other animal tissues (see references given in figure legends for this chapter). Examples from the existing data on the frequency and temperature dependence of the relaxation times of water protons and the temperature dependence of the self-diffusion coefficients of water in tissues will be described in the following sections.

**6.2.A.*1*.** *Blood.* Blood consists of a fluid portion (plasma) and a cellular portion. The cellular portion consists of erythrocytes (red blood cells) and leukocytes (white blood cells), while the plasma consists of approximately 91% water, 7% proteins, and 2% other electrolytes and biochemicals. The most important constituents of erythrocytes, from an NMR perspective, are hemoglobin and water. Despite the fact that blood is a liquid tissue, the spin–lattice ($T_1$) and spin–spin ($T_2$) relaxation times of protons in blood are frequency dependent. The frequency dependence demonstrated by existing data on $T_1$ of blood from various sources under different experimental conditions is shown in Figure 6.1. The data shown for different frequencies were not measured at the same temperature (range 25–37°C). Because of the various sources of data, differences among different groups' data must be considered carefully. Generally, however, $T_1$ values increase with increasing frequency.

In addition to temperature and frequency, blood pH can influence relaxation times. $T_1$ data at different frequencies and temperatures (range 22–33°C) are shown in Figure 6.2. Part of the differences at the same temperature and frequency can be attributed to differences in pH, and part of the variation might be due to differences in hematocrit (the ratio of cell volume to total blood volume). All the relevant parameters that could affect the relaxation times of protons in blood are shown in Tables 6.1–6.3.

As a rough guide for estimating $T_1$ changes in blood

**Table 6.3. Inorganic constitutents of blood**

| Description of constituent | Mean | 95% Range |
|---|---|---|
| Water (g/L) | | |
| Whole blood | 850 | 830–865 |
| Plasma | 945 | 930–955 |
| Water (g/kg) | | |
| Whole blood | 681.3 | 654–709 |
| Total cations (mEq/L) | | |
| Arterial blood (plasma) | 152.9 | 149–157 |
| Venous blood (plasma) | 154.1 | 149–159 |
| Potassium (mEq/L) | | |
| Umbilical vein blood | 99.6 | — |
| 1-day-old newborn | 105 | — |
| 2-day-old newborn | 107 | — |
| Adult | 81.7 | — |
| Adult | 88 | — |
| Sodium (mEq/L) | | |
| Adult | 16.4 | — |
| Adult | 8.7 | — |
| Cobalt (μg/L) | 0.35 | — |
| Iron (mg/L) | | |
| Male | — | 440–560 (whole blood) |
| Female | — | 420–480 (whole blood) |
| Copper (mg/L) | | |
| Cord blood | 1.03 | — |
| Male | 0.89 | — |
| Female | 0.89 | — |
| Manganese (μg/L) | 200 | |

Data from ref. 42.

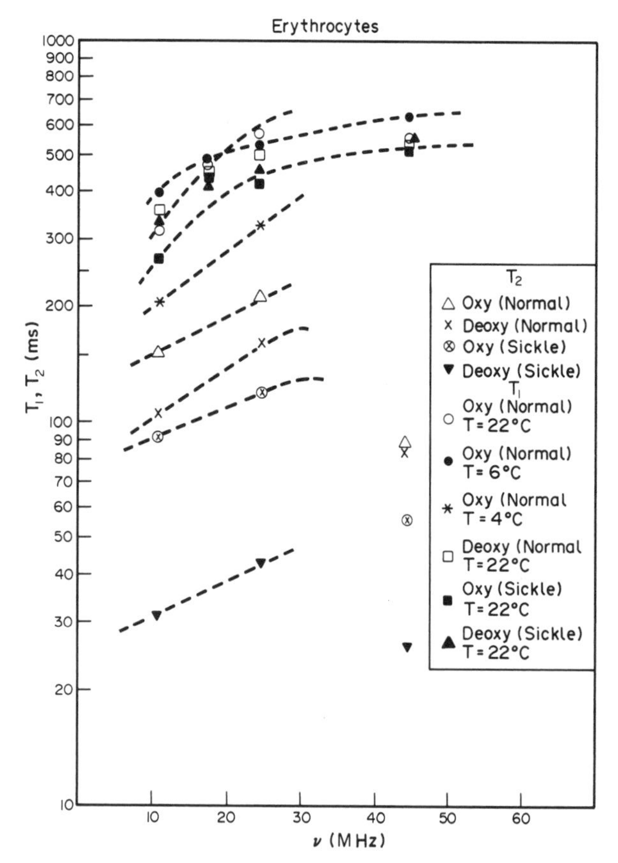


Fig. 6.3. Proton spin–lattice relaxation times ($T_1$) and spin–spin relaxation times ($T_2$) at various frequencies and three temperatures are shown for normal and sickle cell erythrocytes. Data for both the oxy and deoxy forms are shown. Data at frequencies of 10.7 and 24.3 MHz and at temperatures of 22 and 4°C are from the work of Thompson et al., 1973[9]; data at 6°C for all frequencies are from Lindstrom et al., 1974[10]; all the rest of the data are from Zipp et al., 1976.[8]

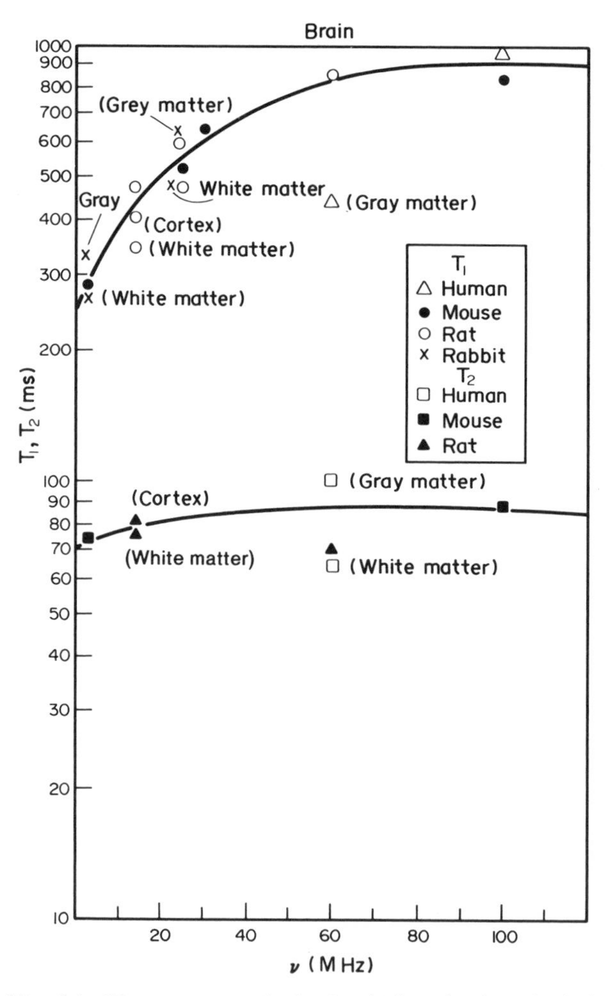


Fig. 6.4. Water proton spin–lattice ($T_1$) and spin–spin ($T_2$) relaxation times at various frequencies for brain tissue measured *in vitro* by different laboratories.[1,20,23,35,44–52]

due to temperature, a 2% increase in $T_1$ for an increase of 1°C is expected[7] at low frequencies. Similarly, an approximate 25% increase in $T_1$ per pH unit decrease is expected at low frequencies.

$T_1$ and $T_2$ data measured at different frequencies[8–10] and at three different temperatures for normal and sickle cell erythrocytes are shown in Figure 6.3. Again, in the case of erythrocytes, while comparing $T_1$ and $T_2$ values from different studies, even for the same temperature and frequencies, some differences should be expected due to hemoglobin concentration, the duration and *g*-force of centrifugation, and the extent of cell washing; all of these affect the hematocrit. Note in Figure 6.3 that at 44.4 MHz, $T_2$ decreased by a factor of 2 for sickle cells but remained relatively constant for normal erythrocytes upon deoxygenation of the cells. The situation was different at lower frequencies. There was no change in $T_1$ observed upon deoxygenation for normal or sickle cells at 44.4 MHz, while there were small changes observed at lower frequencies for both types of cells.

**6.2.A.2.** *Other tissues. In vitro* values in the literature for $T_1$ and $T_2$ of water protons at different frequencies have been collected and presented as plots of $T_1$ and $T_2$ versus frequency (on semilog scales) in Figures 6.4–6.11 for different tissues. There is some scatter in the data because the $T_1$ and $T_2$ values have been measured by different workers at slightly different temperatures (23–28°C) and the tissues were from various sources—human, mouse, rat, rabbit, etc. There are two things that are apparent in these plots: (a) The $T_1$ is strongly dependent on frequency, especially in the frequency range of 1–40 MHz, while $T_2$ is not so dependent on frequency in this region. Also, little $T_2$ data exist as a

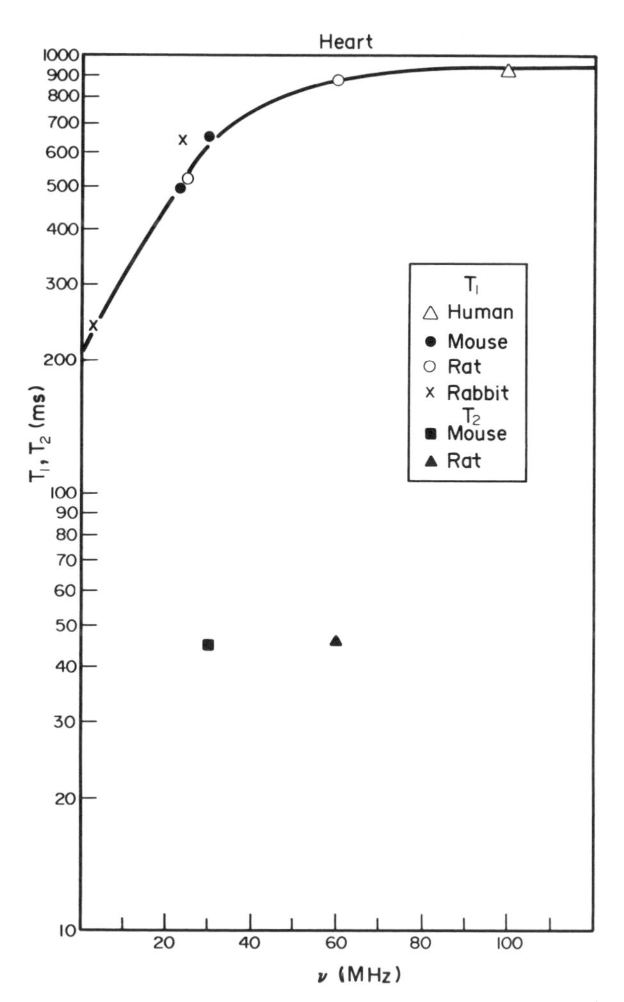


Fig. 6.5. Water proton $T_1$ and $T_2$ at various frequencies for heart tissue measured *in vitro* by different laboratories.[23,35,45–47,51]

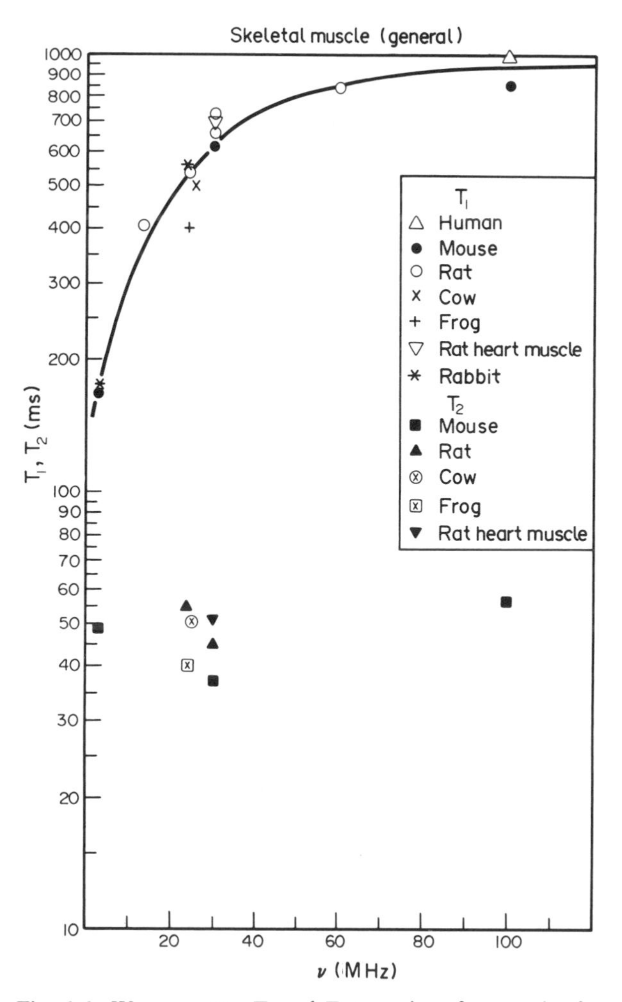


Fig. 6.6. Water proton $T_1$ and $T_2$ at various frequencies for skeletal muscle (general) measured *in vitro* by different laboratories.[11,20,23,25,40,46,47,50,53–58]

function of frequency for most of the tissues. (b) It is possible to estimate the approximate $T_1$ or $T_2$ expected at any frequency using these plots, irrespective of the tissue source. We hope the plots will be useful for this purpose. They are not intended to show any theoretical dependence of $T_1$ or $T_2$ on frequency. Very often, if a $T_1$ or $T_2$ is to be measured at a different frequency or on a new instrument, it may be helpful to use such plots to check if one is getting values in the same range as other investigators.

As mentioned earlier in this section, theoretically, $T_1$ is expected to be directly proportional to $\nu^2$. Studies at various frequencies of the same tissues and at the same temperatures have been made by several workers. An example of such a study[11] for muscle tissue in the frequency range of 14–45 MHz (at 25°C) is shown in Figure 6.12. Note that the $1/T_1$ versus $1/\nu^2$ plot is not linear; it shows only an approximate dependence of $T_1$ on $\nu^2$. Also, $1/T_1$ does not extrapolate to zero at the high end of the frequency range, thus suggesting a contribution to $T_1$ from frequency-independent molecular motions. The frequency dependence of the water proton relaxation time in the low frequency range (kHz) is obtained by measuring the relaxation time in the rotating frame ($T_{1\rho}$) (see ref. 12 for details about $T_{1\rho}$) as a function of the strength of the radiofrequency (rf) field ($B_1$). Although $T_{1\rho}$ is not commonly used for *in vivo* or imaging experiments, it is interesting to note the frequency dependence of $T_{1\rho}$ in the kHz range. An example of such a study for muscle tissue is shown in Figure 6.13 (from the work of Finch and Homer[13]) as plots of $1/T_{1\rho}$ versus $1/\nu_1^2$, where $\nu_1 = \omega_1/2\pi = \gamma B_1/2\pi$ ($\gamma$ is the magnetogyric ratio and $B_1$ is the strength of the rf field used in $T_{1\rho}$ measurement). The clear deviation of $1/T_{1\rho}$ versus $1/\nu_1^2$ from linearity, even in the low frequency range, implies that the

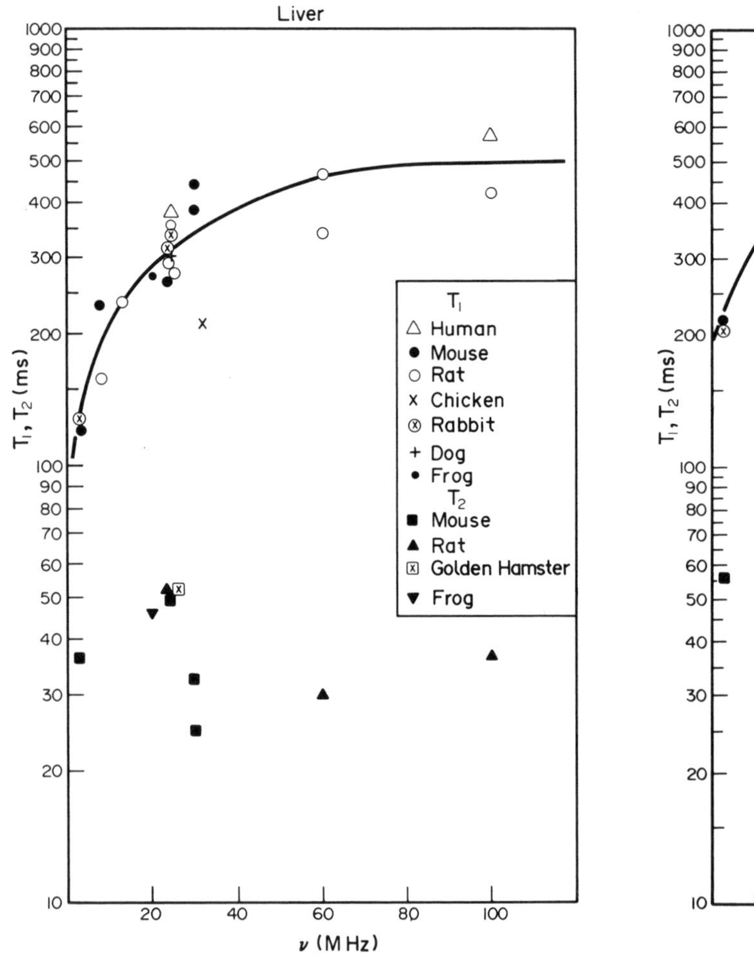


Fig. 6.7. Water proton $T_1$ and $T_2$ at various frequencies for liver tissue measured *in vitro* by different laboratories.[1,11,20,23,25,27,35,43,44,45,47,50,51,56,58–63]

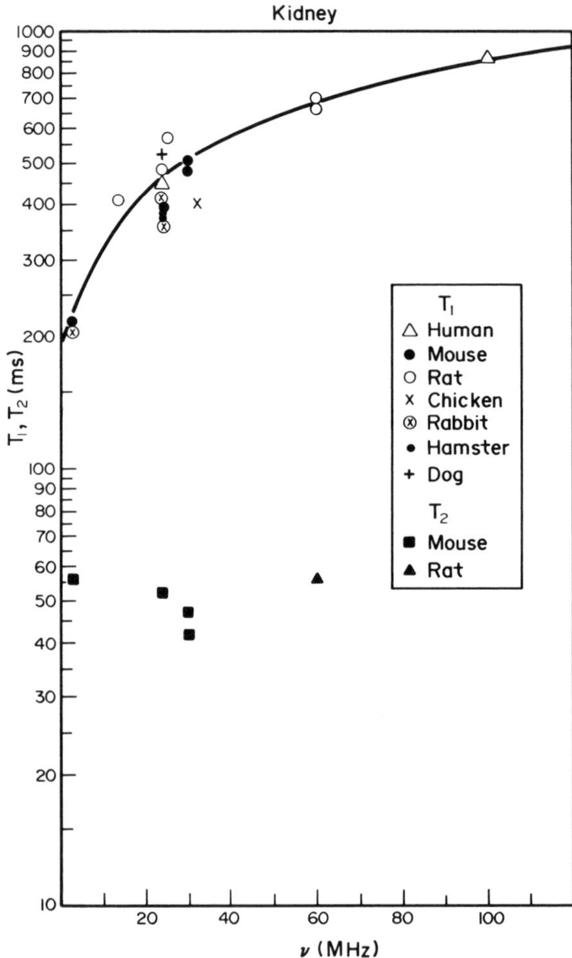


Fig. 6.8. Water proton $T_1$ and $T_2$ at various frequencies for kidney tissue measured *in vitro* by different laboratories.[1,11,20,23,25–27,35,43,44–47,51,58]

relaxation times of water protons in muscle tissue cannot be described by a single mechanism.

$T_1$, $T_{1\rho}$, and $T_2$ for water in tissues are temperature dependent. In the temperature range of 0–40°C, they decrease with temperature, approximately linearly on a log $T_1$ versus $10^3/T$ plot. As an example of how $T_1$, $T_{1\rho}$, and $T_2$ vary with temperature, data for frog gastrocnemius muscle[13] are shown in Figure 6.14. Notice that $T_1$ and $T_{1\rho}$ change with temperature more rapidly than does $T_2$ at this frequency (23 MHz). It is interesting to compare the temperature dependence of muscle water with that of pure water, which is shown in Figure 6.15. In the same temperature range, the $T_1$ for pure water changes more rapidly then does that of muscle water and is slightly nonlinear. This is because of the contribution to the relaxation of water protons by two processes. Over the temperature range of $-16$ to 145°C, the variation of $T_1$ for water as a function of temperature can be represented by a double-exponential form.[14]

Although the *in vitro* data on the frequency and temperature dependence of the water proton relaxation times presented here have been limited, extensive data exist in the literature on the relaxation times (mostly $T_1$) as a function of wide ranges of temperatures and frequencies. Some particular examples are given here: Thompson et al.[15] (1973) measured the spin–lattice relaxation time in the rotating frame ($T_{1\rho}$) as a function of $\nu$ at room temperature for water in mouse muscle and spleen tissues. Outhred and George[16] (1973) measured the water proton $T_1$ for toad muscle in the frequency range of 2–40 MHz at 25° and 2°C. Duff and Derbyshire[17] (1974) measured water proton $T_1$ at 10.7, 30, 60, and 90 MHz as a function of temperature in the range of $-70$ to $-10$°C. They also measured $T_{1\rho}$ and $T_2$ over this range. Fung and

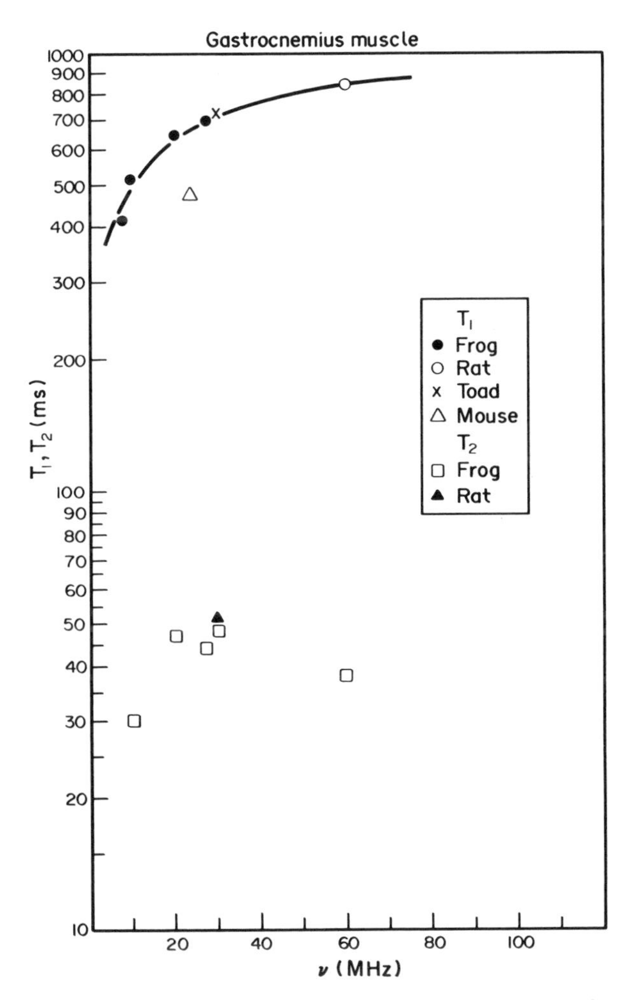


Fig. 6.9. Water proton $T_1$ and $T_2$ at various frequencies for gastrocnemius muscle measured *in vitro* by different laboratories.[13,16,18,26,27,35,54,57,59,64]

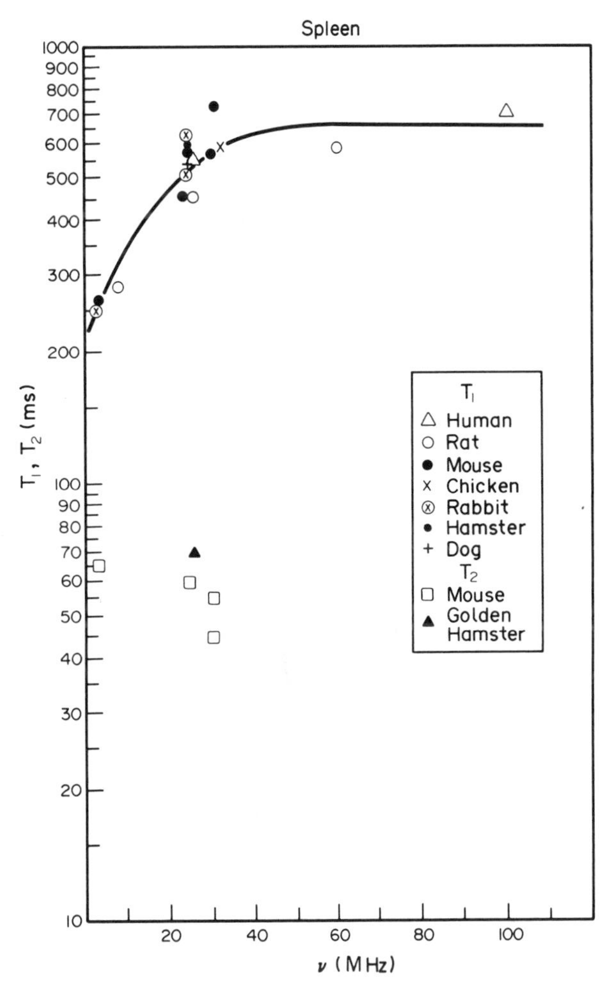


Fig. 6.10. Water proton $T_1$ and $T_2$ at various frequencies for spleen tissue measured *in vitro* by different laboratories.[11,20,23,25,27,35,43,45–47,56,58,62,63,66]

McGaughy[18] (1974) measured water proton $T_1$ at 4.5, 7.5, 10.5, 16.5, 33.5, and 60 MHz in the temperature range of +37 to −70°C. Knispel et al.[11] (1974) measured $T_2$ as a function of frequency in the range of 17–45 MHz and $T_{1\rho}$ in the range of 1–100 kHz at 25°C for mouse muscle (healthy and tumorous), as well as for mammary adenocarcinoma tumor. Finch and Homer[13] (1974) measured $T_1$, $T_{1\rho}$, and $T_2$ for frog muscle in the temperature range of 5–30°C. Fung et al.[19] (1975) measured $T_1$ of water protons in mouse liver and egg white at 4.5, 7.5, 12, 20, 35, and 60 MHz in the temperature range of −70 to +37°C. Coles[20] (1976) measured $T_1$ and $T_2$ at two frequencies, 2.7 and 15 MHz, for five tissues from normal and tumor-bearing animals. Kasturi et al.[21] (1980) measured the $T_1$ and $T_2$ for muscle water as a function of the orientation of the muscle fibers with respect to the static magnetic field at 28, −5 and −10°C. Burnell et al.[22] (1981) measured $T_1$ and $T_2$ at 30, 60, and 90 MHz and $T_{1\rho}$ as a function of frequency in the range of 1.2–37 kHz for barnacle muscle. Ling et al.[23] (1980) measured water proton $T_1$ for rabbit tissues at 24 and 2.5 MHz and at 20–23°C. Escanye et al.[24] (1982) measured the frequency dependence of water proton $T_1$ in mouse tissues at 20°C in the range of 6.7–90 MHz. The reader is referred to these articles for details.

### 6.2.B. *Dependence on water content of tissue*

Another parameter that significantly affects the relaxation times of water protons in tissues is the water content. Hence, care should be exercised while handling tissue samples to ensure that they do not absorb or lose water before or during measurement (see Chapter 5). Several investigators have studied the relationship between water content and water proton

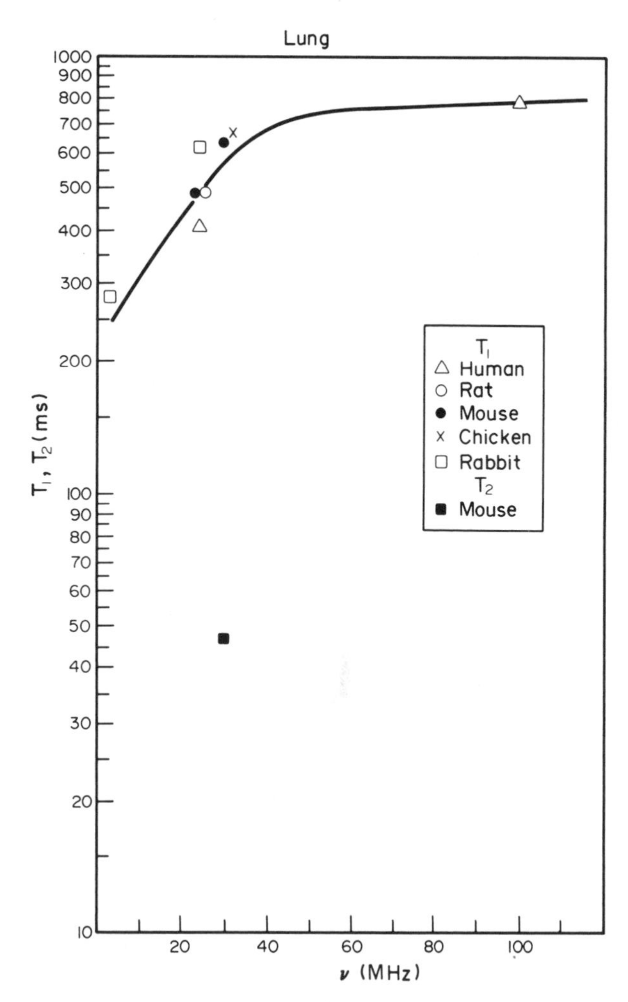


Fig. 6.11. Water proton $T_1$ and $T_2$ at various frequencies for lung tissue measured *in vitro* by different laboratories.[11,23,35,43,45–47,58]

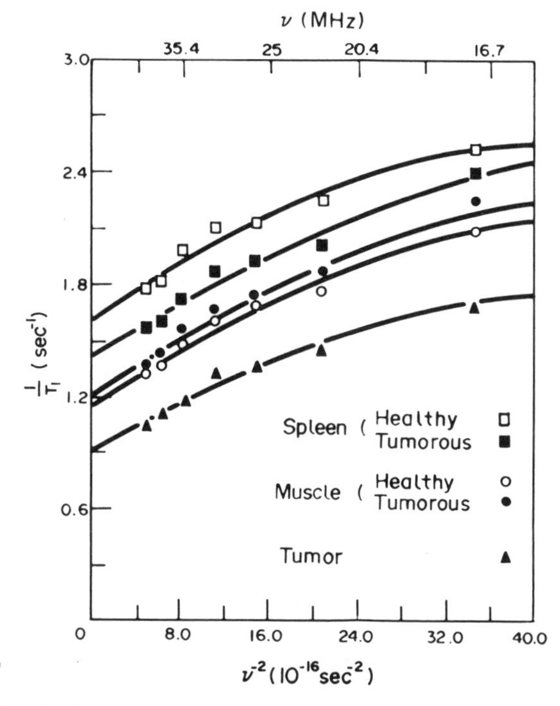


Fig. 6.12. Dependence of the water proton relaxation rate ($1/T_1$) on $\nu^2$ for muscle and spleen tissue from normal and mammary adenocarcinoma tumor-bearing mice at 25°C. Also shown is a similar plot for the adenocarcinoma tumor tissue. (Reproduced Knispel et al., 1974.[11])

relaxation times.[23,25–29] It is well established that alterations in tissue hydration can affect the relaxation times of water protons. Nevertheless, it is clear from the investigations that there are many factors that may be responsible for the relaxation time differences that are observed. Thus, no strictly linear dependence of $T_1$ or ($1/T_1$) on water content is to be expected. An example of the relationship between the $T_1$ and water content is shown in Figure 6.16 for human breast tissues. While there is a definite correlation between $T_1$ and water content, it is not a linear relationship. It should be remembered, however, that the higher the water content, the larger is $T_1$ for water in most tissues.

### 6.2.C. *Dependence on paramagnetic ion concentrations*

In NMR imaging experiments, some paramagnetic species such as molecular oxygen and the $Mn^{2+}$ ion are

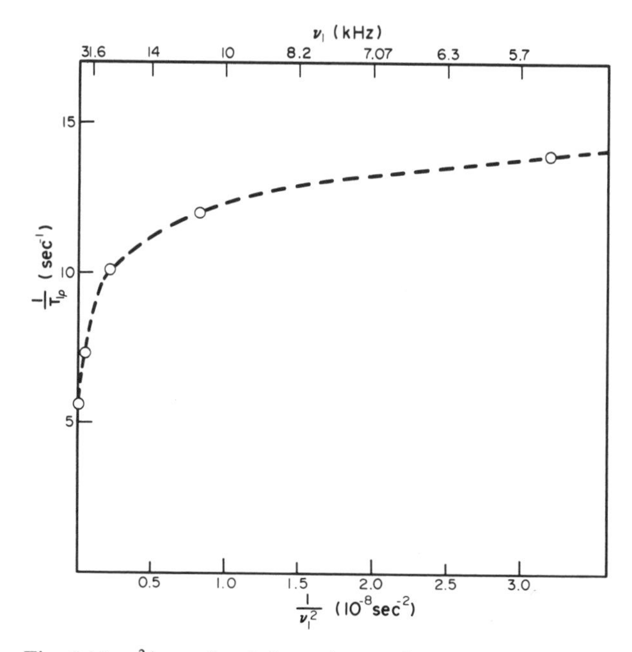


Fig. 6.13. $\nu^2(\omega_1 = 2\pi\nu_1)$ dependence of water proton relaxation rate ($1/T_{1\rho}$) in the kHz frequency range for frog gastrocnemius muscle tissue at 23°C. (Data from Finch and Homer, 1974.[13])

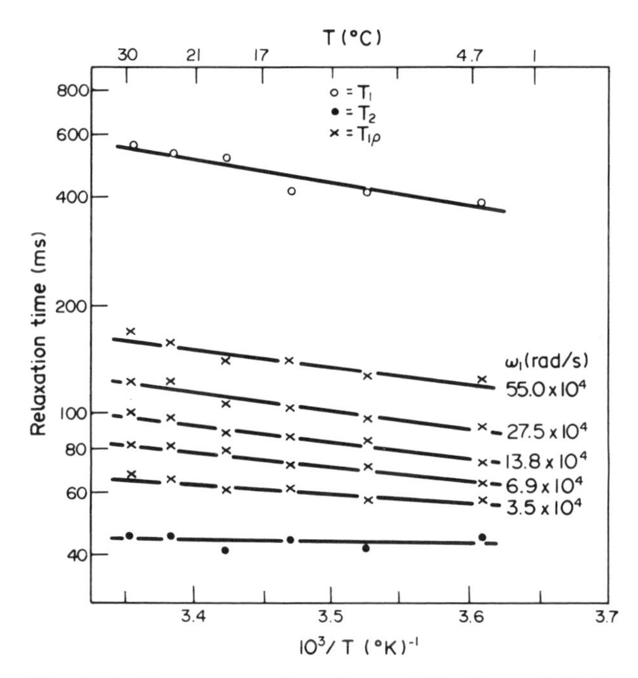


Fig. 6.14. Temperature dependence of $T_1$, $T_{1\rho}$, and $T_2$ for water protons in frog gastrocnemius muscle. $T_1$ and $T_2$ data were measured at 23.3 MHz, while $T_{1\rho}$ was obtained at the frequencies shown in the figure legend. The uncertainty in the data was ±5%. (Reproduced from Finch and Homer, 1974.[13])

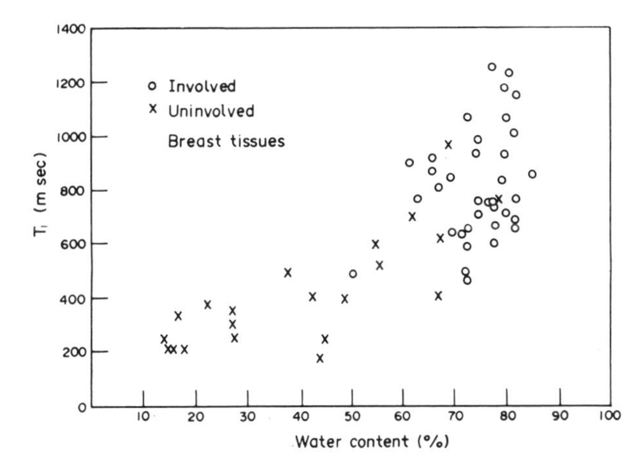


Fig. 6.16. Proton spin–lattice relaxation times ($T_1$) versus percentage water content in human breast tissues (involved and uninvolved tissues). Involved tissues were surgically resected material from the tumor region while the uninvolved tissues were taken from the region which appeared normal on gross examination.[28]

being tested as possible contrast agents because it is known that the addition of paramagnetic ions can shorten the relaxation times of protons.[30–33] Such shortening is produced by the dipolar interactions of the proton spins and the electron spins of the paramagnetic ion. An example of the effect of paramagnetic ions on water proton relaxation times ($T_1$) is shown in Table 6.4. Note that even a concentration as low as 10 $\mu$M of $Mn^{2+}$ can reduce the water relaxation by ~19%. The relaxation rate depends on the number of paramagnetic ions per milliliter. The theoretical equations for the relaxation of water protons in the presence of paramagnetic ions such as $Mn^{2+}$, $V^{2+}$, $Cr^{2+}$, $Ni^{2+}$, $Co^{2+}$, $Fe^{2+}$, and $Cu^{2+}$, and the temperature dependence of these relaxation times at a few frequencies can be found in an article by Bloembergen and Morgan.[33] In the presence of macromolecules and paramagnetic ions, an enhancement of water proton relaxation rate has been observed.[34] It has been suggested by Hollis et al.[35] that such a mechanism could be operative for water proton relaxation in tissues. Neverthe-

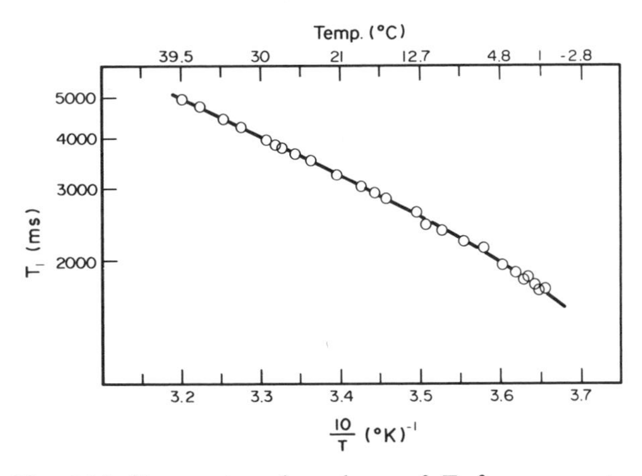


Fig. 6.15. Temperature dependence of $T_1$ for pure water (triply distilled, conductivity-grade). $T_1$ was measured at 60 MHz. The uncertainty in the data was about ±2%. (Data from Hindman et al., 1973.[14])

**Table 6.4. Effect of the paramagnetic ion $Mn^{2+}$ on water proton relaxation time measured at 20 MHz and 36°C**

| Concentration of $MnCl_2$ in water ($\mu$M) | $T_1$ (ms) |
|---|---|
| 0 (pure, double-distilled deionized water) | 3,269 |
| 10 | 2,660 |
| 50 | 1,475 |
| 100 | 982 |
| 250 | 518 |
| 400 | 316 |
| 500 | 262 |
| 800 | 164 |
| 1,000 | 132 |
| 5,000 | 27 |
| 10,000 | 15 |

Solutions were $MnCl_2$ in double-deionized water. Error in the $T_1$ measurement was ~4%.

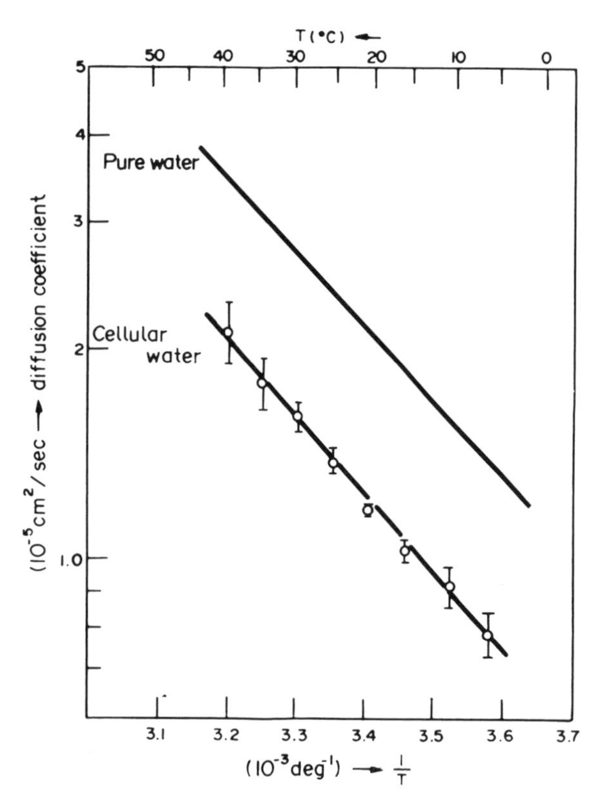


Fig. 6.17. Temperature dependence of the diffusion coefficients for muscle water and pure water. The diffusion coefficient for pure water is $2.4 \times 10^{-5}$ $cm^2/s$ (at 25°C). (Reprinted from Chang et al., 1973,[41] with permission.)

less, to date, no direct evidence exists for such an effect in intact tissues and cells, although it is known that abnormal tissues[36] have a tendency to accumulate trace metal ions, some of which are paramagnetic. The usefulness of paramagnetic ions as contrast agents is being explored in clinical NMR imaging.[37]

### 6.2.D. *Dependence on protein content*

It has been shown in dilute protein solutions that the relaxation times of water protons depend on the protein concentration.[38] It already has been mentioned, with regard to blood, that changes in hemoglobin content can alter relaxation times.[7,39] For dilute protein solutions, the proton relaxation time increases as the protein content decreases. Although the water in cells may not simply be described as a dilute protein solution,[2,3] intuitively one would expect the relaxation time to decrease as the protein content increases.

## 6.3. DIFFUSION COEFFICIENTS

It is possible to measure the self-diffusion coefficient of nuclear spins in general and water in particular by using spin–echo methods.[2,3] Extensive data exist in the literature[2,3] on the diffusion coefficients of water in various tissues. The diffusion coefficients for water in a variety of tissues are also tabulated in this book (see Table 9.7 in Chapter 9). Although no *in vivo* measurements of diffusion coefficients have been made, it is conceivable that *in vivo* water diffusion coefficients will be measured in the future using imaging techniques. From *in vitro* studies, it has been found that the self-diffusion coefficient of water in tissues is reduced by a factor of two, compared with bulk water.[2,3] For muscle water, the diffusion coefficient was found to be anisotropic and is larger for the muscle fiber orientation parallel to the magnetic field ($1.39 \times 10^{-5}$ $cm^2/s$) than for the orientation perpendicular to the field ($1.01 \times 10^{-5}$ $cm^2$ s).[40] The diffusion coefficient of pure water, as well as that of muscle water, is temperature dependent, and follows an Arrhenius-type of temperature dependence in the range of 0–50°C, as shown in Figure 6.17.[41] That is, the diffusion coefficient decreases with decreases in temperature. The explanation(s) for the reduced diffusion coefficient for water in tissues has been discussed by Hazlewood.[2] The interested reader is referred to that article for possible explanations.

## 6.4. REFERENCES

1. Damadian, R. *Science* **171**:1151, 1971.
2. Hazlewood, C.F., In: *Cell Associated Water*. Drost-Hansen, W., and Clegg, J., eds. Academic Press, New York, 1979, pp. 165–260.
3. Hazlewood, C.F., ed. *Ann. N.Y. Acad. Sci.* **204**:1–631, 1973.
4. Cooke, R., and Kuntz, I.D. *Annu. Rev. Biophys. Bioeng.* **3**:95, 1974.
5. De'Vre, R.M. *Prog. Biophys. Mol. Biol.* **35**:103, 1979.
6. Abragam, A. *The Principles of Nuclear Magnetism*. Clarendon Press, Oxford, 1961, Chapter 3.
7. Brooks, R.A., Battocletti, J.H., Sances, A. Jr., Larson, S.J., Bowman, R.L., and Kudravcev, V. *IEEE Trans. Biomed. Eng.* **BME-22**:12, 1975.
8. Zipp, A., James, T.L., Kuntz, I.D., and Shohet, S.B. *Biochim. Biophys. Acta* **428**:291, 1976.
9. Thompson, B.C., Waterman, M.R., and Cottam, G.L. *Arch. Biochem. Biophys.* **166**:193, 1973.
10. Lindstrom, T.R. and Koenig, S.J. *J. Mag. Reson.* **15**:344, 1974.
11. Knispel, R.R., Thompson, R.T., and Pintar, M.M. *J. Magn. Reson.* **14**:44, 1974.
12. Fukushima, E., and Roeder, S.B.W. *Experimental Pulse NMR: A Nuts and Bolts Approach*. Addison-Wesley, Reading, MA, 1981.
13. Finch, E.D., and Homer, L.D. *Biophys. J.* **14**:907, 1974.
14. Hindmann, J.C., Svirmickas, A., and Wood, M. *J. Chem. Phys.* **59**:1517, 1973.
15. Thompson, R.T., Knispel, R.R., and Pintar, M.M. *Chem. Phys. Lett.* **22**:335, 1973.
16. Outhred, R.K. and George, E.P. *Biophys. J.* **13**:97, 1973.
17. Duff, I.D., and Derbyshire, W. *J. Magn. Reson.* **15**:130, 1974.

18. Fung, B.M., and McGaughy, T.W. *Biochim. Biophys. Acta* **343**:663, 1974.
19. Fung, B.M., Durham, D.L., and Wassil, D.A. *Biochim. Biophys, Acta* **399**:191, 1975.
20. Coles, B.A. *J.N.C.I.* **57**:389, 1976.
21. Kasturi, S.R., Chang, D.C., and Hazlewood, C.F. *Biophys, J.* **30**:369, 1980.
22. Burnell, E.E., Clark, M.E., Hinke, J.A.M., and Chapman, N.R. *Biophys. J.* **33**:1, 1981.
23. Ling, C.R., Foster, M.A., and Hutchison, J.M.S. *Phys. Med. Biol.* **25**:748, 1980.
24. Escanye, J.M., Canet, D., and Robert, J. *Biochim. Biophys. Acta* **721**:305, 1982.
25. Boveé, W., Huismann, P., and Smidt, J. *J.N.C.I.* **52**:595, 1974.
26. Hazlewood, C.F., Chang, D.C., Nichols, B.L., and Woessner, D.E. *Biophys. J.* **14**:583, 1974.
27. Hollis, D.P., Saryan, L.A., Economou, J.S., Eggleston, J.C., Czeisler, J.L., and Morris, H.P. *J.N.C.I.* **53**:807, 1974.
28. Kasturi, S.R., Ranade, S.S., and Shah, S. *Proc. Ind. Acad. Sci.* **84B**:60, 1976.
29. Kiricuta, I.C., Jr., and Simplaceanu, V. *Cancer Res.* **35**:431, 1975.
30. Bloembergen, N., Purcell, E.M., and Pound, R.V. *Physiol. Rev.* **73**:679, 1948.
31. Solomon, I. *Physiol. Rev.* **99**:559, 1955.
32. Bloembergen, N. *J. Chem. Phys.* **27**:572, 1957.
33. Bloembergen, N., and Morgan L.O. *J. Chem. Phys.* **34**:842, 1961.
34. Mildvan, A.S., and Cohn, M. *Adv. Enzymol.* **33**:1, 1970.
35. Hollis, D.P., Economou, J.S., Parks, L.C., Eggleston, J.C., Saryan, L.A., and Czeisler, J.L. *Cancer Res.* **33**:2156, 1973.
36. Sissoef, J.G., Grisvard, J., and Guille, E. *Prog. Biophys. Mol. Biol.* **31**:165, 1976.
37. Lauterbur, P.C., Dias, M.H.M., and Rudin, A.M. *Front. Biol. Energetics* **1**:753, 1978.
38. Koenig, S.H., and Schillinger, W.H. *J. Biol. Chem.* **244**:3283, 1969.
39. Koivula, A., Suominen, K., Timonen, J., and Kiviniitty, K. *Phys. Med. Biol.* **7**:937, 1982.
40. Cleveland, G.G., Chang, D.C., Hazlewood, C.F., and Rorschach, H.E. *Biophys. J.* **16**:1043, 1976.
41. Chang, D.C., Rorschach, H.E., Nichols, B.L., and Hazlewood, C.F. *Ann. N.Y. Acad. Sci.* **204**:434, 1973.
42. Diem, K., and Leutner, C., eds. *Documenta Geigy Scientific Tables,* 7th ed., Ciba Geigy, Basel, Switzerland, 1973.
43. Ratkovic, S. and Rusov, C. *Perspect. Biol. Med.* **76**:19, 1974.
44. Block, R.E., and Maxwell, G.P. *J. Magn. Reson.* **14**:329, 1974.
45. Chaughule, R.S., Kasturi, S.R., Vijayaraghavan, R., and Ranade, S.S. *Ind. J. Biochem. Biophys.* **11**:256, 1974.
46. Damadian, R., Zaner, K., Hor, D., and DiMaio, T. *Proc. Natl. Acad. Sci. U.S.A.* **71**:1471, 1974.
47. Frey, H.E., Knispel, R.R., Kruuv, J., Sharp, A.A., Thompson, R.T., and Pintar, M.M. *J.N.C.I.* **49**:903, 1972.
48. Gwango, K., and Edzes, H.T. *Arch. Neurol.* **32**:462, 1975.
49. Hansen, J.R. *Biochim. Biophys. Acta* **230**:482, 1971.
50. Iijima, N., Saitoo, S., Yoshida, Y., Fuji, N., and Koike, T. *Physiol. Chem. Phys.* **5**:431, 1973.
51. Kiricuta, I.C,. Jr., and Simplaceanu, V. *Cancer Res.* **35**:1164, 1975.
52. Parrish, R.G., Kurland, R.J., Janese, W.W., and Bakay, L. *Science* **183**:438, 1974.
53. Chang, D.C., Hazlewood, C.F., Nichols, B.L., and Rorschach, H.E. *Nature* **235**:170, 1972.
54. Finch, E., Harmon, J.F., and Muller, B.H. *Arch. Biochem. Biophys.* **147**:299, 1971.
55. Hazlewood, C.F., Nichols, B.L., Chang, D.C., and Brown, B. *Johns Hopkins Med. J.* **128**:117, 1971.
56. Hazlewood, C.F., Cleveland, G., and Medina, D. *J.N.C.I.* **52**:1849, 1974.
57. Held, C., Noack, F., Pollak, V., and Mellon, B. *Z. Naturforsch.* [*C*] **28**:59, 1973.
58. Inch, W.R., McCredie, J.A., Knispel, R.R., Thompson, R.T., and Pintar, M.M. *J.N.C.I.* **52**:353, 1974.
59. Abetsedarskaya, L.A., Miftakhutdinova, F.G., and Fedotoz, V.C. *Biophys.* **13**:750, 1968.
60. Barroilhet, L.E., and Moran, P.R. *Med. Phys.* **3**:410, 1976.
61. Cooper, R.L., Chang, D.B., Young, A.C., Martin, C.J., and Johnson, B.A. *Biophys. J.* **14**:161, 1974.
62. Floyd, R.A., Yoshida, T., and Leigh, J.S. *Proc. Natl. Acad. Sci. U.S.A.* **72**:56, 1975.
63. Lewa, C.J., and Baczkowski, A. *Acta Physiol. Pol.* **A50**:865, 1976.
64. Civan, M.M., and Shporer, M. *Biophys. J.* **15**:299, 1975.
65. Lewa, C.J., and Zbytniewski, Z. *Bull. Cancer (Paris)* **63**:69, 1976.
66. Eggleston, J.C., Sarayan, L.A. and Hollis, D.P. *Cancer Res.* **35**:1326, 1975.

CHAPTER 7

# INTRODUCTION TO ELEMENTARY STATISTICS FOR NMR DATA HANDLING



## 7.1. BASIC DEFINITIONS OF STATISTICAL TERMS

*Alternative hypothesis:* A hypothesis acceptable as an alternative to the null hypothesis.

*Chi-square distribution:* The distribution of the sum of the squares of $n$ independent normal variates, where each normal variate emanates from a normal population with mean of 0 and standard deviation of 1. The parameter $n$ is the number of degrees of freedom.

*Coefficient of variation:* The standard deviation ($\sigma$) of a distribution divided by the mean ($\bar{x}$) of the distribution, i.e., $\sigma/\bar{x} \times 100$.

*Continuous variable:* A variable that can assume values in a continuous range.

*Correlation coefficient:* Measures the interdependence of two variables. It takes values between $-1$ and $+1$, with the intermediate value of 0 indicating the absence of association and with the $+1$ and $-1$ values indicating perfect positive association and perfect negative association, respectively.

*Degrees of freedom (DF):* Term for a parameter indexing a family of distributions.

*Dependent variable:* Term used in contradistinction to "independent variable" in regression analysis. A random variable $y$ is called a dependent variable if it is expressed as a function of a controlled variable $x$.

*Discrete variable:* Variable that assumes a finite or at most a countable number of possible values.

*Distribution:* Description of a set of measurement values along with their frequencies of occurrence. A distribution of measurements is illustrated with a histogram or a cumulative, relative-frequency polygon.

F-*distribution:* The distribution of the statistic $F = S_2^2/S_1^2$, where $S_1^2$ is the variance of a sample of size $m$

from a normal population having a variance $\sigma^2$ and $S_2^2$ is the variance of an independent sample of size $n$ from a normal population with variance $\sigma^2$. $F$ is said to have an $F$ distribution with $n - 1$ degrees of freedom in the numerator and $m - 1$ degrees of freedom in the denominator. (The $F$-distribution is ascribed to R. F. Fisher.)

*Finite population:* A finite collection of individuals.

*Finite population correction:* If a sample of $n$ values is drawn without replacement from a finite population of size $N$, then the standard deviation of the sample mean $\bar{x}$ can be written as

$$\sigma_{\bar{x}} = \sqrt{\frac{N-n}{N-1}} \cdot \frac{\sigma}{\sqrt{n}} \quad ,$$

where $\sigma$ is the population standard deviation. The factor $(N - n)/(N - 1)$ is sometimes called the finite population correction.

*Frequency:* The number of times a given type of event occurs, or the number of members of a population that fall into a specified class.

*Independent variable:* Compare "dependent variable" above. When a random variable $y$ is expressed as a function of a controlled variable $x$, $x$ is known as the independent variable.

*Linear regression* (of $y$ on $x$): When the regression curve of $y$ on $x$ is a straight line, that is, when $u_{yx} = A + Bx$, then we say there is a linear regression of $y$ on $x$.

*Normal distribution:* A symmetric, bell-shaped probability distribution, given by

$$f(x) = \frac{1}{\sigma\sqrt{2\pi}} e^{-(x-\mu)^2/2\sigma^2}, \qquad -\infty < x < +\infty \quad .$$

The value $f(x)$ is the height of the normal curve at point $x$. The parameters $\mu$ and $\sigma$ are the mean and standard deviation, respectively. $\mu$ can be any number and $\sigma$ can be any positive number.

*Null hypothesis:* In general, this refers to a particular hypothesis being tested that contrasts with the alternative hypotheses under consideration.

p *value:* See "significance probability" below.

*Parameter:* In statistics, a parameter is a quantity associated with a conceptual population. For example, the mean $\mu$ of a normal population is a population parameter.

*Population:* A term used to describe any finite or infinite collection of individuals.

*Population mean:* For a finite population with measurement values $x_1, x_2, x_3, \ldots, x_N$, the population mean is defined as

$$\mu = \frac{x_1 + x_2 + x_3 + \cdots + x_N}{N} \quad .$$

*Population median:* That value which divides the total population into two halves.

*Population standard deviation:* The most widely used measure of dispersion of a population. For a finite population with measurement values $x_1, x_2, \ldots, x_N$, the population standard deviation is defined as

$$\sigma = \sqrt{\frac{\Sigma(x-\mu)^2}{N}} \quad ,$$

where $\mu$ is the population mean.

*Population variance:* A parameter that measures dispersion in a population. It is given by the standard deviation of the population.

*Probability (of an event):* A number between 0 and 1 indicating how likely it is that the event will occur.

*Random sample* (of $n$ objects): A sample of $n$ objects (each with an associated measurement) such that for any given interval, each object's measurement has an equal and independent chance of falling in that interval. For example, the chance is indicated by the area of a histogram.

*Regression curve* (of $y$ on $x$): A curve that represents the regression equation. For a particular point on a given curve, the abscissa is the value of $x$ and the ordinate is $u_{yx}$ (the mean of the distribution of $y$ for that specified fixed value of $x$).

*Residual:* The difference between an observed value $y_i$ and the predicted value $a + bx_i$ is called the residual, in linear regression analyses.

*Sample:* A subset of a population used to investigate properties of the parent population.

*Sample mean:* The sample mean is

$$\bar{x} = \frac{x_1 + x_2 + x_3 + \cdots + x_n}{n}$$

for a sample of size $n$ with $x_1, x_2, \ldots, x_n$ as the measurement values.

*Sample median:* That value that divides the sample into two halves.

*Sample standard deviation:* The standard deviation of a sample is

$$S = \sqrt{\frac{\Sigma(x_i - \overline{x})^2}{n - 1}} \quad ,$$

where $\overline{x}$ is the sample mean for a sample of size $n$, and with $x_i$ taking the measurement values of $x_1, x_2, \ldots, x_n$.

*Scatter plot:* A diagram for examining the relationship between two variables $x$ and $y$.

*Significance probability* (p): The lowest value of $\alpha$ (the probability of a Type I error) that can be used and still be capable of rejecting the null hypothesis.

*Standard deviation:* See "population standard deviation" and "sample standard deviation."

*Standard error:* The standard error of a statistic $T$ is an estimator of the standard deviation of $T$. The *true* standard deviation SD($T$) is estimated from the data. Such an estimate is referred to as standard error and is denoted by SD($T$). It is represented by SD when the statistic under consideration can be clearly understood from context.

*Standard normal distribution:* A normal distribution with mean $\mu = 0$ and standard deviation $\sigma = 1$.

t *distribution:* The distribution of $T = (\overline{x} - \mu)/(S/\sqrt{n})$ where $\overline{x}$ is the sample mean of a random sample of size $n$ from a normal population with mean $\mu$ and S is the sample standard deviation. In this case, the distribution has $n - 1$ degrees of freedom. The distribution is used for making inferences about the unknown population mean $\mu$. (This distribution is ascribed to Student. "Student" was the pen name of British statistician W.S. Gossett.)

*Test statistic:* A function of a sample of observations that provides a basis for testing a statistical hypothesis.

*Type I error:* A false rejection of the null hypothesis, i.e., a hypothesis is rejected when, in fact, it is true.

*Type II error:* A false acceptance of the null hypothesis, i.e., a null hypothesis is accepted when, in fact, an alternative hypothesis is true.

## 7.2. FORMULAE

Sum of squares:

$$\Sigma(x_i - \overline{x})^2$$

Variance ($S^2$):

$$S^2 = \frac{\Sigma(x_i - \overline{x})^2}{n - 1}$$

Coefficient of variance (CV):

$$CV = \frac{S}{\overline{x}} \times 100$$

Proportion percent ($P$):

$$P = \frac{\text{number with the characteristic}}{\text{total number in the sample}} \times 100$$

Standard error of the mean (SEM):

$$SEM = \frac{S}{\sqrt{n}}$$

SE of the difference between two means ($S_{\overline{d}}$):

$$S_{\overline{d}} = \sqrt{\frac{S_1^2}{n_1} + \frac{S_2^2}{n_2}}$$

Level of significance: Expressed as $p = 0.05$ for 5%, $p = 0.10$ for 10%, etc.

$t$ test:

$$t = \frac{\text{observed difference between means}}{\text{SE of difference between means}}$$

Degrees of freedom (DF) for $S_{\overline{d}}$:

$$DF = n_1 + n_2 - 2 \quad ,$$

where $n_1$ and $n_2$ are the numbers in two samples.

$\chi^2$ test:

$$\text{If DF} > 1, \chi^2 = \Sigma \frac{(O - E)^2}{E} \quad ,$$

where $O$ is the observed value and $E$ is the expected value.

$$\text{If DF} = 1, \ \chi^2 = \Sigma \frac{[(O - E) - \tfrac{1}{2}]^2}{E} \quad .$$

Degrees of freedom (DF) in $\chi^2$ test (for independence of two classifications):

$$DF = (r - 1)(c - 1) \quad ,$$

where $r$ and $c$ are the number of rows and columns in a table.

Correlation coefficient ($r$):

$$r = \frac{\Sigma \Delta x \Delta y}{\sqrt{\Sigma(\Delta x)^2 \Sigma(\Delta y)^2}} ,$$

where $\Delta x$ and $\Delta y$ are deviations from the mean, that is

$$\Delta x = x - \bar{x} \quad \text{and} \quad \Delta y = y - \bar{y}$$

and

$$r = \frac{\Sigma(x_i - \bar{x})(y_i - \bar{y})}{\sqrt{\Sigma(x_i - \bar{x})^2 \, \Sigma(y_i - \bar{y})^2}} .$$

An alternative formula for $r$, which does not require the calculation of the mean, is

$$r = \frac{n\Sigma x_i y_i - \Sigma x_i \Sigma y_i}{\sqrt{[n\Sigma x_i^2 - (\Sigma x_i)^2][n\Sigma y_i^2 - (\Sigma y_i)^2]}} .$$

Degrees of freedom (DF):

$$\text{DF} = n - 2$$

(where $n$ is the number of paired measurements), used to find $p$ from tables of correlation coefficients.

Regression coefficient:

$$b_{yx} = \frac{\Sigma \Delta x \Delta y}{\Sigma(\Delta x)^2} = \frac{\Sigma(x_i - \bar{x})(y_i - \bar{y})}{\Sigma(x_i - \bar{x})^2} ,$$

where $b_{yx}$ is the value of $y$ for 1 unit of $x$. Similarly,

$$b_{xy} = \frac{\Sigma \Delta x \Delta y}{\Sigma(\Delta y)^2} = \frac{\Sigma(x_i - \bar{x})(y_i - \bar{y})}{\Sigma(y_i - \bar{y})^2} .$$

Regression equation for $y$ on $x$ or $x$ on $y$:

$$y_c = \bar{y} + b_{yx}(x_i - \bar{x}) = \bar{y} + b(x_i - \bar{x}) = a + bx_i ,$$

where $y_c$ are calculated values of $y$ for $x$ and $a$ denotes $\bar{y} = b\bar{x}$, a constant.

## 7.3. METHODS OF FINDING THE SIZE OF A REQUIRED SAMPLE

As in any other experiment, in NMR experiments one faces the problem of determining the number of samples required for investigation that will minimize error. The larger the sample, the less will be the error due to chance. To choose a suitable number to reduce sampling error to a minimum, the following methods are recommended.

*For quantitative data,* e.g., $T_1$ and $T_2$ values of protons of tissues: Assume that variations in a population indicated by S are known. If S and SE are known from previous measurements, these values can be used to determine the size of the sample ($n$). Otherwise, these variations have to be estimated by preliminary investigations.

$$\text{SE of mean} = \frac{\text{S}}{\sqrt{n}}, \text{ i.e.,}$$

$$\sqrt{n} = \frac{\text{S}}{\text{SE}} \text{ or } n = \left(\frac{\text{S}}{\text{SE}}\right)^2 \text{ for 68\% confidence limits.}$$

The critical level of significance is delineated by the 95% confidence limits. Therefore,

$$n = \left(\frac{2\text{S}}{\text{SE}}\right)^2$$

and

$$n = \left(\frac{3\text{S}}{\text{SE}}\right)^2 \text{ for the 99\% confidence limits.}$$

(A sample size of 1 is enough if there is no variation.)

*For qualitative data:* Qualitative data, such as morbidity rate, cure rate, etc., have to be treated in a different way. The size should be such that the error is within a sampling error of ±10%, and it can be calculated by the formula for SE of proportion:

$$\text{SE} = \sqrt{\frac{Pq}{n}} ,$$

where $P$ = proportion percent = number with the characteristic/total number in sample × 100, and $q$ = complement of $P = 100 - P$ (i.e., proportion without characteristic). *Example:* If the anticipated cure rate is 5%, the SE should be within 10% of this, i.e., 0.5%. Therefore,

$$\text{SE} = \sqrt{\frac{Pq}{n}}, \quad \text{i.e., } 0.5 = \sqrt{\frac{5 \times 95}{n}}$$

and

$$n = \frac{5 \times 95}{(0.5)^2} = 1{,}900 \quad .$$

Thus, 1,900 people have to be examined.

## 7.4. HYPOTHESIS TESTING

Hypothesis tests are used in experiments to help determine if small differences between two samples of data are more likely to be due to random sampling variation or to known differences in particular experimental variables. Here the two commonly used methods, particularly with regard to NMR data in medicine, are briefly discussed.

### 7.4.A. *Student's* t *test*

For small samples, the ratio of observed difference between two means to the standard error of the difference follows a distribution called the $t$ distribution, with $n - 1$ degree of freedom. It is denoted by the letter $t$.

The $t$ test is an accurate method for deciding whether or not the difference between two means of small samples is significant. Table 7.1 gives the values of $t$ that are numerically exceeded with given probabilities for different degrees of freedom. The probabilities ($p$) are given in decimals as 0.01 (1%), 0.05 (5%), etc. The probability of 0.05 is regarded as the critical level of significance.

As mentioned above, the $t$ test is used for small samples, i.e., those with fewer than 30 members. There are basically two types of $t$ test: unpaired $t$ test and paired $t$ test.

*Unpaired* t *test:* The unpaired test is used more often and is applied to means of two *different or separate groups of samples* drawn from two populations to determine if the difference between means is real or can be attributed to sampling variability.

**Table 7.1. Student's $t$ test values, exceeded with probability ($p$)**

| DF | $p = 0.1$ | 0.05 | 0.02 | 0.01 | 0.002 | 0.001 |
|---|---|---|---|---|---|---|
| 1 | 6.314 | 12.706 | 31.821 | 63.657 | 318.31 | 636.62 |
| 2 | 2.920 | 4.303 | 6.965 | 9.925 | 22.327 | 31.598 |
| 3 | 2.353 | 3.182 | 4.541 | 5.841 | 10.214 | 12.924 |
| 4 | 2.132 | 2.776 | 3.747 | 4.604 | 7.173 | 8.610 |
| 5 | 2.015 | 2.571 | 3.365 | 4.032 | 5.893 | 6.869 |
| 6 | 1.932 | 2.447 | 3.143 | 3.707 | 5.208 | 5.959 |
| 7 | 1.895 | 2.365 | 2.998 | 3.499 | 4.785 | 5.408 |
| 8 | 1.860 | 2.306 | 2.896 | 3.355 | 4.501 | 5.041 |
| 9 | 1.833 | 2.262 | 2.821 | 3.250 | 4.297 | 4.781 |
| 10 | 1.812 | 2.228 | 2.764 | 3.169 | 4.144 | 4.587 |
| 11 | 1.796 | 2.201 | 2.718 | 3.106 | 4.025 | 4.437 |
| 12 | 1.782 | 2.179 | 2.681 | 3.055 | 3.930 | 4.318 |
| 13 | 1.771 | 2.160 | 2.650 | 3.012 | 3.852 | 4.221 |
| 14 | 1.761 | 2.145 | 2.624 | 2.977 | 3.787 | 4.140 |
| 15 | 1.753 | 2.131 | 2.602 | 2.947 | 3.733 | 4.073 |
| 16 | 1.746 | 2.120 | 2.583 | 2.921 | 3.686 | 4.015 |
| 17 | 1.740 | 2.110 | 2.567 | 2.898 | 3.646 | 3.965 |
| 18 | 1.734 | 2.101 | 2.552 | 2.878 | 3.610 | 3.922 |
| 19 | 1.729 | 2.093 | 2.539 | 2.861 | 3.579 | 3.883 |
| 20 | 1.725 | 2.086 | 2.528 | 2.845 | 3.552 | 3.850 |
| 21 | 1.721 | 2.080 | 2.518 | 2.831 | 3.527 | 3.819 |
| 22 | 1.717 | 2.074 | 2.508 | 2.819 | 3.505 | 3.792 |
| 23 | 1.714 | 2.069 | 2.500 | 2.807 | 3.485 | 3.767 |
| 24 | 1.711 | 2.064 | 2.492 | 2.797 | 3.467 | 3.745 |
| 25 | 1.708 | 2.060 | 2.485 | 2.787 | 3.450 | 3.725 |
| 26 | 1.706 | 2.056 | 2.479 | 2.779 | 3.435 | 3.707 |
| 27 | 1.703 | 2.052 | 2.473 | 2.771 | 3.421 | 3.690 |
| 28 | 1.701 | 2.048 | 2.467 | 2.763 | 3.408 | 3.674 |
| 29 | 1.699 | 2.045 | 2.462 | 2.756 | 3.396 | 3.659 |
| 30 | 1.697 | 2.042 | 2.457 | 2.750 | 3.385 | 3.646 |
| 40 | 1.684 | 2.021 | 2.423 | 2.704 | 3.307 | 3.551 |
| 60 | 1.671 | 2.000 | 2.390 | 2.660 | 3.232 | 3.460 |
| 120 | 1.658 | 1.980 | 2.358 | 2.617 | 3.160 | 3.373 |
| $P$ | 1.645 | 1.960 | 2.326 | 2.576 | 3.090 | 3.291 |

*Example:* The probability of observing a value of $t$ greater than 3.169 with 10 degrees of freedom is 0.01 or 1%.

Let us start with the hypothesis that there is no real difference between the means of the two samples, i.e., the samples are drawn from the same population. We can then proceed as follows to test the significance of the difference:

1. Calculate the difference between two means $(\bar{x}_1 - \bar{x}_2)$.

2. Calculate the standard error of difference between the two means, i.e.

$$S_{\bar{d}} = \sqrt{\frac{S_1^2}{n_1} + \frac{S_2^2}{n_2}} \quad .$$

This gives a variation which can occur purely by chance between the means of two random samples drawn from the same population.

3. Find $t$ by using the formula:

$$\pm t = \frac{\bar{x}_1 - \bar{x}_2}{\text{SE of difference between two means}}$$

$$= \frac{\bar{x}_1 - \bar{x}_2}{\sqrt{\frac{S_1^2}{n_1} + \frac{S_2^2}{n_2}}} \quad .$$

4. Calculate the pooled degrees of freedom using the formula: DF $= n_1 + n_2 - 2$.

5. Now refer to the $t$ table (Table 7.1) and find the probability of the calculated $t$ corresponding to the degrees of freedom for the sample under study.

*Paired* t *test:* The paired test is applied to a paired sample, i.e., every individual of the sample gives a pair of readings. The observations are paired, i.e., they are made on the same sample (e.g., the $T_1$ values of a tissue before and after treatment).

Again, we start with the null hypothesis, i.e., we assume that there is no real difference between the two paired observations. We then proceed as follows to test the significance:

1. Calculate the difference in each set of paired observations: $(\Delta x = x_1 - x'_1)$.

2. Calculate the mean difference $(\overline{\Delta x})$, i.e., $\overline{\Delta x} = \Sigma \Delta x_i / n$ and calculate $\Sigma(\Delta x_i - \overline{\Delta x})^2$.

3. Find the standard deviation (S) and the standard error (SE) of the above difference from the mean:

$$S = \sqrt{\frac{\Sigma(\Delta x_i - \overline{\Delta x})^2}{n-1}} \text{ and SE} = \frac{S}{\sqrt{n}}$$

4. Find $t$ by substituting the above values in the formula $\pm t = (m - \overline{\Delta x})/\text{SE}$, where $m = 0$ by null hypothesis. Thus, $\pm t = \Delta x/\text{SE}$. An assumption is made that the mean difference is zero, i.e., the treatment has no effect on $T_1$.

5. Calculate the degrees of freedom. Since the same sample is being used, it is $n - 1$.

6. Refer to the $t$ table (Table 7.1) to find the probability of the calculated $t$ corresponding to the degrees of freedom used.

7. If the probability ($p$) is more than 5%, or 0.05, then the difference observed is not significant, but if it is less than 0.05, the difference observed is significant (because such a difference is less likely to occur by chance).

The $t$ table, therefore, gives the significant limit of any value, i.e., the highest that can be obtained by chance, for a particular probability or level of significance. The $t$ value estimated from the sample is significant and cannot be obtained by chance if it is higher than the value listed in the $t$ table for a particular probability.

**7.4.B.** *The chi square* $(\chi^2)$ *test*

The chi square $(\chi^2)$ test is another probability test frequently used in medical statistics. It is used when a series of observed values are to be compared with corresponding, matched, expected values to determine if any overall dissimilarities can be explained by the randomness of natural sampling fluctuations. It is also sometimes used as a "goodness of fit" test.

$$\chi^2 = \sum_i \frac{(O_i - E_i)^2}{E_i}$$

if degrees of freedom (DF) > 1, and

$$\chi^2 = \sum_i \frac{[(O_i - E_i)^{-1/2}]^2}{E_i} \qquad \text{if DF} = 1,$$

where $O$ is the observed value and $E$ is the expected value. $\chi^2$ is therefore a measure of the overall differences between the observed and expected frequencies.

When the $\chi^2$ value is calculated, the $\chi^2$ distribution table (Table 7.2) has to be consulted to determine whether or not the value of $\chi^2$ obtained is significant at the 5% or 1% level, or any other level. Whether or not

**Table 7.2. Chi square ($\chi^2$) values**

| DF | $p$ = 0.900 | 0.500 | 0.100 | 0.050 | 0.010 | 0.001 |
|---|---|---|---|---|---|---|
| 1 | 0.016 | 0.455 | 2.71 | 3.84 | 6.63 | 10.83 |
| 2 | 0.211 | 1.39 | 4.61 | 5.99 | 9.21 | 13.82 |
| 3 | 0.584 | 2.37 | 6.25 | 7.81 | 11.34 | 16.27 |
| 4 | 1.06 | 3.36 | 7.78 | 9.49 | 13.28 | 18.47 |
| 5 | 1.61 | 4.35 | 9.24 | 11.07 | 15.09 | 20.52 |
| 6 | 2.20 | 5.35 | 10.64 | 12.59 | 16.81 | 22.46 |
| 7 | 2.83 | 6.35 | 12.02 | 14.07 | 18.48 | 24.32 |
| 8 | 3.49 | 7.34 | 13.36 | 15.51 | 20.09 | 26.13 |
| 9 | 4.17 | 8.34 | 14.68 | 16.92 | 21.67 | 27.88 |
| 10 | 4.87 | 9.34 | 15.99 | 18.31 | 23.21 | 29.59 |
| 11 | 5.58 | 10.34 | 17.28 | 19.68 | 24.73 | 31.26 |
| 12 | 6.30 | 11.34 | 18.55 | 21.03 | 26.22 | 32.91 |
| 13 | 7.04 | 12.34 | 19.81 | 22.36 | 27.69 | 34.53 |
| 14 | 7.79 | 13.34 | 21.06 | 23.68 | 29.14 | 36.12 |
| 15 | 8.55 | 14.34 | 22.31 | 25.00 | 30.58 | 37.70 |
| 16 | 9.31 | 15.34 | 23.54 | 26.30 | 32.00 | 39.25 |
| 17 | 10.09 | 16.34 | 24.77 | 27.59 | 33.41 | 40.79 |
| 18 | 10.86 | 17.34 | 25.99 | 28.87 | 34.81 | 42.31 |
| 19 | 11.65 | 18.34 | 27.20 | 30.14 | 36.19 | 43.82 |
| 20 | 12.44 | 19.34 | 28.41 | 31.41 | 37.57 | 45.32 |
| 21 | 13.24 | 20.34 | 29.62 | 32.67 | 38.93 | 46.80 |
| 22 | 14.04 | 21.34 | 30.81 | 33.92 | 40.29 | 48.27 |
| 23 | 14.85 | 22.34 | 32.01 | 35.17 | 41.64 | 49.73 |
| 24 | 15.66 | 23.34 | 33.20 | 36.42 | 42.98 | 51.18 |
| 25 | 16.47 | 24.34 | 34.38 | 37.65 | 44.31 | 52.62 |
| 26 | 17.29 | 25.34 | 35.56 | 38.89 | 45.64 | 54.05 |
| 27 | 18.11 | 26.34 | 36.74 | 40.11 | 46.96 | 55.48 |
| 28 | 18.94 | 27.34 | 37.92 | 41.34 | 48.28 | 56.89 |
| 29 | 19.77 | 28.34 | 39.09 | 42.56 | 49.59 | 58.30 |
| 30 | 20.60 | 29.34 | 40.26 | 43.77 | 50.89 | 59.70 |
| 40 | 29.05 | 39.34 | 51.81 | 55.76 | 63.69 | 73.40 |
| 50 | 37.69 | 49.33 | 63.17 | 67.50 | 76.15 | 86.66 |
| 60 | 46.46 | 59.33 | 74.40 | 79.08 | 88.38 | 99.61 |
| 70 | 55.33 | 69.33 | 85.53 | 90.53 | 100.43 | 112.32 |
| 80 | 64.28 | 79.33 | 96.58 | 101.88 | 112.33 | 124.84 |
| 90 | 73.29 | 89.33 | 107.57 | 113.15 | 124.12 | 137.21 |
| 100 | 82.36 | 99.33 | 118.50 | 123.34 | 135.81 | 149.45 |

the difference between the observed and expected frequencies is significant depends not only on the value of $\chi^2$ but also on the number of degrees of freedom (DF). The value of the calculated $\chi^2$ is compared with the $\chi^2$ value at a particular level of significance in the table. If the value of $\chi^2$ calculated is greater than the $\chi^2$ value obtained from the $\chi^2$ table, then the difference between the observed and expected frequencies is significant.

## 7.5. CORRELATION COEFFICIENT

The correlation coefficient ($r$) is a measure of the degree of linear association or relationship found between two series of measurements, such as $T_1$ and water content. The $\chi^2$ test for qualitative data does not give the degree of association between two characteristics.

There are two ways of expressing the correlation coefficient: mathematical and graphic.

*Mathematical:* There are basically five types of correlation, depending on the degree and direction of the correlation.

1. Perfect positive correlation: In this case, the two variables denoted by the letters $x$ and $y$ are directly proportional and completely dependent on each other, and the coefficient is $+1$, i.e., both rise or fall proportionally.
2. Perfect negative correlation: In this case, the two variables are inversely proportional, i.e., one increases and the other decreases, and the coefficient of correlation is $-1$.
3. Absolutely no correlation: In this case, the two

variables are not related; they are independent of each other.

4. Partial positive correlation: the coefficient of correlation lies between 0 and +1.
5. Partial negative correlation: the coefficient of correlation lies between 0 and −1.

A partial correlation between the two variables may be significant if it is large, irrespective of its sign. Otherwise it can be considered as having arisen by chance. The significance of the calculated value of the correlation coefficient can be evaluated by finding the standard error of correlation. When the sample is small, the $t$ test can be applied. The probability ($p$) can be obtained by looking at the correlation coefficient table (Table 7.3), using the correct number of degrees of freedom (DF). DF $= n - 2$, where $n$ is the number of paired measurements.

*Graphic:* Correlation between two parameters can also be shown graphically by plotting the two parameters $x$ and $y$ on graph paper. Such a diagram is called a scatter diagram. As in the case of mathematical expression, there can be five types of correlation. For partial positive correlations or partial negative correlations, one line can be drawn through the mean values, called the regression line and indicating a positive or negative association; the correlation coefficient is obtained from this. The regression line can be $y$ on $x$ if the corresponding values of $y$ (i.e., $y_c$) are calculated for $x$ from the regression coefficient $b_{yx}$. The regression coefficient $b_{yx}$ is the value of $y_c$ (i.e., $y_c - \bar{y}$) for one unit of $x$ beyond $\bar{x}$, that is:

$$y_c = \bar{y} + b_{yx}(x - \bar{x})$$

and

$$b_{yx} = \frac{\Sigma \Delta x_i \Delta y_i}{\Sigma(\Delta x_i)^2} = \frac{\Sigma(x_i - \bar{x})(y_i - \bar{y})}{\Sigma(x_i - \bar{x})^2} .$$

**Table 7.3. The correlation coefficient ($r$)**

| DF | Probability ($p$) | | | | |
|---|---|---|---|---|---|
| | 0.10 | 0.05 | 0.02 | 0.01 | 0.001 |
| 1 | 0.9877 | 0.9969 | 0.9995 | 0.99988 | 0.99999 |
| 2 | 0.9000 | 0.9500 | 0.9800 | 0.9900 | 0.9990 |
| 3 | 0.805 | 0.878 | 0.9343 | 0.9587 | 0.9912 |
| 4 | 0.729 | 0.811 | 0.882 | 0.9172 | 0.9741 |
| 5 | 0.669 | 0.754 | 0.833 | 0.875 | 0.9509 |
| 6 | 0.621 | 0.707 | 0.789 | 0.834 | 0.9249 |
| 7 | 0.582 | 0.666 | 0.750 | 0.798 | 0.898 |
| 8 | 0.549 | 0.632 | 0.715 | 0.765 | 0.872 |
| 9 | 0.521 | 0.602 | 0.685 | 0.735 | 0.847 |
| 10 | 0.497 | 0.576 | 0.658 | 0.708 | 0.823 |
| 11 | 0.476 | 0.553 | 0.634 | 0.684 | 0.801 |
| 12 | 0.457 | 0.532 | 0.612 | 0.661 | 0.780 |
| 13 | 0.441 | 0.514 | 0.592 | 0.641 | 0.760 |
| 14 | 0.426 | 0.497 | 0.574 | 0.623 | 0.742 |
| 15 | 0.412 | 0.482 | 0.558 | 0.606 | 0.725 |
| 16 | 0.400 | 0.468 | 0.543 | 0.590 | 0.708 |
| 17 | 0.389 | 0.456 | 0.529 | 0.575 | 0.693 |
| 18 | 0.378 | 0.444 | 0.516 | 0.561 | 0.679 |
| 19 | 0.369 | 0.433 | 0.503 | 0.549 | 0.665 |
| 20 | 0.360 | 0.423 | 0.492 | 0.537 | 0.652 |
| 25 | 0.323 | 0.381 | 0.445 | 0.487 | 0.597 |
| 30 | 0.296 | 0.349 | 0.409 | 0.449 | 0.554 |
| 35 | 0.275 | 0.325 | 0.381 | 0.418 | 0.519 |
| 40 | 0.257 | 0.304 | 0.358 | 0.393 | 0.490 |
| 45 | 0.243 | 0.288 | 0.338 | 0.372 | 0.465 |
| 50 | 0.231 | 0.273 | 0.322 | 0.354 | 0.443 |
| 60 | 0.211 | 0.250 | 0.295 | 0.325 | 0.408 |
| 70 | 0.195 | 0.232 | 0.274 | 0.302 | 0.380 |
| 80 | 0.183 | 0.217 | 0.257 | 0.283 | 0.357 |
| 90 | 0.173 | 0.205 | 0.242 | 0.267 | 0.338 |
| 100 | 0.164 | 0.195 | 0.230 | 0.254 | 0.321 |

## 7.6. CURVE FITTING

### 7.6.A. *Least-squares fit to a straight line*

Let us consider a set of observations $(x_i, y_i)$ that are related by a linear relation of the form: $y = ax + b$.

Let us assume that we know the $x$ values accurately and that the error is contained in the $y$ values. Let us also assume that the weights of the $y$ values are equal:

$$\delta y_i = \text{the difference between } y \text{ and } y_i$$
$$= y_i - ax_i + b \quad .$$

The sum of squares of these differences is to be minimized. Therefore:

$$(\delta y_i)^2 = [\,y_i - (ax_i + b)]^2$$
$$= y_i^2 + a^2x_i^2 + b^2 + 2ax_ib - 2ax_iy_i - 2y_ib \quad .$$

If there are $n$ pairs of observations, then the sum $M$ of the squares of these differences is

$$M = \Sigma\,(\delta y_i^2) = \Sigma y_i^2 + a^2\,\Sigma x_i^2 + nb^2 + 2ab\,\Sigma x_i - 2a\,\Sigma x_iy_i - 2b\,\Sigma y_i \quad .$$

$\Sigma(\delta y_i)^2$ should be a minimum for the best fit of the values of $a$ and $b$. Therefore:

$$\frac{\partial M}{\partial a} = 0 \quad \text{and} \quad \frac{\partial M}{\partial b} = 0 \quad .$$

From the first condition, we get

$$2a\,\Sigma x_i^2 + 2b\,\Sigma x_i - 2\,\Sigma(x_iy_i) = 0 \quad . \qquad (7.1)$$

The second condition gives

$$2nb + 2a\,\Sigma x_i - 2\,\Sigma y_i = 0 \quad . \qquad (7.2)$$

Eqs. 7.1 and 7.2 can be solved to give

$$a = \frac{n\Sigma(x_iy_i) - (\Sigma x_i)(\Sigma y_i)}{n\Sigma x_i^2 - (\Sigma x_i)^2}$$

and

$$b = \frac{\Sigma x_i^2\Sigma y_i - \Sigma x_i\Sigma(x_iy_i)}{n\Sigma x_i^2 - (\Sigma x_i)^2} \quad .$$

Standard deviations for $a$ and $b$ can be calculated as follows:

**Table 7.4. NMR parameters for mouse mammary tissue samples**

| | Normal glands | | | Preneoplastic tissue | | | Adenocarcinoma cell line CDH-3 | | |
|---|---|---|---|---|---|---|---|---|---|
| $n$ | $T_1$ | $T_2$ | % $H_2O$ | $T_1$ | $T_2$ | % $H_2O$ | $T_1$ | $T_2$ | % $H_2O$ |
| 1 | 365 | 77 | 34.9 | 437 | 74 | 18.2 | 436 | 91 | 53.5 |
| 2 | 334 | 71 | 33.3 | 382 | 74 | 19.7 | 636 | 93 | 80.7 |
| 3 | 273 | 69 | 24.7 | 449 | 70 | 20.2 | 722 | 132 | 63.3 |
| 4 | 311 | 73 | 26.9 | 428 | 64 | 18.3 | 409 | 157 | 42.1 |
| 5 | 323 | 68 | 30.3 | 482 | 62 | 18.9 | 902 | 94 | 90.1 |
| 6 | 343 | 85 | 32.5 | 336 | 89 | 31.6 | 815 | 98 | 84.9 |
| 7 | 306 | 84 | 42.7 | 375 | 88 | 29.8 | 847 | 113 | 81.1 |
| 8 | 290 | 84 | 31.9 | 299 | 88 | 31.9 | 789 | 85 | 77.3 |
| 9 | 273 | 88 | 28.9 | 338 | 142 | 65.3 | 707 | 91 | 78.9 |
| 10 | 299 | 87 | 29.6 | 213 | 90 | 17.2 | 777 | 85 | 80.1 |
| 11 | 232 | 94 | 22.6 | 306 | 84 | 40.0 | 815 | 112 | 82.5 |
| 12 | 207 | — | 23.4 | 262 | 80 | 31.9 | 871 | 101 | 72.5 |
| 13 | 200 | — | 16.4 | 315 | 77 | 52.7 | 752 | 103 | 79.1 |
| 14 | 216 | 96 | 12.9 | 360 | 90 | 33.9 | 697 | 73 | 72.2 |
| 15 | 274 | 89 | 26.4 | 231 | 96 | 25.8 | 740 | 95 | 70.4 |
| 16 | 252 | 87 | 30.8 | 358 | 92 | 35.0 | 700 | 89 | 79.0 |
| 17 | 275 | 91 | 21.8 | 376 | 93 | 38.4 | 677 | 71 | — |
| 18 | 283 | 85 | 40.3 | | | | 746 | 96 | 83.0 |
| 19 | 244 | 90 | 19.6 | | | | 666 | 85 | 73.3 |
| 20 | 283 | 90 | 33.6 | | | | | | |
| 21 | 209 | 93 | 10.5 | | | | | | |
| 22 | 254 | 84 | 15.8 | | | | | | |
| Mean | 275 | 84 | 26.8 | 350 | 85 | 31.1 | 721 | 98 | 74.7 |
| ± SD | 46 | 8 | 8.4 | 75 | 18 | 13.1 | 127 | 20 | 11.7 |
| ± SE | 10 | 2 | 1.8 | 18 | 4 | 3.2 | 29 | 5 | 2.7 |
| $n$ | 22 | 20 | 22 | 17 | 17 | 17 | 19 | 19 | 19 |

$$S_y = \text{standard deviation of } y \text{ values} = \sqrt{\frac{\Sigma(\delta y_i)^2}{n-2}} .$$

The standard deviation of a computed function is given by

$$S^2 = \left(\frac{\partial f}{\partial x}\right)^2 S_x^2 + \left(\frac{\partial f}{\partial y}\right)^2 S_y^2 + \cdots .$$

In our sample, $x$ and $y$ are $y_1$, $y_2$, etc., which are part of our set of observations. Therefore, the function for $a$ is

$$a = \frac{1}{n\Sigma x_i^2 - (\Sigma x_i)^2} [nx_1y_1 - y_1\Sigma x_i + nx_2y_2 - y_2\Sigma x_i + \cdots] .$$

Therefore

$$\frac{\partial a}{\partial y_k} = \frac{1}{n\Sigma x_i^2 - (\Sigma x_i)^2} [nx_k - \Sigma x_i]$$

and

$$\left(\frac{\partial a}{\partial y_k}\right)^2 = \frac{1}{[n\Sigma x_i^2 - (\Sigma x_i)^2]^2} [n^2x_k^2 + (\Sigma x_i)^2 - 2nx_k\Sigma x_i] .$$

$S_y$ is common to all the contributions. Therefore, we can sum $(\partial a/\partial y_k)^2$ directly to obtain

$$\sum_k \left(\frac{\partial a}{\partial y_k}\right)^2 = \frac{1}{[n\Sigma x_i^2 - (\Sigma x_i)^2]^2} \times [n^2\Sigma x_i^2 + n(\Sigma x_i)^2 - 2n(\Sigma x_i)^2] .$$

In the above, use is made of the fact that $\Sigma x_k = \Sigma x_i$, etc. Therefore:

$$\sum_k \left(\frac{\partial a}{\partial y_k}\right)^2 = \frac{1}{[n\Sigma x_i^2 - (\Sigma x_i)^2]^2} [n^2\Sigma x_i^2 - n(\Sigma x_i)^2 = \frac{n}{n\Sigma x_i^2 - (\Sigma x_i)^2} ,$$

or $S_a$ = standard deviation of $a$, and is equal to

$$S_y \sqrt{\frac{n}{n\Sigma x_i^2 - (\Sigma x_i)^2}} .$$

Similarly, the standard deviation for $S_b$ can be obtained by the same procedure.

The coefficient of correlation, $r^2$, which determines the quality of the fit obtained by regression is given by

$$r^2 = \frac{[n\Sigma x_iy_i - \Sigma x_i\Sigma y_i]^2}{[n\Sigma x_i^2 - (\Sigma x_i)^2][n\Sigma y_i^2 - (\Sigma y_i)^2]} .$$

**7.6.B.** *Exponential function:* $y = ae^{bx}$ $(a > 0)$

Using the procedures described above, equations for the best fit values of $a$ and $b$ can be obtained:

$$b = \frac{n\Sigma x_i \ln y_i - (\Sigma x_i)(\Sigma \ln y_i)}{n\Sigma x_i^2 - (\Sigma x_i)^2}$$

$$a = \exp\left[\frac{\Sigma \ln y_i}{n} - \frac{b\Sigma x_i}{n}\right]$$

$$r^2 = \frac{[n\Sigma x_i \ln y_i - \Sigma x_i \Sigma \ln y_i]^2}{[n\Sigma x_i^2 - (\Sigma x_i)^2]\,[n\Sigma(\ln y_i)^2 - (\Sigma \ln y_i)^2]}$$

In NMR, one comes across the problem of fitting the data to multiexponentials of the form:

$$y = a_1e^{-b_1x} + a_2e^{-b_2x} + a_3e^{-b_3x}$$

There are several methods available for fitting the data to multiexponential forms, and the reader is referred to the report of Simon[1] and references therein for methods of carrying out such an analysis.

## 7.7. EXAMPLE CALCULATIONS FOR STATISTICAL HANDLING OF A SET OF NMR DATA

The following set of experimental data (in Table 7.4) was generated by Beall and colleagues during their investigations of the distinction on NMR between normal mouse mammary gland and preneoplastic and neoplastic mammary tissues. The raw data, measured at 30 MHz and 25°C, are presented to give an idea of the range and variation found in a reasonably large number of real tissue samples. These data will be used to demonstrate the calculation of standard statistical parameters, for testing by the Student's $t$ test, and for determining the correlation coefficient between $T_1$ and % $H_2O$.

1. Calculate the mean ($\overline{x}$), standard deviation (SD), and standard error of the mean (SE) for each group and each variable by the following formulae:

   Mean:

$$\overline{x} = \frac{x_1 + x_2 + x_3 + \ldots + x_n}{n} .$$

Standard deviation for group 1:

$$S_1 = \sqrt{\frac{\Sigma(x_i - \bar{x})^2}{n - 1}} \quad .$$

Standard error of the mean:

$$SE = \frac{S}{\sqrt{n}} \quad .$$

2. Using these data, compare the normal samples with the adenocarcinoma group to test for significant differences in $T_1$ values, using the unpaired Student's $t$ test (Table 7.5):
   a. Calculate the difference between the means of the two groups as

$$721 - 275 = 446 \quad .$$

   (A large difference between the means offers hope of statistical significance, but is not conclusive.)
   b. Calculate the standard error of the difference between the two means:

$$S_{\bar{d}} = \sqrt{\frac{S_1^2}{n_1} + \frac{S^2}{n_2}}$$

$$S_{\bar{d}} = \sqrt{\frac{(127)^2}{19} + \frac{(46)^2}{22}} = \sqrt{849 + 96} = \sqrt{945} = 31$$

   c. Calculate the $t$ value:

$$t = \frac{\bar{x}_1 - \bar{x}_2}{S_{\bar{d}}} = \frac{446}{31} = 14.4 \quad .$$

   d. Calculate the pooled degrees of freedom (DF):

$$DF = n_1 + n_2 - 2 = 19 + 22 - 2 = 39 \quad .$$

   e. Refer to the $t$ table and find the $p$ value for $t$ = 14.4 and DF = 39 ($p << 0.001$ for the distinction of cancers from normal tissues).

3. Calculate the correlation coefficient between two variables (Table 7.5), in this case the $T_1$ value and % $H_2O$ value, for the set of normal samples:

$$r = \frac{\Sigma \Delta x \Delta y}{\sqrt{\Sigma(\Delta x)^2 \Sigma(\Delta y)^2}} \quad ,$$

where $\Delta x_1 = x_1 - \bar{x}$ (mean of the group).

$$\Delta x_2 = x_2 - \bar{x}$$

$$\Delta x_3 = x_3 - \bar{x}$$

$$\Delta x_n = x_n - \bar{x} \quad .$$

**Table 7.5. Calculation of $r$ between $T_1$ and % $H_2O$ for normal subjects ($n$ = 22)**

| $T_1$ | $-\bar{x}$ | $\Delta x$ | $(\Delta x)^2$ | $\Delta xy$ | % $H_2O$ ($y$) | $-\bar{y}$ | $\Delta y$ | $(\Delta y)^2$ |
|---|---|---|---|---|---|---|---|---|
| 365 | −275 | 90 | 8,100 | 729 | 34.9 | −26.8 | 8.1 | 65.6 |
| 334 | | 59 | 3,481 | 383 | 33.3 | | 6.5 | 42.3 |
| 273 | | −2 | 4 | 379 | 24.7 | | −2.1 | 4.4 |
| 311 | | 36 | 1,296 | 3.6 | 26.9 | | 0.1 | 0.01 |
| 323 | | 48 | 2,304 | 168 | 30.3 | | 3.5 | 12.3 |
| 343 | | 68 | 4,624 | 388 | 32.5 | | 5.7 | 32.5 |
| 306 | | 31 | 961 | 493 | 42.7 | | 15.9 | 252.8 |
| 290 | | 15 | 225 | 77 | 31.9 | | 5.1 | 26.0 |
| 273 | | −2 | 4 | −4.2 | 28.9 | | 2.1 | 4.4 |
| 299 | | 24 | 576 | 67 | 29.6 | | 2.8 | 7.8 |
| 232 | | −43 | 1,849 | 180 | 22.6 | | −4.2 | 17.6 |
| 207 | | −68 | 4,624 | 231 | 23.4 | | −3.4 | 11.6 |
| 200 | | −75 | 5,625 | 780 | 16.4 | | −10.4 | 108.2 |
| 216 | | −59 | 3,481 | 820 | 12.9 | | −13.9 | 193.2 |
| 274 | | −1 | 1 | 0.4 | 26.4 | | −0.4 | 0.1 |
| 252 | | −23 | 529 | 92 | 30.8 | | 4.0 | 16.0 |
| 275 | | 0 | 0 | 0 | 21.8 | | −5.0 | 25.0 |
| 283 | | 8 | 64 | 108 | 40.3 | | 13.5 | 182.3 |
| 244 | | −31 | 961 | 223 | 19.6 | | −7.2 | 51.8 |
| 283 | | 8 | 64 | 54 | 33.6 | | 6.8 | 46.2 |
| 209 | | −66 | 4,356 | 1,076 | 10.5 | | −16.3 | 265.7 |
| 254 | | −21 | 441 | 231 | 15.8 | | −11.0 | 121.0 |
| | | | Sum = 43,570 | 6,479 | | | | Sum = 1,486.9 |

Do the same calculations for the other variable ($y$).
Calculate:

$$r = \frac{\Sigma \Delta x \Delta y}{\sqrt{\Sigma(\Delta x)^2 \Sigma(\Delta y)^2}} = \frac{6,479}{\sqrt{43,570 \times 1486.9}}$$

$$= \frac{6,479}{8,049} = 0.805 \quad .$$

This suggests a strong, but not perfect, linear correlation between $T_1$ and % $H_2O$ in these samples. The calculation of $r$ is usually possible on most hand-held calculators today, which store $\Delta x$ and $(\Delta x)^2$ as they calculate SD and SE.

4. Fit the multicomponent relaxation behavior of a biological sample using multiexponentials: An example of the fitting of multiexponential data taken for spin–spin relaxation times ($T_2$) in rat skeletal muscle is provided in the report of Hazlewood et al.[2]

Figure 7.1 shows the amplitude of the spin–echoes of a $T_2$ measurement on rat gastrocnemius muscle as a function of time ($2\tau$). The amplitude has been normalized by taking the amplitude of the free induction decay signal at 100 $\mu$s after the 90° pulse to be unity. From Figure 7.1 it is apparent that the amplitude decay is not a simple exponential, and thus the spin–spin relaxation rate cannot be described by a single $T_2$ value. After correction for the diffusional decay, the typical result of a spin–echo series is shown in Figure 7.2. It clearly indicates that there are multiple frac-

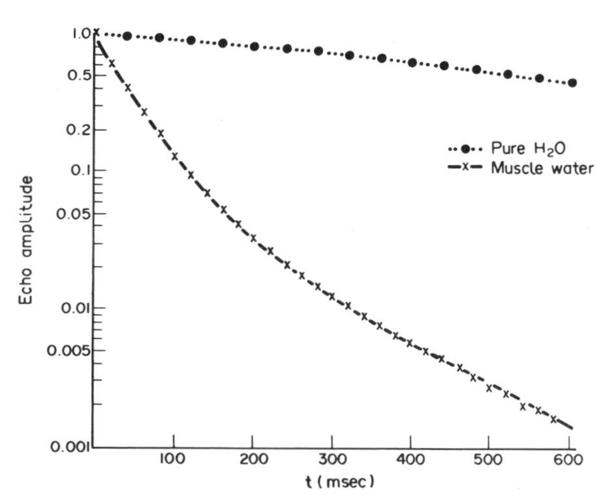


Fig. 7.1. The initial portion of the echo-decay curve in a 90°–$\tau$–180° spin–echo relaxation time ($T_2$) measurement on rat leg muscle. Echoes were measured at very small intervals and demonstrate at least two short-relaxing fractions of protons in muscle.

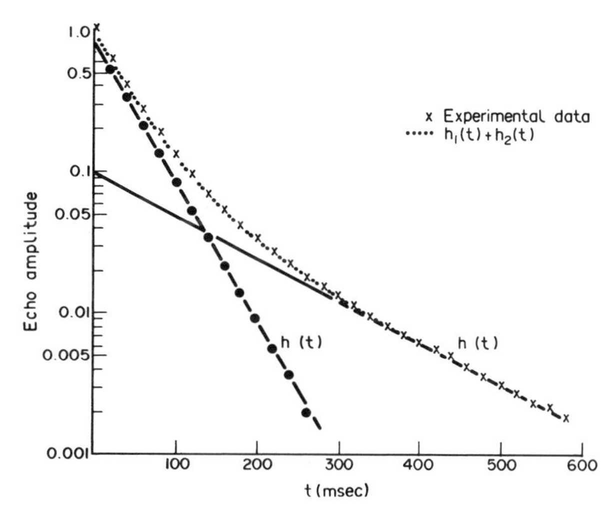


Fig. 7.2. Echo-decay curve in a 90°–$\tau$–180° sequence for rat muscle, taken over a span of 500 ms. A slow-relaxing fraction of protons can be identified.

tions of tissue water which are not in rapid exchange and which involve distinct spin–spin relaxation times. One may assume that the measured signal is a summation of the signals from several non- (or slowly) exchanging fractions, i.e.:

$$M(t) = \sum_i M_i(t)$$

and

$$M_i(t) = M_{oi} \exp(-t/T_{2,i}) \quad ,$$

where $M(t)$ is the normalized echo amplitude at time $t = 2\tau$, and $M_{oi}$ and $T_{2,i}$ are the relative population and the transverse relaxation time of the *ith* fraction. This model does not exclude the possibility that within each $M(t)$ there may be several fast-exchanging subfractions, but those fractions which are in fast exchange will show a single $T_2$ and will appear as one fraction. It is important to know the minimum number of fractions required to fit the data.

First we attempt to fit the data with two fractions. We follow the usual procedures for curve decomposition by finding the slowest-decaying fraction and subtracting this fraction from the observed data. As can be seen from Figure 7.2 the observed $M(t)$ approaches a straight line when $t$ is larger than 350 ms. This suggests that $M(t)$ is composed predominately of the slow-relaxing fraction for $t > 350$ ms. The straight line which fits this part of the curve is labeled $h_1(t)$ or $M_1(t)$. The points resulting from the subtraction of $M_1(t)$ from the decay curve can be fit with another straight line on the semilog plot and are called $h_2(t)$ or $M_2(t)$. The sum of $M_1(t)$ and $M_2(t)$ closely approxi-

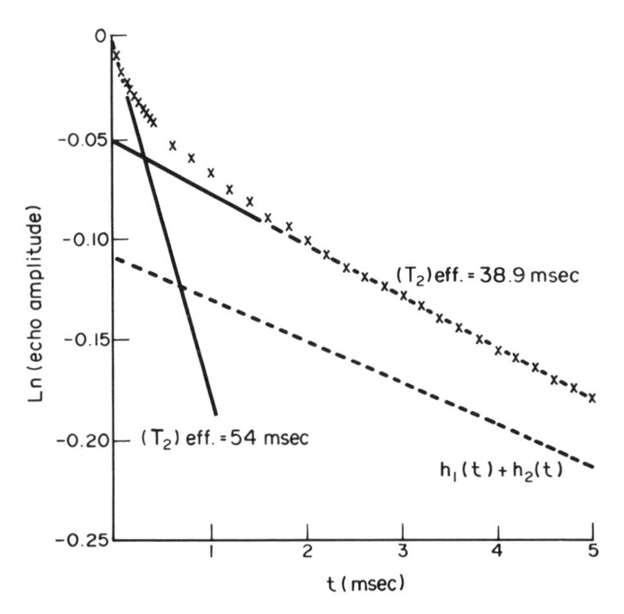


Fig. 7.3. Echo-decay curve in a 90°–$\tau$–180° sequence for pure water and rat muscle. Note the characteristic nonexponential decay of the echo series in a biological sample.

mates the observed data, except where it is very close to zero. This indicates that there may be more than two fractions of tissue water. The initial part of the decay curve has been studied closely, and suggests an initial fast-relaxing fraction of protons with a $T_2$ of 5.4 ms. This short relaxation time is evidence that a fast-relaxing fraction of muscle protons must exist, but whether these are water protons or protons on other molecules remains to be proven. Therefore, using a curve-stripping method such as this, it is possible to demonstrate and characterize multiple populations of protons in tissues. Such behavior is occasionally seen for $T_1$ data as well, especially when the data are extended over a long time scale, on the order of $5T_1$ (Fig. 7.3). In this tissue, three populations of protons are demonstrable, with $T_2$ values of 155, 44, and $<5$ ms. Some authors have related them to anatomical water compartments. When this procedure is used in the reporting of NMR data, it is important to describe the method of fitting and the range of $t$ for which each population was fit. Only if this is done can the data be compared with those from other laboratories studying the same tissues.

## 7.8. PROPAGATION OF ERRORS

In many experiments, the final quantity is obtained by measuring one or more different quantities, each measured individually and subject to individual uncertainties. The uncertainty in the final quantity is thus dependent on the uncertainty in the individual quantities. For NMR, the uncertainty is related to the uncertainties stemming from both the biological material and the machinery. As another example, if we are trying to measure the area of a plot, there is uncertainty in the measurement of its area due to uncertainty in the measurement of the length and width.

The uncertainty $\delta x$ in $x$ represents the outer limits of confidence within which we are almost certain that the measurement lies. The uncertainty $\delta x$ is absolute uncertainty, while the ratio $\delta x/x_0$ represents the relative uncertainty in the measured value of $x$, which is $x_0$. That is, $\delta x/x_0$ represents the precision of the measurement. When a measurement is repeated many times to get statistical significance in the result, then the standard deviation for the sample or the standard error of the mean is quoted.

We will discuss below the method of computing the resultant uncertainty ($\delta z$) and standard deviation $\delta_z$ in $z$, which is a function of two variables $x$ and $y$ with uncertainties of $\delta x$ and $\delta y$, respectively.

### 7.8.A. *Propagation of uncertainty*

1. Sum of two or more variables.
   Consider $z = x + y$. Then the uncertainty ($\delta z$) in $z$ will be obtained from

$$z_0 \pm \delta z = x_0 \pm \delta x + y_0 \pm \delta y \quad ,$$

   where $x_0$, $y_0$, and $z_0$ are the measured values of $x$, $y$, and $z$, and $\delta x$, $\delta y$, $\delta z$ are the corresponding uncertainties of each variable.
2. Differences of two variables.
   Consider $z = x - y$. Then $\delta z$ will be obtained from

$$z_0 \pm \delta z = (x_0 \pm \delta x) - (y_0 \pm \delta y) \quad .$$

3. General method for uncertainty in functions of two or more variables.
   If we have $z = f(x,y)$, then the uncertainty $\delta z$ in $z$ is given by

$$\delta z = \left(\frac{\partial f}{\partial x}\right)\delta x + \left(\frac{\partial f}{\partial y}\right)\delta y \quad ,$$

   where the derivatives $\partial f/\partial x$ and $\partial f/\partial y$ are normally evaluated from the values of $x_0$ and $y_0$ at which $\delta z$ is required. Special cases are discussed below.
4. Product of two or more variables.
   Consider $z = xy$. Using the equation above we need:

$$\frac{\partial z}{\partial x} = y \quad \text{and} \quad \frac{\partial z}{\partial y} = x \quad .$$

   Therefore, $\delta z = y\delta x + x\delta y$.

5. Quotient of two or more variables.
Consider a function of the form $z = x^a y^b$, where $a$ and $b$ may be positive or negative, integral or fractional, powers. The formula can be simplified by taking logarithms of both sides before differentiating. Therefore, $\log z = a \log x + b \log y$ and

$$\frac{\delta z}{z} = a\frac{\delta x}{x} + b\frac{\delta y}{y} \quad .$$

The quotient $z = x/y$ is a special case of the above, for which $a = 1$ and $b = 1$. Thus,

$$\frac{\delta z}{z} = \frac{\delta x}{x} + \frac{\delta y}{y} \quad .$$

**7.8.B.** *Standard deviation of computed values*

In the section above, we considered the outer limits of possibility for the computed value of $z$. However, the more useful quantity is the *probable* value for $z$. The limits given by this quantity will be smaller than $\pm\delta z$ calculated above. Let $x$ and $y$ be a pair of observations with standard deviations $S_x$ and $S_y$. Each pair defines a value for $z$. If the repetition had yielded $n$ pairs, then we have a set of $n$ values of $z$ showing statistical fluctuations. The quantity $S_z$ is the standard deviation of this set of $z$ values.

Let $z = f(x,y)$.

Then

$$\delta_z = \left(\frac{\partial z}{\partial x}\right) S_x + \left(\frac{\partial z}{\partial y}\right) S_y$$

$S_z$ = standard deviation for the $n$ different $z$-values, or

$$\sqrt{\frac{\Sigma(\delta z)^2}{n}} \quad .$$

$$S_z^2 = \frac{1}{n}\Sigma\left[\left(\frac{\partial z}{\partial x}\right)\delta x + \left(\frac{\partial z}{\partial y}\right)\delta y\right]^2$$

$$= \frac{1}{n}\Sigma\left[\left(\frac{\partial z}{\partial x}\right)^2(\delta x)^2 + \left(\frac{\partial z}{\partial y}\right)^2(\delta y)^2 + 2\left(\frac{\partial z}{\partial x}\right)\left(\frac{\partial z}{\partial y}\right)\delta x\delta y\right]$$

$$= \left(\frac{\partial z}{\partial x}\right)^2\frac{\Sigma(\delta x)^2}{n} + \left(\frac{\partial z}{\partial y}\right)^2\frac{\Sigma(\delta y)^2}{n} + \frac{2}{n}\left(\frac{\partial z}{\partial x}\right)\left(\frac{\partial z}{\partial y}\right)\Sigma\delta x\delta y$$

$\delta x$ and $\delta y$ are considered to be independent perturbations. That is, $\Sigma\delta x\delta y = 0$. Therefore

$$S_z = \sqrt{\left(\frac{\partial z}{\partial x}\right)^2 S_x^2 + \left(\frac{\partial z}{\partial y}\right)^2 S_y^2} \quad ,$$

where

$$S_x^2 = \frac{\Sigma(\delta x)^2}{n} \quad \text{and} \quad S_y^2 = \frac{\Sigma(\delta y)^2}{n} \quad .$$

If $z$ is a function of more than two variables, then the equation is extended by adding similar terms.

## 7.9. REFERENCES

1. Simon, W. A method of exponential separation applicable to small computers. *Phys. Med. Biol.* **15:**355–360, 1970.
2. Hazlewood, C.F., Chang, D.C., Nichols, B.L., and Woessner, D.E. Nuclear magnetic resonance transverse relaxation times of water protons in skeletal muscle. *Biophys. J.* **14:**583–606, 1974.

## 7.10. BIBLIOGRAPHY

Baird, D.C. *Experimentation—An Introduction to Measurement Theory and Experimental Design.* Prentice-Hall, Englewood Cliffs, NJ, 1962.

Brown, B.W., and Hollander, M. *Statistics—A Biomedical Introduction.* John Wiley and Sons, New York, 1977.

Mahajan, B.K. *Methods in Biostatistics for Medical Students.* Kothari Book Depot, Bombay, India, 1970.

Snedecor, G.W., and Cochran, W.G. *Statistical Methods.* University of Iowa Press, Ames, IA, 1976.

CHAPTER 8

# *PHYSICAL DATA TABLES AND FORMULAE*

**Table 8.1. The electromagnetic spectrum**

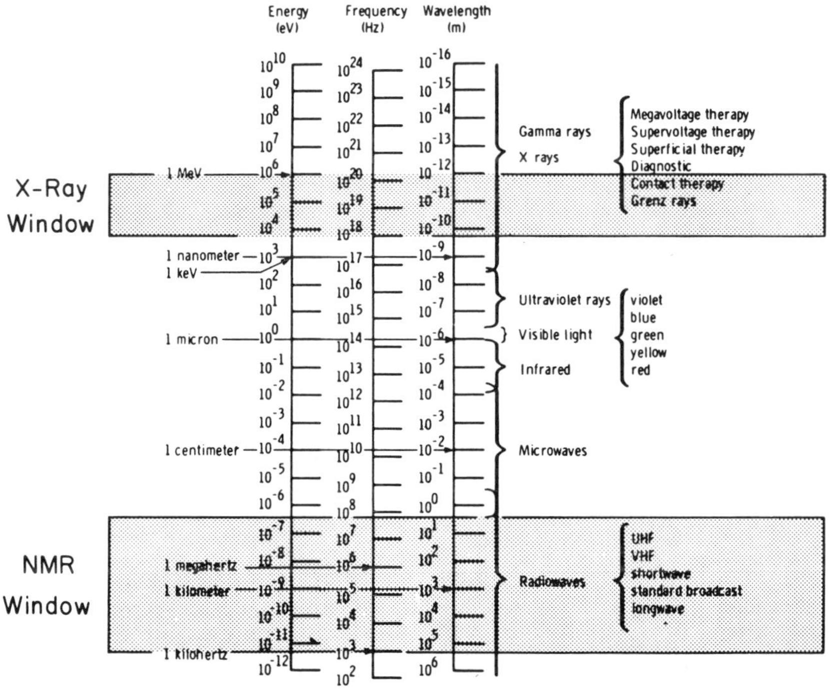

Energy (eV) | Frequency (Hz) | Wavelength (m)

X-Ray Window

NMR Window

1 MeV
1 nanometer
1 keV
1 micron
1 centimeter
1 megahertz
1 kilometer
1 kilohertz

Gamma rays
X rays
Megavoltage therapy
Supervoltage therapy
Superficial therapy
Diagnostic
Contact therapy
Grenz rays
Ultraviolet rays
Visible light
Infrared
violet
blue
green
yellow
red
Microwaves
Radiowaves
UHF
VHF
shortwave
standard broadcast
longwave

**Table 8.2. Typical ranges for electromagnetic radiation**

| Type of radiation | Wavelength* (m) | Frequency range (Hz) | Energy range (eV)† |
|---|---|---|---|
| Electricity | $\infty$–$3 \times 10^5$ | $0$–$10^3$ | $0$–$4.1 \times 10^{-1}$ |
| Radio waves | $3 \times 10^4$–$3 \times 10^{-4}$ | $10^4$–$10^{12}$ | $4.1 \times 10^{-11}$–$4.1 \times 10^{-7}$ |
| NMR | $6 \times 10^2$–$0.43$ | $0.5 \times 10^6$–$7 \times 10^8$ | $2.0 \times 10^{-4}$–$29 \times 10^{-7}$ |
| Electron spin resonance | 0.3–0.003 | $10^9$–$10^{11}$ | $4.1 \times 10^{-6}$–$4.1 \times 10^{-4}$ |
| Infrared | $3 \times 10^{-3}$–$7.6 \times 10^{-7}$ | $10^{11}$–$10^{14}$ | $4.1 \times 10^{-4}$–$1.6$ |
| Visible light | $7.6 \times 10^{-7}$–$3.8 \times 10^{-7}$ | $4 \times 10^{14}$–$7.9 \times 10^{14}$ | 1.6–3.3 |
| Ultraviolet | $3.8 \times 10^{-7}$–$3 \times 10^{-9}$ | $7.9 \times 10^{14}$–$10^{17}$ | 3.3–410 |
| X-rays | $1.2 \times 10^{-7}$–$4.1 \times 10^{-17}$ | $2.5 \times 10^{15}$–$7.3 \times 10^{24}$ | $10$–$3 \times 10^7$ |
| Gamma rays | $1.5 \times 10^{-17}$–$1.2 \times 10^{-13}$ | $2 \times 10^{18}$–$2.5 \times 10^{21}$ | $8 \times 10^3$–$10^7$ |
| Cosmic rays | $1.2 \times 10^{-7}$–. . . . | $2.5 \times 10^{15}$–. . . . . | 10–. . . . . |

*Ranges are approximate.
†The energy $E$ of a photon of frequency $\nu$ is calculated by Planck's equation: $E = h\nu$ where $h$ is Planck's constant.

**Table 8.3. Units and definitions**

| Quantity | SI unit | cgs unit | British unit |
|---|---|---|---|
| Length | meter (m) | centimeter (cm) | foot (ft) |
| Time | second (s) | second (s) | second |
| Mass | kilogram (kg) | gram (g) | slug |
| Velocity | m $s^{-1}$ | cm $s^{-1}$ | ft $s^{-1}$ |
| Acceleration | m $s^{-2}$ | cm $s^{-2}$ | ft $s^{-2}$ |
| Force | newton (N) [kg-m $s^{-2}$] | dyne [g-cm $s^{-2}$] | pound (lb) [slug-ft $s^{-2}$] |
| Work (energy) | joule (J) [N-m] | erg [dyne-cm] | ft-lb |
| Power | watt (W) | erg $s^{-1}$ | ft-lb $s^{-1}$ |
| Torque | N-m | dyne-cm | lb-ft |
| Pressure | N $m^{-2}$ | dyne $cm^{-2}$ | lb $ft^{-2}$ |

SI, Système International d'Unités, or International System of Units; cgs, centimeter-gram-second system of units.

*(Continued)*

**Table 8.3. Units and definitions** *(Continued)*

The SI (International System) units use meter for length, kilogram for mass, second for time, and ampere for electrical current. SI units are widely used in the scientific literature.

*Length (l):* The meter (m) used to be based on the platinum–iridium standard kept in Paris, France, but it is now defined as the length equal to 1,650,763.73 wavelengths in vacuum of the radiation corresponding to the transition between the levels $2p_{10}$ and $5d_5$ of the krypton-86 atom.

*Mass (m):* The kilogram (kg) is the mass of the platinum–iridium prototype kept at Sevres, France. It was originally based on the meter and was intended to be the mass of a cubic decimeter of pure water at 4°C.

*Time (t):* The second (s) is no longer based on the mean solar day; it is now defined as the duration of 9,192,631,770 periods of the radiation corresponding to the transition between two hyperfine levels of the ground state of the cesium-133 atom.

*Acceleration* ($a$ or $l\ t^{-2}$): The rate of change of velocity. When the velocity of a point changes at the rate of 1 meter per second it is said to experience 1 unit of acceleration.

*Angular velocity* ($\omega$ or radian $t^{-1}$) The rate of change of angle. A rate of change of 1 radian per second is the unit.

*Density* ($\rho$ or $m\ l^{-3}$): The mass of 1 unit volume of a substance.

*Electrical current* ($I$): An ampere (A) is the constant current which produces a force of $2 \times 10^{-7}$ newtons per meter between two parallel conductors, which are infinitely long, of negligible cross section, and spaced in vacuum 1 meter apart.

*Energy or Work* ($W$ or $l^2\ m\ t^{-2}$): When a force moves its point of application in the direction in which the force is acting, it is said to do work.

*Force* ($F$ or $l\ m\ t^{-2}$): A mass which moves with acceleration is said to be acted on by a force equal to mass times acceleration. The unit is the newton (N), or that force which acts on a mass of 1 kilogram to produce an acceleration of 1 meter per second per second.

*Frequency* ($v$ or $f$). The unit is the hertz (Hz), equivalent to 1 cycle per second.

*Length* ($l$): Basic units are kilometer, centimeter, and millimeter. The term micron has been abandoned; the unit is the micrometer ($\mu$m). 1 $\mu$m = $10^{-6}$ m, and 1 Angstrom (Å) = $10^{-10}$ m. It may also be called nanometer (nm).

*Mole* (mol): The mole is the amount of substance which contains as many elementary particles as there are atoms in 0.012 kilograms of carbon-12.

*Power* ($P$ or $l^2\ m\ t^{-3}$): The rate at which work is done. The unit is 1 joule (J) per second, equivalent to the watt (W).

*Pressure* ($p$ or $l^{-1}\ m\ t^{-2}$): When a force is transmitted to act over an area it is said to exert a pressure; unit pressure is defined as unit force over unit area. In SI units it is the Pascal (Pa), equivalent to 1 newton per square meter.

*Radian:* The radian is the plane angle subtended at the center of a circle by an arc of its circumference equal in length to the radius (1 radian = 57.296°).

*Steradian:* The steradian is the solid angle subtended at the center of a sphere by an area of its surface equal in magnitude to the square of its radius.

*Thermodynamic temperature* ($T$): The unit of temperature is the kelvin (K), or the fraction of 1/273.16 of the thermodynamic temperature of the triple point of water.

*Velocity* ($v$ or $l/t$): A point which moves in a straight line with a uniform rate of change of position is said to have uniform velocity.

***Electrical and magnetic units***

*Capacitance* ($C$): The unit is the farad (F) or coulomb per volt; it is possessed by a conductor if a charge increase of 1 coulomb raises its potential by 1 volt. 1 F = 1 $m^{-2}\ kg^{-1}\ s^{-4}\ A^{-2}$.

*Charge* ($Q$): The unit is the total charge transferred when a current of 1 ampere flows for 1 second, called the coulomb (C).

*Electrical current* ($I$): The ampere (A) is the basic unit.

*Magnetic field strength* ($H$): The unit is ampere per meter (A $m^{-1}$) and it exists at the center of a one turn coil having a diameter of 1 meter, carrying a current of 1 ampere.

*Magnetic flux* ($\Phi$): The magnetic flux threading an area $ds$, sufficiently small to render the flux density ($B$) uniform over it, is defined to be the scalar product $B \cdot ds$.

*(Continued)*

**Table 8.3. Units and definitions** *(Continued)*

*Magnetic induction or field density* ($B$): The sum of the vector intensity of magnetization (**M**) and the product of the magnetic field strength ($H$) and permeability ($\mu$).

*Mutual inductance* ($M$): The unit is possessed by two circuits if the unit current in either causes a total flux of 1 weber (Wb) to thread the other. The unit is the Henry (H).

*Potential* ($V$): The unit is the volt (V). A difference of potential of 1 volt exists between two points if 1 joule of work is done when 1 coulomb is transferred between them. $1\ V = 1\ m^2\ kg\ s^{-3}\ A^{-1}$.

*Resistance* ($R$): The resistance of a conductor which passes 1 ampere when a difference of potential of 1 volt exists between its ends is 1 ohm ($\Omega$).
$1\ \Omega = 1\ m^2\ kg\ s^{-3}\ A^{-2}$.

**Table 8.4. Common conversion factors**

| To convert: | To: | Multiply by: |
|---|---|---|
| inches (in) | centimeters (cm) | 2.54 cm/in |
| feet (ft) | cm | 30.48 cm/ft |
| meters (m) | feet | 3.28 ft/m |
| miles | feet | 5280 ft/mile |
| miles | kilometers (km) | 1.61 km/mile |
| pounds (lb) | newtons (N) | 4.45 N/lb |
| slugs | kilograms (kg) | 14.59 kg/slug |
| atmospheres (atm) | $lb/in^2$ | $14.7\ lb/in^2/atm$ |
| $lb/in^{-2}$ | mm Hg | $51.7\ mm\ Hg/lb/in^2$ |
| mm Hg | mm $H_2O$ | 13.6 mm $H_2O$/mm Hg |
| cm $H_2O$ | dyne $cm^{-2}$ | 980 dyne/cm/cm $H_2O$ |
| mm Hg | dyne $cm^{-2}$ | $1333\ dyne/cm^2/mm\ Hg$ |
| ft-lb | joules (J) | 1.356 J/ft-lb |
| ft-lb/s | watts (W) | 1.356 W/ft-lb/s |
| horsepower (HP) | watts | 746 W/HP |
| calories | joules | 4.186 J/calories |

At sea level, where $g$ (gravity) $= 980.665\ cm\ s^{-2} = 32.174\ ft\ s^{-2}$, 1 kg weighs 2.2 lb, 1 slug weighs 32.174 lb, 453.6 g weighs 1.0 lb.

Conversion formulas for different temperature scales:

Degrees Fahrenheit (°F) = 9/5 × (Degrees Celsius + 32)

Degrees Celsius (°C) = 5/9 × (Degrees Fahrenheit − 32)

Temperature on Kelvin (absolute) scale = Degrees Celsius + 273

°C −40 −20 0 +20 37 40 +60

°F −40 0 32 60 98.6 120 140

**Table 8.5. Temperature conversions (Celsius to Fahrenheit)**

| $T$(°C) | $T$(°F) | $T$(°C) | $T$(°F) | $T$(°C) | $T$(°F) | $T$(°C) | $T$(°F) |
|---|---|---|---|---|---|---|---|
| 0 | 32 | 26 | 78.8 | 51 | 123.8 | 76 | 168.8 |
| 1 | 33.8 | 27 | 80.6 | 52 | 125.6 | 77 | 170.6 |
| 2 | 35.6 | 28 | 82.4 | 53 | 127.4 | 78 | 172.4 |
| 3 | 37.4 | 29 | 84.2 | 54 | 129.2 | 79 | 174.2 |
| 4 | 39.2 | 30 | 86.0 | 55 | 131.0 | 80 | 176.0 |
| 5 | 41.0 | 31 | 87.8 | 56 | 132.8 | 81 | 177.8 |
| 6 | 42.8 | 32 | 89.6 | 57 | 134.6 | 82 | 179.6 |
| 7 | 44.6 | 33 | 91.4 | 58 | 136.4 | 83 | 181.4 |
| 8 | 46.4 | 34 | 93.2 | 59 | 138.2 | 84 | 185.0 |
| 9 | 48.2 | 35 | 95.0 | 60 | 140.0 | 85 | 185.0 |
| 10 | 50.0 | 36 | 96.8 | 61 | 141.8 | 86 | 186.8 |
| 11 | 51.8 | 37 | 98.6 | 62 | 143.6 | 87 | 188.6 |
| 12 | 53.6 | 38 | 100.4 | 63 | 145.4 | 88 | 190.4 |
| 13 | 55.4 | 39 | 102.2 | 64 | 147.2 | 89 | 192.2 |
| 14 | 57.2 | 40 | 104.0 | 65 | 149.0 | 90 | 194.0 |
| 15 | 59.0 | 41 | 105.8 | 66 | 150.8 | 91 | 195.8 |
| 16 | 60.8 | 42 | 107.6 | 67 | 152.6 | 92 | 197.6 |
| 17 | 62.6 | 43 | 109.4 | 68 | 154.4 | 93 | 199.4 |
| 18 | 64.4 | 44 | 111.2 | 69 | 156.2 | 94 | 201.2 |
| 19 | 66.2 | 45 | 113.0 | 70 | 158.0 | 95 | 203.0 |
| 20 | 68.0 | 46 | 114.8 | 71 | 159.8 | 96 | 204.8 |
| 21 | 69.8 | 47 | 116.6 | 72 | 161.6 | 97 | 206.6 |
| 22 | 71.6 | 48 | 118.4 | 73 | 163.4 | 98 | 208.4 |
| 23 | 73.4 | 49 | 120.2 | 74 | 165.2 | 99 | 201.2 |
| 24 | 75.2 | 50 | 122.0 | 75 | 167.0 | 100 | 212.0 |
| 25 | 77.0 | | | | | | |

**Table 8.6. Conversion of gaussian units to SI units for electromagnetic quantities**

| Fundamental quantity | Gaussian units | | SI units | | Multiplication factor |
|---|---|---|---|---|---|
| | Symbol | Unit | Symbol | Unit | |
| Magnetization | $4\pi M$ | gauss (G) | $M$ | $\mathrm{A \cdot m^{-1}}$ | $10^3/4\pi$ |
| Permeability of free space | — | — | $\mu_0 = 4\pi \times 10^{-7}$ | $\mathrm{Wb/A \cdot m}$ | $4\pi \times 10^{-7}$ |
| Anisotropy | K | $\mathrm{erg\ cm^{-3}}$ | K | $\mathrm{J\ m^{-3}}$ | $10^{-1}$ |
| Exchange | A | erg cm | A | J m | $10^{-5}$ |
| Gyromagnetic ratio | $\gamma$ | $(\mathrm{s\ Oe})^{-1}$ | $\gamma$ | $(\mathrm{s\ A \cdot m^{-1}})$ | $4\pi/10^3$ |
| **Parameter** | **Formula** | **Unit** | **Formula** | **Unit** | **Factor** |
| Magnetic field strength (**H**) | $\int \mathbf{H} dl = 4\pi NI/10$ | Oe | $\int \mathbf{H} dl = NI$ | $\mathrm{A \cdot m}$ | $10^3/4\pi$ |
| Magnetic induction (**B**) | $\mathbf{B} = H + 4\pi\mathbf{M}$ | G | $\mathbf{B} = \mu_0(\mathbf{H} + \mathbf{M})$ | $\mathrm{Wb\ m^{-2}}$ | $10^{-4}$ |
| Flux ($\Phi$) | $\Phi = \int \mathbf{B} dA$ | Mx | $\Phi = \int \mathbf{B} dA$ | Wb | $10^{-8}$ |
| Volume susceptibility ($x$, $\kappa$) | $x = d\mathbf{M}/d\mathbf{H}$ | — | $\kappa = d\mathbf{M}/d\mathbf{H}$ | — | $4\pi$ |
| Permeability ($\mu$) | $\mu = d\mathbf{B}/d\mathbf{H}$ $= 1 + 4\pi x$ | — | $\mu = d\mathbf{B}/d\mathbf{H}$ $= \mu_0(1 + \kappa)$ | — | $4\pi \times 10^{-7}$ |

m = meter; s = second; A = ampere; Oe = Oersted; G = gauss; cm = centimeter; N = number of turns; J = joule; Wb = Weber; Mx = Maxwell; M = magnetization.

**Table 8.7. Energy conversion factors**

1 erg = 1 dyne-centimeter
1 joule = 1 newton-meter = $10^7$ ergs
1 thermochemical calorie = 4.184 joules
1 electronvolt (eV) = $1.602 \times 10^{-19}$
1 atomic mass unit ($u$) = 931.478 MeV
rest mass of the electron ($m_e$) = 0.511 MeV
rest mass of the proton ($m_p$) = 938.256 MeV
rest mass of the neutron ($m_n$) = 939.550 MeV

**Table 8.8. Physical constants (SI units)**

| | |
|---|---|
| Speed of light ($c$) | $2.998 \times 10^8$ m/s |
| Acceleration of gravity ($g$) | 9.80 m/s$^2$ |

Earth's magnetic field:

| Locale | Horizontal component | Dip angle |
|---|---|---|
| El Paso, TX | $\mathbf{B} = 2.2 \times 10^{-5}$ weber m$^{-2}$ | 60° |
| Cambridge, MA | $\mathbf{H} = 17.5$ amp-turn m$^{-1}$ | |
| | $\mathbf{B} = 1.7 \times 10^{-5}$ weber m$^{-2}$ | 73° |
| | $\mathbf{H} = 13.5$ amp-turn m$^{-1}$ | |

| | |
|---|---|
| Electronic charge | $1.602 \times 10^{-19}$ coulombs |
| Mass of the electron | $9.108 \times 10^{-31}$ kilograms |
| Mass of the proton | $1.673 \times 10^{-27}$ kilograms |
| Mass of the hydrogen atom | $1.674 \times 10^{-27}$ kilograms |
| Planck's constant ($h$) | $6.625 \times 10^{-34}$ joule second |
| Permittivity of free space | $8.854 \times 10^{-12}$ farad meter$^{-1}$ |
| Permeability of free space | $4\pi \times 10^{-7}$ henry meter$^{-1}$ |
| Conductivity of copper | $5.7 \times 10^{-7}$ mhos meter$^{-1}$ |
| Conductivity of sea water | 5 mhos meter$^{-1}$ |
| Relative permittivity of water | 80 farad meter$^{-1}$ |
| Resistivity of copper | $1.724 \times 10^{-8}$ ohm-meter |
| Dielectric strength of air | $3 \times 10^{-6}$ volts meter$^{-1}$ |
| Magnetic moment of electron | $9.29 \times 10^{-24}$ joule tesla$^{-1}$ |
| Magnetic moment of proton | $1.41 \times 10^{-26}$ joule tesla$^{-1}$ |
| Faraday constant | $9.64870 \times 10^{-4}$ coulombs mole$^{-1}$ |
| Coulomb force constant | $8.98755 \times 10^{-9}$ newtons-meters$^{-2}$ coulombs$^{-2}$ |
| Volume of an ideal gas | 22413.6 cm$^{-3}$ mole |
| Gas constant | 8.31434 joules mole$^{-1}$ K$^{-1}$ |
| Boltzmann constant | $1.38054 \times 10^{-23}$ joules K$^{-1}$ |
| Bohr radius | $5.29167 \times 10^{-11}$ meters |
| Bohr magneton | $9.2732 \times 10^{-24}$ joules tesla$^{-1}$ |
| Nuclear magneton | $5.05050 \times 10^{-27}$ joules tesla$^{-1}$ |
| Gyromagnetic ratio of proton | $2.675192 \times 10^{-8}$ radians second$^{-1}$ tesla$^{-1}$ |

**Table 8.9. Table of spins, moments, and NMR frequencies of nuclei**

| Isotope | NMR frequency (MHz) at 14,092 G | NMR frequency (MHz) at 23,490 G | NMR field values (kG) at 60 MHz | NMR field values (kG) at 90 MHz | Magnetic moment ($\mu$) | Spin ($I$) | Electric quadrupole moment ($Q$) |
|---|---|---|---|---|---|---|---|
| $^{1}n$ | (29.167)* | | | | −1.91315 | $-\frac{1}{2}$ | |
| $^{1}$H | 60.000 | 100.00 | 14.09 | 21.06 | 2.79268 | $\frac{1}{2}$ | |
| $^{2}$H | 9.2101 | 15.352 | (45.5)† | | 0.85738 | 1 | $2.73 \times 10^{-3}$ |
| $^{3}$H | (45.414)* | | | | 2.9788 | $\frac{1}{2}$ | |
| $^{3}$He | (32.435)* | | 18.49 | 27.75 | −2.1274 | $-\frac{1}{2}$ | |
| $^{6}$Li | (6.265)* | | (47.88)† | | 0.82192 | 1 | $4.6 \times 10^{-4}$ |
| $^{7}$Li | 23.317 | 38.867 | 36.26 | 54.39 | 3.2560 | $\frac{3}{2}$ | −0.03 |
| $^{9}$Be | 8.4318 | 14.055 | (50.14)† | | −1.1773 | $-\frac{3}{2}$ | 0.052 |
| $^{10}$B | 6.4477 | 10.748 | | | 1.8005 | 3 | 0.074 |
| $^{11}$B | 19.250 | 32.087 | 43.92 | | 2.6880 | $\frac{3}{2}$ | 0.0355 |
| $^{13}$C | 15.086 | 25.147 | 56.05 | | 0.70220 | $\frac{1}{2}$ | |
| $^{14}$N | 4.3341 | 7.2246 | | | 0.40358 | 1 | 0.071 |
| $^{15}$N | 6.0796 | 10.134 | | | −0.28304 | $-\frac{1}{2}$ | |
| $^{17}$O | 8.134 | 13.56 | (51.97)† | | −1.8930 | $-\frac{5}{2}$ | −0.026 |
| $^{19}$F | 56.444 | 94.087 | 14.98 | 22.47 | 2.6273 | $\frac{1}{2}$ | |
| $^{21}$Ne | (3.363)* | | | | −0.66176 | $-\frac{3}{2}$ | |
| $^{22}$Na | (4.434)* | | | | 1.746 | 3 | |
| $^{23}$Na | 15.870 | 26.454 | 53.28 | | 2.2161 | $\frac{3}{2}$ | 0.14–0.15 |
| $^{24}$Na | (3.22)* | | | | 1.69 | 4 | |
| $^{25}$Mg | (2.606)* | | | | −0.85471 | $-\frac{5}{2}$ | |
| $^{27}$Al | 15.634 | 26.060 | 54.08 | | 3.6385 | $\frac{5}{2}$ | 0.149 |
| $^{29}$Si | 11.919 | 19.867 | (35.47)† | | −0.55477 | $-\frac{1}{2}$ | |
| $^{31}$P | 24.288 | 40.485 | 34.81 | 52.22 | 1.1305 | $\frac{1}{2}$ | |
| $^{33}$S | 4.6016 | 7.6704 | | | 0.64274 | $\frac{3}{2}$ | −0.064 |
| $^{35}$Cl | 5.8788 | 9.7993 | | | 0.82091 | $\frac{3}{2}$ | −0.0789 |
| $^{37}$Cl | 4.893 | 8.156 | | | 0.68330 | $\frac{3}{2}$ | −0.0621 |
| $^{39}$K | (1.987)* | | | | 0.39094 | $\frac{3}{2}$ | −0.07 |
| $^{41}$K | (1.092)* | | | | 0.21488 | $\frac{3}{2}$ | |
| $^{45}$Sc | (10.344)* | | 58.00 | | 4.7492 | $\frac{7}{2}$ | −0.22 |
| $^{47}$Ti | (2.400)* | | | | −0.78711 | $-\frac{5}{2}$ | |
| $^{49}$Ti | (2.401)* | | | | −1.1022 | $-\frac{7}{2}$ | |
| $^{51}$V | 15.77 | 26.29 | 53.60 | | 5.1392 | $\frac{7}{2}$ | −0.04 |
| $^{55}$Mn | 14.798 | 24.667 | 56.85 | | 3.4611 | $\frac{5}{2}$ | 0.55 |
| $^{59}$Co | 14.168 | 23.617 | 59.39 | | 4.6388 | $\frac{7}{2}$ | 0.40 |
| $^{63}$Cu | 15.903 | 26.508 | 53.17 | | 2.2206 | $\frac{3}{2}$ | −0.16 |
| $^{65}$Cu | 17.036 | 28.397 | 49.63 | | 2.3790 | $\frac{3}{2}$ | −0.15 |
| $^{69}$Ga | (10.219)* | | 58.71 | | 2.0108 | $\frac{3}{2}$ | 0.178 |
| $^{71}$Ga | (12.984)* | | 46.21 | | 2.5549 | $\frac{3}{2}$ | 0.112 |
| $^{75}$As | 10.276 | 17.129 | (41.14)† | | 1.4349 | $\frac{3}{2}$ | 0.3 |
| $^{77}$Se | 11.44 | 19.07 | | | 0.5325 | $\frac{1}{2}$ | |
| $^{79}$Br | 15.032 | 25.057 | 56.25 | | 2.0991 | $\frac{3}{2}$ | 0.33 |
| $^{81}$Br | 16.203 | 27.009 | 52.18 | | 2.2626 | $\frac{3}{2}$ | 0.28 |
| $^{83}$Kr | (1.64)* | | | | −0.96705 | $\frac{9}{2}$ | 0.15 |
| $^{85}$Rb | (4.111)* | | | | 1.3482 | $\frac{5}{2}$ | 0.28 |
| $^{87}$Rb | 19.632 | 32.724 | | | 2.7414 | $\frac{3}{2}$ | 0.14 |
| $^{89}$Y | (2.086)* | | | | −0.13682 | $-\frac{1}{2}$ | |
| $^{91}$Zr | (3.958)* | | | | −1.298 | $-\frac{5}{2}$ | |
| $^{93}$Nb | 14.666 | 24.446 | 57.65 | | 6.1135 | $\frac{9}{2}$ | −0.2 |
| $^{95}$Mo | (2.774)* | | | | −0.9099 | $-\frac{5}{2}$ | |
| $^{101}$Ru | (2.1)* | | | | −0.69 | $-\frac{5}{2}$ | |
| $^{103}$Rh | (1.340)* | | | | −0.0879 | $-\frac{1}{2}$ | |

*continued*

**Table 8.9. Table of spins, moments, and NMR frequencies of nuclei (*Continued*)**

| Isotope | NMR frequency (MHz) at 14,092 G | NMR frequency (MHz) at 23,490 G | NMR field values (kG) at 60 MHz | NMR field values (kG) at 90 MHz | Magnetic moment ($\mu$) | Spin ($I$) | Electric quadrupole moment ($Q$) |
|---|---|---|---|---|---|---|---|
| $^{105}$Pd | (1.74)* | | | | −0.57 | −5/2 | |
| $^{107}$Ag | (1.723)* | | | | −0.1130 | −1/2 | |
| $^{109}$Ag | (1.981)* | | | | −0.1299 | −1/2 | |
| $^{111}$Cd | (9.028)* | | (33.23)† | | −0.5922 | −1/2 | |
| $^{113}$Cd | (9.444)* | | (31.77)† | | −0.6195 | −1/2 | |
| $^{115}$In | (9.329)* | | (32.16)† | | 5.5073 | 9/2 | 0.761; 1.16 |
| $^{117}$Sn | 21.375 | 35.630 | 39.55 | 59.33 | −0.9949 | −1/2 | |
| $^{119}$Sn | 22.363 | 37.276 | 37.81 | 56.71 | −1.0409 | −1/2 | |
| $^{121}$Sb | 14.358 | 23.934 | | | 3.3417 | 5/2 | −0.53; −1.2 |
| $^{123}$Sb | (5.518)* | | | | 2.5334 | 7/2 | −0.68; −1.5 |
| $^{125}$Te | (13.45)* | | | | −0.8824 | −1/2 | |
| $^{127}$I | 12.004 | 20.009 | (35.21)† | | 2.7937 | 5/2 | −0.69 |
| $^{129}$Xe | (11.78)* | | | | −0.77255 | −1/2 | |
| $^{131}$Xe | (3.490)* | | | | 0.68680 | 3/2 | −0.12 |
| $^{133}$Cs | 7.8699 | 13.118 | (53.71)† | | 2.5642 | 7/2 | −0.004 |
| $^{137}$Ba | (4.732)* | | | | 0.93107 | 3/2 | |
| $^{139}$La | (6.014)* | | | | 2.7615 | 7/2 | 0.5 |
| $^{141}$Pr | (11.95)* | | | | 3.92 | 5/2 | 0.054 |
| $^{143}$Nd | (2.72)* | | | | −1.25 | −7/2 | −0.57 |
| $^{145}$Nd | (1.7)* | | | | −0.78 | 7/2 | −0.30 |
| $^{147}$Sm | (1.5)* | | | | −0.68 | 7/2 | 0.72 |
| $^{149}$Sm | (1.2)* | | | | −0.55 | 7/2 | 0.72 |
| $^{151}$Eu | (10.49)* | | | | 3.441 | 5/2 | 1.2 |
| $^{153}$Eu | (4.638)* | | | | 1.521 | 5/2 | 2.5 |
| $^{155}$Gd | (1.2)* | | | | −0.25 | 3/2 | 1.1 |
| $^{157}$Gd | (1.7)* | | | | −0.34 | 3/2 | 1.0 |
| $^{159}$Tb | (7.72)* | | | | 1.52 | 3/2 | |
| $^{161}$Dy | (1.2)* | | | | −0.38 | 5/2 | |
| $^{163}$Dy | (1.6)* | | | | −0.53 | 5/2 | |
| $^{165}$Ho | (7.22)* | | | | 3.31 | 7/2 | 2 |
| $^{167}$Er | (1.04)* | | | | 0.48 | 7/2 | (10) |
| $^{169}$Tm | (3.05)* | | | | −0.20 | 1/2 | |
| $^{171}$Yb | (7.51)* | | | | 0.4926 | 1/2 | |
| $^{173}$Yb | (2.1)* | | | | −0.677 | 5/2 | 3.9 |
| $^{175}$Lu | (6.3)* | | | | 2.9 | 7/2 | 5.5 |
| $^{177}$Hf | (1.3)* | | | | 0.61 | 7/2 | 3 |
| $^{179}$Hf | (0.80)* | | | | −0.47 | 9/2 | 3 |
| $^{181}$Ta | (5.09)* | | | | 2.340 | 7/2 | 4.0 |
| $^{183}$W | (1.75)* | | | | 0.115 | 1/2 | |
| $^{185}$Re | (9.586)* | | | | 3.1437 | 5/2 | 2.8 |
| $^{187}$Re | (9.684)* | | | | 3.1760 | 5/2 | 2.6 |
| $^{189}$Os | (3.307)* | | | | 0.6507 | 3/2 | 2.0 |
| $^{191}$Ir | (0.813)* | | | | 0.16 | 3/2 | 1.5 |
| $^{193}$Ir | (0.86)* | | | | 0.17 | 3/2 | 1.5 |
| $^{195}$Pt | 12.90 | 21.50 | (32.78)† | | 0.6004 | 1/2 | |
| $^{197}$Au | | | | | 0.1439 | 3/2 | 0.56 |
| $^{199}$Hg | 10.696 | 17.829 | | | 0.4979 | 1/2 | |
| $^{201}$Hg | 3.9597 | 6.6005 | | | −0.5513 | −3/2 | 0.50 |
| $^{203}$Tl | 34.289 | 57.156 | | | 1.5960 | 1/2 | |
| $^{205}$Tl | 34.624 | 57.715 | | | 1.6115 | 1/2 | |
| $^{207}$Pb | 12.553 | 20.924 | (33.71)† | | 0.5837 | 1/2 | |
| $^{209}$Bi | 9.6414 | 16.071 | | | 4.0389 | 9/2 | −0.4 |

In this table the magnetic moment $\mu$ is in multiples of the nuclear magneton ($eh/4\pi Mc$), the spin $I$ is in multiples of $h/2\pi$, and the electric quadrupole moment $Q$ is in multiples of $e \times 10^{-24}$ cm$^2$. *At 10,000 gauss. †At 30 MHz.

**Table 8.10. NMR properties of some nuclei used in biology**

| Nuclide | Natural abundance (%) | Spin ($I$) | Gyromagnetic ratio (radians s $T$) ($\gamma$) | Relative sensitivity | Resonance frequency (MHz) at 1.4090 tesla | Resonance frequency (MHz) at 2.3010 tesla |
|---|---|---|---|---|---|---|
| $^{1}$H | 99.98 | 1/2 | $2.676 \times 10^{-8}$ | 1.000 | 60.00 | 100.00 |
| $^{2}$H | 0.016 | 1 | $4.106 \times 10^{-7}$ | 0.0097 | 9.21 | 15.35 |
| $^{13}$C | 1.11 | 1/2 | $6.725 \times 10^{-7}$ | 0.016 | 15.09 | 25.14 |
| $^{14}$N | 99.64 | 1 | $1.933 \times 10^{-7}$ | 0.001 | 4.33 | 7.22 |
| $^{15}$N | 0.36 | 1/2 | $-2.711 \times 10^{-7}$ | 0.001 | 6.08 | 10.13 |
| $^{17}$O | 0.037 | 5/2 | $-3.628 \times 10^{-7}$ | 0.029 | 8.13 | 7.22 |
| $^{19}$F | 100.0 | 1/2 | $2.517 \times 10^{-8}$ | 0.830 | 56.45 | 94.08 |
| $^{23}$Na | 100.0 | 3/2 | $7.075 \times 10^{-7}$ | 0.093 | 15.86 | 24.78 |
| $^{31}$P | 100.0 | 1/2 | $1.082 \times 10^{-8}$ | 0.066 | 24.29 | 40.48 |
| $^{35}$Cl | 75.53 | 3/2 | $2.6198 \times 10^{-7}$ | 0.047 | 5.88 | 9.60 |
| $^{39}$K | 93.10 | 3/2 | $1.2477 \times 10^{-7}$ | 0.0005 | 2.80 | 4.57 |

**Table 8.11. List of elements**

| Atomic No. | Symbol | Name | Atomic No. | Symbol | Name | Atomic No. | Symbol | Name |
|---|---|---|---|---|---|---|---|---|
| 0 | n | Neutron | 35 | Br | Bromine | 70 | Yb | Ytterbium |
| 1 | H | Hydrogen | 36 | Kr | Krypton | 71 | Lu | Lutetium |
| 2 | He | Helium | 37 | Rb | Rubidium | 72 | Hf | Hafnium |
| 3 | Li | Lithium | 38 | Sr | Strontium | 73 | Ta | Tantalum |
| 4 | Be | Beryllium | 39 | Y | Yttrium | 74 | W | Tungsten |
| 5 | B | Boron | 40 | Zr | Zirconium | 75 | Re | Rhenium |
| 6 | C | Carbon | 41 | Nb | Niobium | 76 | Os | Osmium |
| 7 | N | Nitrogen | 42 | Mo | Molybdenum | 77 | Ir | Iridium |
| 8 | O | Oxygen | 43 | Tc | Technetium | 78 | Pt | Platinum |
| 9 | F | Fluorine | 44 | Ru | Ruthenium | 79 | Au | Gold |
| 10 | Ne | Neon | 45 | Rh | Rhodium | 80 | Hg | Mercury |
| 11 | Na | Sodium | 46 | Pd | Palladium | 81 | Tl | Thallium |
| 12 | Mg | Magnesium | 47 | Ag | Silver | 82 | Pb | Lead |
| 13 | Al | Aluminum | 48 | Cd | Cadmium | 83 | Bi | Bismuth |
| 14 | Si | Silicon | 49 | In | Indium | 84 | Po | Polonium |
| 15 | P | Phosphorus | 50 | Sn | Tin | 85 | At | Astatine |
| 16 | S | Sulfur | 51 | Sb | Antimony | 86 | Rn | Radon |
| 17 | Cl | Chlorine | 52 | Te | Tellurium | 87 | Fr | Francium |
| 18 | Ar | Argon | 53 | I | Iodine | 88 | Ra | Radium |
| 19 | K | Potassium | 54 | Xe | Xenon | 89 | Ac | Actinium |
| 20 | Ca | Calcium | 55 | Cs | Cesium | 90 | Th | Thorium |
| 21 | Sc | Scandium | 56 | Ba | Barium | 91 | Pa | Protactinium |
| 22 | Ti | Titanium | 57 | La | Lanthanum | 92 | U | Uranium |
| 23 | V | Vanadium | 58 | Ce | Cerium | 93 | Np | Neptunium |
| 24 | Cr | Chromium | 59 | Pr | Praseodymium | 94 | Pu | Plutonium |
| 25 | Mn | Manganese | 60 | Nd | Neodymium | 95 | Am | Americium |
| 26 | Fe | Iron | 61 | Pm | Promethium | 96 | Cm | Curium |
| 27 | Co | Cobalt | 62 | Sm | Samarium | 97 | Bk | Berkelium |
| 28 | Ni | Nickel | 63 | Eu | Europium | 98 | Cf | Californium |
| 29 | Cu | Copper | 64 | Gd | Gadolinium | 99 | Es | Einsteinium |
| 30 | Zn | Zinc | 65 | Tb | Terbium | 100 | Fm | Fermium |
| 31 | Ga | Gallium | 66 | Dy | Dysprosium | 101 | Md | Mendelevium |
| 32 | Ge | Germanium | 67 | Ho | Holmium | 102 | No | Nobelium |
| 33 | As | Arsenic | 68 | Er | Erbium | 103 | Lw | Lawrencium |
| 34 | Se | Selenium | 69 | Tm | Thulium | | | |

**Table 8.12. Magnetic susceptibility ($x_m$) of some common substances**

| | |
|---|---|
| Aluminum | 0.81 |
| Benzene | −0.89 |
| Bismuth | −1.70 |
| Cadmium | −0.23 |
| Copper | −0.107 |
| Diamond | −0.62 |
| Ethanol | −1.00 |
| Glass | −1.25 |
| Gold | −0.19 |
| Sulfur | −0.62 |
| Tin | 0.031 |
| Tungsten | 0.35 |
| Water | −0.905 |
| Zinc | −0.197 |

Values are per kilogram at 18°C. In every case, the unit is $10^{-8}$ for the right column.

**Table 8.13. Resistivity ($\rho$) of some common substances**

| Substance | $\rho(\times 10^{-8}$ ohm-m) | Temperature coefficient (20°C) |
|---|---|---|
| Aluminum | 2.67 | 0.0045 |
| Antimony | 44 | 0.0051 |
| Bismuth | 117 | 0.0046 |
| Brass | 8 | 0.0015 |
| Bronze (phosphor) | 8 | 0.0035 |
| Copper | 1.72 | 0.0042 |
| German silver | 28 | 0.0003 |
| Gold | 2.20 | 0.0040 |
| Iridium | 5.2 | 0.0045 |
| Iron | 10.3 | 0.0065 |
| Lead | 20.63 | 0.0042 |
| Magnesium | 4.24 | 0.0043 |
| Manganese | 44 | 0.00001 |
| Mercury | 95.93 | 0.0089 |
| Nickel | 6.94 | 0.0070 |
| Palladium | 10.7 | 0.0039 |
| Platinum | 10.5 | 0.0039 |
| Silver | 1.63 | 0.0038 |
| Steel (hardened) | 45 | 0.0015 |
| Tantalum | 13.4 | 0.0033 |
| Tin | 11.4 | 0.0044 |
| Tungsten | 5.5 | 0.0050 |
| Zinc | 5.92 | 0.0038 |

**Table 8.14. Dielectric constant ($\epsilon_r$) and dielectric strength ($S$) of some common substances**

| Substance | $\epsilon_r$ | $S$ (kV/mm) |
|---|---|---|
| Solids | | |
| Amber | 2.8 | |
| Chlornapth | 2.8 | |
| Ebonite | 2.8 | 30–110 |
| Glass (crown) | 5–7 | 30–150 |
| Glass (flint) | 7–10 | 30–150 |
| Mica | 6 | 80–200/0.05 |
| Paper (paraffined) | | 40–60 |
| Paraffin wax | 2.2 | 15–50 |
| Porcelain | 5.6 | |
| Quartz | 4.5 | |
| Rubber | 2.2 | 16–40 |
| Shellac | 3.1 | |
| Sulfur | 4–4.2 | |
| Wood | 2–8 | 0.4–0.6/25 |
| Liquids (18°C) | | |
| Ethanol | 26 | |
| Methanol | 32 | |
| Aniline | 7.3 | |
| Benzene | 2.3 | |
| Oil | | |
| Olive | 3.1 | |
| Paraffin | 4.7 | |
| Transformer | 2.2 | |
| Vaseline | 1.9 | |
| Water | 81 | |

**Table 8.15. Standard wire gauge**

| | 0 | 10 | 20 | 30 | 40 | 50 |
|---|---|---|---|---|---|---|
| 0 | 8.23 | 3.25 | 0.914 | 0.315 | 0.122 | 0.025 |
| 1 | 7.62 | 2.95 | 0.813 | 0.295 | 0.112 | |
| 2 | 7.01 | 2.64 | 0.711 | 0.274 | 0.102 | |
| 3 | 6.40 | 2.34 | 0.610 | 0.254 | 0.091 | |
| 4 | 5.89 | 2.03 | 0.559 | 0.234 | 0.081 | |
| 5 | 5.39 | 1.83 | 0.508 | 0.213 | 0.071 | |
| 6 | 4.88 | 1.63 | 0.457 | 0.193 | 0.061 | |
| 7 | 4.47 | 1.42 | 0.417 | 0.173 | 0.051 | |
| 8 | 4.06 | 1.22 | 0.376 | 0.132 | 0.031 | |
| 9 | 3.66 | 1.02 | 0.345 | 0.132 | 0.031 | |

Diameters are in millimeters.

**Table 8.16. Resistor and capacitor color code**

| Resistor | | Color code | | | | | | | | | | | | |
|---|---|---|---|---|---|---|---|---|---|---|---|---|---|---|
| Band | Function | Black | Brown | Red | Orange | Yellow | Green | Blue | Violet | Gray | White | Silver | Gold | Band |
| | | −90 | −90 | −30 | −80 | −150 | −220 | −330 | −470 | −750 | +30 | — | +100 | 1 |
| 1 | First digit | 0 | 1 | 2 | 3 | 4 | 5 | 6 | 7 | 8 | 9 | — | — | 2 |
| 2 | Second digit | 0 | 1 | 2 | 3 | 4 | 5 | 6 | 7 | 8 | 9 | — | — | 3 |
| 3 | Multiplier | 1 | 10 | $10^2$ | $10^3$ | $10^4$ | $10^5$ | $10^6$ | $10^7$ | $10^8$ | — | — | — | |
| 4 | % Tolerance | | | | | | | | | | | | | |
| | Capacitor multiplier | 1 | 10 | $10^2$ | $10^3$ | $10^4$ | | | | $10^{-2}$ | $10^{-1}$ | $10^{-2}$ | $10^{-1}$ | 4 |
| | Capacitor tolerance (% above 10 pF) | 20 | 1 | 2 | 3 | | 5 | | | | 10 | 10 | | 5 |
| | Capacitor tolerance (% below 10 pF) | 2 | 0.1 | | | | 0.5 | | | 0.25 | 1 | | | 5 |

## 8.2. MATHEMATICAL FORMULAE

**Table 8.17. Greek alphabet**

| | | | | | | | | |
|---|---|---|---|---|---|---|---|---|
| A | $\alpha$ | alpha | I | $\iota$ | iota | P | $\rho$ | rho |
| B | $\beta$ | beta | K | $\kappa$ | kappa | $\Sigma$ | $\sigma$ | sigma |
| $\Gamma$ | $\gamma$ | gamma | $\Lambda$ | $\lambda$ | lambda | T | $\tau$ | tau |
| $\Delta$ | $\delta$ | delta | M | $\mu$ | mu | $\Upsilon$ | $\upsilon$ | upsilon |
| E | $\epsilon$ | epsilon | N | $\nu$ | nu | $\Phi$ | $\phi$ | phi |
| Z | $\zeta$ | zeta | $\Xi$ | $\xi$ | xi | X | $\chi$ | chi |
| H | $\eta$ | eta | O | $o$ | omicron | $\Psi$ | $\psi$ | psi |
| $\Theta$ | $\theta$ | theta | $\Pi$ | $\pi$ | pi | $\Omega$ | $\omega$ | omega |

**Table 8.18. Signs and symbols**

| | |
|---|---|
| $\div$ / − division | $\sqrt{\ }$ square root |
| $\times$ · ( ) multiplication | $x^n$ *nth* power |
| ( ) [ ] collection | $a^n$ *nth* power of *a* |
| = equal to | $a^{-r}$ reciprocal of the *nth* power |
| $\neq$ not equal to | log $\log_{10}$ common logarithm |
| $\equiv$ identical to | ln $\log_e$ natural logarithm |
| $\simeq$ equals approximately | $e$ $\epsilon$ base of natural logarithms |
| > greater than | $\pi$ pi (3.14159265) |
| $\not>$ not greater than | $n$ any number |
| $\geq$ $\geqq$ greater than or equal to | $\mid n \mid$ absolute value of $n$ |
| < less than | $\bar{n}$ average value of $n$ |
| $\not<$ not less than | $n°$ $n$ degrees |
| $\leq$ $\leqq$ less than or equal to | $n'$ $n$ minutes or $n$ feet |
| :: proportional to | $n''$ $n$ seconds or $n$ inches |
| : ratio | $f(x)$ function of $x$ |
| ~ similar to | $\Delta x$ increment of $x$ |
| $\propto$ varies as, proportional to | $dx$ differential of $x$ |
| $\rightarrow$ approaches | sin sine |
| $\infty$ infinity | cos cosine |
| $\therefore$ therefore | tan tangent |

**Table 8.19. Algebraic expressions**

1. Factors and expansions

$$(a + b)^2 = a^2 + 2ab + b^2$$
$$(a \pm b)^3 = a^3 \pm 3a^2b + 3ab^2 \pm b^3$$
$$(a \pm b)^4 = a^4 \pm 4a^3b + 6a^2b^2 \pm 4ab^3 + b^4$$
$$a^2 - b^2 = (a - b)(a + b)$$
$$a^3 - b^3 = (a - b)(a^2 + ab + b^2)$$
$$a^3 + b^3 = (a + b)(a^2 - ab + b^2)$$

2. Powers and roots

$$a^n \times a^y = a^{n+y}$$
$$\frac{a^n}{a^y} = a^{n-y}$$
$$(a^n)^y = a^{n \cdot y}$$
$$(ab)^n = a^n b^n$$
$$(a/b)^n = a^n/b^n$$
$$a^0 = 1$$
$$a^{-n} = 1/a^n$$
$$a^{1/n} = \sqrt[n]{a}$$
$$\sqrt[n]{ab} = \sqrt[n]{a}\,\sqrt[n]{b}$$

3. Logarithmic expressions

log $N$ = the exponent or power to which the base 10 must be raised to obtain a value $N$ (the common logarithm of $N$)

ln $N$ = the power to which the base 2.718 . . . ($e$) must be raised to obtain a value $N$ (the natural logarithm of $N$)

(1) $\log N = 0.4343 \ln N$
(2) $\ln N = 2.3026 \log N$
(3) $\log MN = \log M + \log N$
(4) $\log M/N = \log M - \log N$
(5) $\log N^a = a \log N$
(6) $\log \sqrt[a]{N} = (\log N)/a$

**Table 8.20. Trigonometric relationships**

If $A$, $B$, and $C$ are the vertices of a right triangle ($C$ is the right angle), and $a$, $b$, and $c$ are the sides opposite the vertices, then:

$$\text{sine } A = \sin A = a/c$$
$$\text{cosine } A = \cos A = b/c$$
$$\text{tangent } A = \tan A = a/b$$
$$\text{cosecant } A = \text{cosec } A = c/a$$
$$\text{secant } A = \sec A = c/b$$
$$\text{cotangent } A = \cot A = b/a$$

Value of the functions of various angles

| | 0° | 30° | 45° | 60° | 90° | 180° | 270° |
|---|---|---|---|---|---|---|---|
| sin | 0 | ½ | ½ $\sqrt{2}$ | ½ $\sqrt{3}$ | 1 | 0 | −1 |
| cos | 1 | ½ $\sqrt{3}$ | ½ $\sqrt{2}$ | ½ | 0 | −1 | 0 |
| tan | 0 | ⅓ $\sqrt{3}$ | 1 | $\sqrt{3}$ | ∞ | 0 | ∞ |
| cot | ∞ | $\sqrt{3}$ | 1 | ⅓ $\sqrt{3}$ | 0 | ∞ | 0 |
| sec | 1 | $\frac{2\sqrt{3}}{3}$ | $\sqrt{2}$ | 2 | ∞ | −1 | ∞ |
| cosec | ∞ | 2 | $\sqrt{2}$ | $\frac{2\sqrt{3}}{3}$ | 1 | ∞ | −1 |

**Table 8.21. Mensuration formulae**

1. The area of a triangle whose base is $b$ and whose height is $h = 1/2\ bh$.
2. The area of a triangle whose angles are $A$, $B$, and $C$ and whose sides opposite those angles are $a$, $b$, and $c$, respectively, is:

$$\text{Area} = 1/2\ ab \sin C,$$
$$\text{or } \sqrt{s(s-a)(s-b)(s-c)},$$
$$\text{where } s = 1/2\,(a + b + c).$$

3. The area of a rectangle with sides $a$ and $b$ is $ab$.
4. The circumference of a circle with radius $r$ is $2\pi r$.
5. The area or a circle with radius $r$ is $\pi r^2$.
6. The length of an arc at $\Theta$ degrees of a circle with radius $r$ is:

$$\frac{\pi r \Theta}{180°}.$$

7. The circumference (C) of an ellipse with semiaxes $a$ and $b$ is:

$$C = 2\pi \sqrt{\frac{a^2 + b^2}{2}}.$$

8. The surface area of a sphere of radius $r$ is $4\pi r^2$.
9. The volume of a sphere of radius $r$ is $4/3\ \pi r^3$.
10. The curved surface of a right cylinder of radius $r$ and height $h$ is $2\pi rh$. The volume of the cylinder is $\pi r^2 h$.

**Table 8.22. Space coordinates**

1. Rectangular system: $x$, $y$, and $z$
2. Cylindrical system: $r$, $\Theta$, and $z$
3. Spherical system: $\rho$, $\Theta$, and $\phi$
4. Polar space system: $\rho$, $\alpha$, $\beta$, and $\gamma$

**Table 8.23. Calculus**

Differential calculus is the study of the variation of a function with respect to changes in the independent variable or variables, such as time ($t$). The derivative is defined as the instantaneous rate of change of a function with respect to a variable.

Let $y = f(x)$ be a function of one variable, $x$, and let $\Delta x$ denote a number to be added to the number $n$. Let $\Delta y$ denote the corresponding increment of $y$.

$$\Delta y = f(x + \Delta x) - f(x)$$

The ratio $\Delta y/\Delta x = [f(x + \Delta x) - f(x)]/\Delta x$ is studied as $\Delta x$ approaches zero. If the ratio approaches a limit as $\Delta x$ approaches zero, the limit is called the derivative of $f(x)$ at point $x$. It is denoted by

$$y' \text{ or } dy/dx \text{ or } f'(x) \text{ or } d/dx\,\mathrm{f}(x) \text{ or } df(x)/dx.$$

Two special interpretations of the derivative should be noted:

(i) The derivative of a function at a particular point is the slope of a tangent to that function at that point. If the function is increasing at that point, the slope and the derivative are positive. If the function is decreasing at that point, the slope and the derivative are negative. If the derivative is zero, the function may have a maximum or minimum at that point.

(ii) The derivative may be expressed as speed or rate if the independent variable is time and the dependent variable is distance travelled in that time. The *integral* is the limit, as $\Delta x$ approaches zero, of the sum of areas of the rectangles, $\Delta x$ in width, formed with successive ordinates taken $\Delta x$ apart throughout the interval between two values of $x$:

$$\int_a^b f(x)dx\,.$$

a. *Differentiation formulae.* In the following formulae, $u$, $v$, and $y$ are functions of $x$ which possess derivatives with respect to $x$; the other letters are constants and $\log u = \log_{10} u$.

$$\frac{dc}{dx} = 0$$

$$\frac{d}{dx}x = 1$$

$$\frac{d}{dx}(cv) = c\frac{dv}{dx}$$

$$\frac{d}{dx}x^n = nx^{n-1}$$

$$\frac{dy}{dx} = \frac{dy}{du}\cdot\frac{du}{dx}$$

$$\frac{d}{dx}(u + v) = \frac{du}{dx} + \frac{dv}{dx}$$

$$\frac{d}{dx}(uv) = u\frac{dv}{dx} + v\frac{du}{dx}$$

$$\frac{d}{dx}\left(\frac{u}{v}\right) = \frac{v\dfrac{du}{dx} - u\dfrac{dv}{dx}}{v^2}$$

$$\frac{du^n}{dx} = nu^{n-1}\frac{du}{dx}$$

$n$ = any number (positive or negative, integral or fractional),

$$\frac{d}{dx}(\sin u) = \cos u \cdot \frac{du}{dx}$$

$$\frac{d}{dx}(\cos u) = -\sin u\frac{du}{dx}$$

$$\frac{d}{dx}(\tan u) = \sec^2 u\frac{du}{dx}$$

$$\frac{d}{ax}(\log u) = \frac{\dfrac{du}{dx}}{u}$$

$$\frac{d}{dx}(\log_a u) = \log_a e \cdot \frac{\dfrac{du}{dx}}{u}$$

$$\frac{d}{dx}(e^u) = e^u \cdot \frac{du}{ax}$$

$$\frac{d}{dx}(a^u) = \log a \cdot a^u \cdot \frac{du}{dx}$$

(*Continued*)

**Table 8.23. Calculus** (*Continued*)

b. *Integral table.* In the following table, the constant of integration, $C$, is omitted, but it should be added to the result of every integration. The letter $x$ represents any variable; $u$ represents any function of $x$; the remaining letters represent arbitrary constants. Unless otherwise indicated, all angles are in radians.

$$\int df(x) = f(x)$$

$$d \int f(x)dx = f(x)dx$$

$$\int 0 \cdot dx = C$$

$$\int aj(x)dx = a \int f(x)dx$$

$$\int (u = v)dx = \int udx = \int vdx$$

$$\int u\, dv = uv - \int v\, du$$

$$\int \frac{u\, dv}{dx} dx = uv - \int v \frac{du}{dx} dx$$

$$\int f(y)dx = \int \frac{f(y)dy}{\frac{dy}{dx}}$$

$$\int u^n\, du = \frac{u^{n+1}}{n+1}, n \neq -1$$

$$\int \frac{du}{u} = \log u$$

$$\int e^u\, du = e^u$$

$$\int b^u\, du = \frac{b^u}{\log b}$$

$$\int \sin u\, du = -\cos u$$

$$\int \cos u\, du = \sin u$$

$$\int \tan u\, du = \log \sec u = -\log \cos u$$

**Table 8.24. Vector definitions, identities, and theorems**

1. **Definitions**

1.1. Rectangular coordinates

$$\nabla f = \mathbf{i}\frac{\partial f}{\partial x} + \mathbf{j}\frac{\partial f}{\partial y} + \mathbf{k}\frac{\partial f}{\partial z}$$

$$\nabla \cdot \mathbf{A} = \frac{\partial A_x}{\partial x} + \frac{\partial A_y}{\partial y} + \frac{\partial A_z}{\partial z}$$

$$\nabla \times \mathbf{A} = \mathbf{i}\left(\frac{\partial A_z}{\partial y} - \frac{\partial A_y}{\partial z}\right) + \mathbf{j}\left(\frac{\partial A_x}{\partial z} - \frac{\partial A_z}{\partial x}\right) + \mathbf{k}\left(\frac{\partial A_y}{\partial x} - \frac{\partial A_x}{\partial y}\right)$$

$$\nabla^2 f = \frac{\partial^2 f}{\partial x^2} + \frac{\partial^2 f}{\partial y^2} + \frac{\partial^2 f}{\partial z^2}$$

$$\nabla^2 \mathbf{A} = \mathbf{i}\,\nabla^2 A_x + \mathbf{j}\,\nabla^2 A_y + \mathbf{k}\,\nabla^2 A_z$$

1.2. Cylindrical coordinates

$$\nabla f = \mathbf{i}\frac{\partial f}{\partial \rho} + \mathbf{j}\frac{1}{\rho}\frac{\partial f}{\partial \phi} + \mathbf{k}\frac{\partial f}{\partial z}$$

$$\nabla \cdot \mathbf{A} = \frac{1}{\rho}\frac{\partial}{\partial \rho}(\rho A_\rho) + \frac{1}{\rho}\frac{\partial A_\phi}{\partial \phi} + \frac{\partial A_z}{\partial z}$$

$$\nabla \times \mathbf{A} = \mathbf{i}\left(\frac{1}{\rho}\frac{\partial A_z}{\partial \phi} - \frac{\partial A_\phi}{\partial z}\right) + \mathbf{j}\left(\frac{\partial A_\rho}{\partial z} - \frac{\partial A_z}{\partial \rho}\right) + \mathbf{k}\frac{1}{\rho}\left(\frac{\partial}{\partial \rho}(\rho A_\phi) - \frac{\partial A_\rho}{\partial \phi}\right)$$

$$\nabla^2 f = \frac{1}{\rho}\frac{\partial}{\partial \rho}\left(\rho\frac{\partial f}{\partial \rho}\right) + \frac{1}{\rho^2}\frac{\partial^2 f}{\partial \phi^2} + \frac{\partial^2 f}{\partial z^2}$$

(*Continued*)

**Table 8.24. Vector definitions, identities, and theorems** *(Continued)*

1.3. Spherical coordinates

$$\nabla f = \mathbf{i}\frac{\partial f}{\partial r} + \mathbf{j}\frac{1}{r}\frac{\partial f}{\partial \theta} + \mathbf{k}\frac{1}{r\sin\theta}\frac{\partial f}{\partial \phi}$$

$$\nabla \cdot \mathbf{A} = \frac{1}{r^2}\frac{\partial}{\partial r}(r^2 A_r) + \frac{1}{r\sin\theta}\frac{\partial}{\partial \theta}(A_\theta \sin\theta) + \frac{1}{r\sin\theta}\frac{\partial A_\phi}{\partial \phi}$$

$$\nabla \times \mathbf{A} = \mathbf{i}\frac{1}{r\sin\theta}\left[\frac{\partial}{\partial \theta}(A_\phi \sin\theta) - \frac{\partial A_\theta}{\partial \phi}\right] + \mathbf{j}\frac{1}{r}\left[\frac{1}{\sin\theta}\frac{\partial A_r}{\partial \phi} - \frac{\partial (rA_\phi)}{\partial r}\right] + \mathbf{k}\frac{1}{r}\left[\frac{\partial (rA_\theta)}{\partial r} - \frac{\partial A_r}{\partial \theta}\right]$$

$$\nabla^2 f = \frac{1}{r^2}\frac{\partial}{\partial r}\left(r^2\frac{\partial f}{\partial r}\right) + \frac{1}{r^2\sin\theta}\frac{\partial}{\partial \theta}\left(\sin\theta\frac{\partial f}{\partial \theta}\right) + \frac{1}{r^2\sin^2\theta}\frac{\partial^2 f}{\partial \phi^2}$$

2. **Identities**

$\nabla \cdot (f\mathbf{A}) = f(\nabla \cdot \mathbf{A}) + \mathbf{A} \cdot \nabla f$

$\nabla'(1/r) = \mathbf{r}_1/r^2$, where $\nabla'(1/r)$ is calculated at the source point $(x', y', z')$ and $\mathbf{r}_1$ is the unit vector from the source point $(x', y', z')$ to the field point $(x, y, z)$.

$\nabla(1/r) = -\mathbf{r}_1/r^2$ when the gradient is calculated at the field point, with the same unit vector.

$\nabla \times f\mathbf{A} = f(\nabla \times \mathbf{A}) - \mathbf{A} \times \nabla f$

$\nabla \cdot (\mathbf{A} \times \mathbf{B}) = \mathbf{B} \cdot (\nabla \times \mathbf{A}) - \mathbf{A} \cdot (\nabla \times \mathbf{B})$

$\nabla \times \nabla \times \mathbf{A} = \nabla(\nabla \cdot \mathbf{A}) - \nabla^2 \mathbf{A}$ in rectangular coordinates only.

$$(\mathbf{A} \cdot \nabla)\mathbf{B} = \mathbf{i}\left[A_x\frac{\partial B_x}{\partial x} + A_y\frac{\partial B_x}{\partial y} + A_z\frac{\partial B_x}{\partial z}\right] + \mathbf{j}\left[A_x\frac{\partial B_y}{\partial x} + A_y\frac{\partial B_y}{\partial y} + A_z\frac{\partial B_y}{\partial z}\right] + \mathbf{k}\left[A_x\frac{\partial B_z}{\partial x} + A_y\frac{\partial B_z}{\partial y} + A_z\frac{\partial B_z}{\partial z}\right]$$

3. **Theorems**

*Divergence theorem:* $\int_S \mathbf{A} \cdot d\mathbf{a} = \int_\tau \nabla \cdot \mathbf{A}\, d\tau$,
where $S$ is the surface which bounds the volume $\tau$.

*Stokes's theorem:* $\int_C \mathbf{A} \cdot d\mathbf{l} = \int_S (\nabla \times \mathbf{A}) \cdot d\mathbf{a}$,
where $C$ is the closed curve which bounds the surface $S$.

$\int_\tau (\nabla \times \mathbf{A})\, d\tau = \int_S \mathbf{A} \times d\mathbf{a}$,
where $S$ is the surface which bounds the volume $\tau$.

4. **Vector operations**

$$\mathbf{A} \cdot (\mathbf{B} \times \mathbf{C}) = \mathbf{B} \cdot (\mathbf{C} \times \mathbf{A}) = \mathbf{C} \cdot (\mathbf{A} \times \mathbf{B}) = \mathbf{ABC}$$

$$\mathbf{A} \times (\mathbf{B} \times \mathbf{C}) = (\mathbf{A} \cdot \mathbf{C})\mathbf{B} - (\mathbf{A} \cdot \mathbf{B})\mathbf{C}$$

$$\begin{aligned}(\mathbf{A} \times \mathbf{B}) \cdot (\mathbf{C} \times \mathbf{D}) &= \mathbf{A} \cdot [\mathbf{B} \times (\mathbf{C} \times \mathbf{D})] \\ &= \mathbf{A} \cdot [(\mathbf{B} \cdot \mathbf{D})\mathbf{C} - (\mathbf{B} \cdot \mathbf{C})\mathbf{D})] \\ &= (\mathbf{A} \cdot \mathbf{C})(\mathbf{B} \cdot \mathbf{D}) - (\mathbf{A} \cdot \mathbf{D})(\mathbf{B} \cdot \mathbf{C})\end{aligned}$$

$$\begin{aligned}(\mathbf{A} \times \mathbf{B}) \times (\mathbf{C} \times \mathbf{D}) &= [(\mathbf{A} \times \mathbf{B}) \cdot \mathbf{D}]\mathbf{C} - [(\mathbf{A} \times \mathbf{B}) \cdot \mathbf{C}]\mathbf{D} \\ &= (\mathbf{ABD})\mathbf{C} - (\mathbf{ABC})\mathbf{D}\end{aligned}$$

## 8.3. SOME USEFUL FORMULAE

### Table 8.25. Logarithmic relationships

log $N$ = the exponent or power to which the base 10 must be raised to obtain a value $N$ (the common logarithm of $N$)

ln $N$ = the power to which the base 2.718 . . . ($e$) must be raised to obtain a value $N$ (the natural logarithm of $N$)

(1) $\log N = 0.4343 \ln N$
(2) $\ln N - 2.3026 \log N$
(3) $\log MN = \log M + \log N$
(4) $\log M/N = \log M - \log N$
(5) $\log N^a = a \log N$
(6) $\log \sqrt[a]{N} = (\log N)/a$

### Table 8.26. Classical physics

Unless otherwise noted, the symbols and dimensions in this table are used consistently as follows:

$m$ = mass (g)
$v$ = velocity (cm $s^{-1}$)
$a$ = acceleration (cm $s^{-2}$)
$F$ = force (g-cm $s^{-2}$, dynes)
$r$ = radius of action (cm)
$s$ = distance (cm)

(1) Linear force ($F$)
$F = m\,a$ = (g) (cm $s^{-2}$) = g-cm $sec^{-2}$ = dynes
(2) Momentum ($p$)
$p = mv$ = (g) (cm $s^{-1}$)
(3) Conservation of momentum (any impact between Body A and Body B)
$m_A v_{A_i} + m_B v_{B_i} = m_A v_{A_f} + m_B v_{B_f}$,
where i = initial and f = final
(4) Work ($W$)
$W = Fs = mas$ = (g) (cm $s^{-2}$) (cm)
= g-$cm^2$ $s^{-2}$ = dyne-cm = erg
(5) Energy ($E$)
$E$ = (work) = $Fs$ = (g-cm $s^{-2}$) (cm)
= g-$cm^{-2}$ $s^{-2}$ = erg
(6) Kinetic energy ($KE$)
$KE = \frac{1}{2} m v^2$ = (g) (cm $s^{-2}$) = g-$cm^{-2}$ $s^{-2}$
= erg
(7) Conservation of kinetic energy (elastic impact between Body A and Body B)
$\frac{1}{2} m_A v_{A_i}^2 + \frac{1}{2} m_B v_{B_i}^2 = \frac{1}{2} m_A v_{A_f}^2 + \frac{1}{2} m_B v_{B_f}^2$
(8) Power ($P$)
$P$ = (work/time) = $F\,s/t$ = (g-cm $s^{-2}$) (cm)/s
= erg/s

### Table 8.27. Wave and quantum relationships

Unless otherwise noted, symbols and dimensions in this table are used consistently as follows:

$v$ = velocity of wave or particle (cm s)
$h$ = Planck's constant ($6.6 \times 10^{-27}$ erg s)
$\nu$ = frequency of wave or quanta (hertz)
$\lambda$ = wavelength (cm)
$\lambda_0$ = wavelength of incident radiation (angstroms)
$\lambda_\theta$ = wavelength of scattered radiation at angle $\theta$ (angstroms)
$E$ = energy (ergs)
$\theta$ = angle between incident and scattered radiation
$c$ = velocity of light ($3 \times 10^{10}$ cm s)
$m$ = mass of particle (g)
$\phi$ = work function (ergs)

(1) Wave equation
Wave velocity ($v$ or $c$) = $\lambda\nu$
(2) Associated wavelength of a particle

$$\text{Wavelength} = \lambda = \frac{h}{mv}$$

(3) Photoelectric equation
$E = \phi + \frac{1}{2}mv^2$
(4) Photon energy
$E = h\nu$

$$E = \frac{hc}{\lambda}$$

Energy in electron volts (eV)

$$= \frac{1.242 \times 10^4}{\text{wavelength in angstroms}}$$

(5) Mass–energy relationship
$E = mc^2$
(6) Momentum of photon
$mv = h/\lambda$
(7) Compton scattering of gamma and X-rays
$\lambda_\theta = \lambda_0 + 0.0242\,(1 - \cos\theta)$

### Table 8.28. Electrostatics

The following units apply in this section:

$F$ = force (dynes)
$Q$ = electrostatic charge (statcoulombs)
$s$ = distance (cm)
$V$ = potential (statvolts)
$C$ = capacitance (statfarads)
$W$ = work (ergs)
$\epsilon$ = dielectric constant

(1) Force between two charges, $a$ and $b$ (Coulomb's Law)

$$F = \frac{Q_a Q_b}{\epsilon s^2}$$

(2) Work
$W = QV$
(3) Capacitance
$C = Q/V$
(4) Potential
$V = Q/s$

## Table 8.29. Formulae used in NMR

1. The Bloch equations:

In the laboratory frame,

$$\frac{d\mathbf{M}_x}{dt} = \gamma(\mathbf{M}_y\mathbf{B}_0 + \mathbf{M}_z\mathbf{B}_1 \sin \omega t) - \mathbf{M}_x/T_2$$

$$\frac{d\mathbf{M}_y}{dt} = \gamma(\mathbf{M}_z\mathbf{B}_1 \cos \omega t - \mathbf{M}_x\mathbf{B}_0) - \mathbf{M}_y/T_2$$

$$\frac{d\mathbf{M}_z}{dt} = -\gamma(\mathbf{M}_x\mathbf{B}_1 \sin \omega t + \mathbf{M}_y\mathbf{B}_1 \cos \omega t) - (\mathbf{M}_z - \mathbf{M}_\infty)/T_1$$

where $\mathbf{M}_x$, $\mathbf{M}_y$, and $\mathbf{M}_z$ are the $x$, $y$, and $z$ components of the magnetization vector; $\mathbf{M}_\infty$ is the equilibrium value of the magnetization; $\mathbf{B}_0$ is the static magnetic field; $\mathbf{B}_1$ is the magnetic field vector; $\gamma$ is the magnetogyric ratio of the nucleus; $T_1$ is the spin–lattice relaxation time or longitudinal relaxation time; $T_2$ is the spin–spin relaxation time or transverse relaxation time; $\omega$ is the angular frequency of rotation of the radio-frequency (rf) magnetic field vector, $\mathbf{B}_1$.

2. $\Delta\nu_{1/2}$ = full line width at half maximum (FWHM) = $(1/\pi T_2)$, where $T_2$ is the decay constant of the free induction decay signal.

3. Formulae for different spin–lattice ($T_1$) and spin–spin ($T_2$) relaxation mechanisms.

*(i) Dipole-dipole interaction:*
For a two-spin system with equal spin, $I$:

$$\left.\frac{1}{T_1}\right)_I = \frac{3}{2}\gamma^4 h^2 I(I+1)[J_1(\omega_I) + J_2(2\omega_I)]$$

$$\left.\frac{1}{T_2}\right)_I = \gamma^4 h^2 I(I+1) \cdot \left[\frac{3}{8}J_0(O) + \frac{15}{4}J_1(\omega_I) + \frac{3}{8}J_2(2\omega_I)\right].$$

For intramolecular dipole-dipole interactions, e.g., rotational motion between like spins, the above become

$$\left.\frac{1}{T_1}\right)_{\text{rotational}} = \frac{2}{5}\frac{\gamma^4 h^2 I(I+1)}{r^6}\left[\frac{\tau_c}{1+\omega_I^2\tau_c^2} + \frac{4\tau_c}{1+4\omega_I^2\tau_c^2}\right]$$

$$\left.\frac{1}{T_2}\right)_{\text{rotational}} = \frac{\gamma^4 h^2 I(I+1)}{5r^6}\left[3\tau_c + \frac{5\tau_c}{1+\omega_I^2\tau_c^2} + \frac{2\tau_c}{1+4\omega_I^2\tau_c^2}\right].$$

In these equations,
$\gamma$ is the nuclear magnetogyric ratio; $h = (h/2\pi)$ where $h$ is Planck's constant; $I$ is the nuclear spin; $J_i(\omega)$ are the spectral density functions used to indicate intensity or probabilities of the molecular motions at the angular frequency, $\omega$; $r$ is the internuclear distance between the spins interacting; $\tau_C$ is the correlation time for the molecular motion; $\omega_I$ is the Larmor frequency of the spin $I$.

*(ii) Chemical shift anisotropy relaxation:*
The spin–lattice relaxation rate for the relaxation caused by chemical shift anisotropy interaction assuming axial symmetry for the molecule:

$$\frac{1}{T_1} = \frac{1}{15}\gamma^2\mathbf{B}_0^2(\sigma_\parallel - \sigma_\perp)^2\left[\frac{2\tau_c}{1+\omega_I^2\tau_c^2}\right]$$

$$\frac{1}{T_2} = \frac{1}{90}\gamma^2\mathbf{B}_0^2(\sigma_\parallel - \sigma_\perp)^2\left[\frac{6\tau_c}{1+\omega_I^2\tau_c^2} + 8\tau_c\right],$$

where $\gamma$ and $\omega_I$ have the usual meanings; $\mathbf{B}_0$ is the static magnetic field; $\tau_c$ is the correlation time for the chemical shift interaction; $\sigma_\parallel$ and $\sigma_\perp$ are the components of the chemical shift tenor, parallel and perpendicular to the symmetry axis of the molecule, respectively, and $\sigma = 1/3\,[\sigma_{xx} + \sigma_{yy} + \sigma_{zz}]$, (where $\sigma_{xx}$, $\sigma_{yy}$, and $\sigma_{zz}$ are the components of the chemical shift tensor and $\sigma$ is the average value of the chemical shift).

*(iii) Scalar relaxation:*
If two spins $I$ and $S$ are coupled by spin–spin (scalar) coupling, then, assuming that the nucleus $S$ has a short relaxation time $T_{1_s}^s$ compared to $1/A$ (where $A$ is the spin–spin coupling constant for this interaction), the spin–lattice relaxation rates are

$$\left(\frac{1}{T_1}\right)_I = \frac{2A^2}{3}S(S+1)\frac{\tau_S}{1+(\omega_I-\omega_S)^2\tau_S^2}$$

$$\left(\frac{1}{T_2}\right)_I = \frac{A^2}{3}S(S+1)\left[\tau_S + \frac{\tau_S}{1+(\omega_I-\omega_S)^2\tau_S^2}\right],$$

where $A$ is the spin–spin coupling constant (radians/s), $\tau_S$ is the relaxation time of the nucleus $S$ $(T_{1S})$, $\omega_I$, and $\omega_S$ are the Larmor frequencies of nuclei $I$ and $S$.

*(iv) Spin-rotation relaxation:*
For liquids undergoing isotropic rapid molecular reorientation, the spin-rotation contribution to the relaxation rate is given by

$$\left.\frac{1}{T_1}\right)_{\text{spin-rotation}} = \frac{2\pi IkT}{h^2}C_{\text{eff}}^2\,\tau_J,$$

where $I$ is the moment of inertia of the molecule: $k$ is the Boltzmann constant; $T$ is the absolute temperature; $h$ is Planck's constant; $C_{\text{eff}}^2 = 1/3\,(2C_\perp^2 + C_\parallel^2)$; ($C_\parallel$ and $C_\perp$ are the components of the spin-rotation tensor parallel and perpendicular, respectively, to the symmetry axis of the molecule); and $\tau_J$ is the angular momentum correlation time. Also, $\tau_c\tau_J = I/6kT$, where $\tau_C$ is the molecular reorientation correlation time. All other parameters have the same meaning as above.

*(v) Quadrupolar relaxation:*
For nuclei with $I > 1/2$, the relaxation is caused by the quadrupolar interaction. In the case of motional narrowing (i.e., $\omega_0\tau_c \ll 1$), for the case of molecular reorientation, we have

$$\frac{1}{T_1} = \frac{1}{T_2} = \frac{3}{40}\frac{2I+3}{I^2(2I-1)}\left(1+\frac{\eta^2}{3}\right)\left(\frac{e^2Qq}{h}\right)^2\tau_c,$$

where $\eta$ is the asymmetry parameter; $e^2Qq/h$ is the quadrupole coupling constant; and $\tau_C$ is the correlation time for quadrupole interaction.

*(Continued)*

**Table 8.29. Formulae used in NMR** (*Continued*)

4. The $\mathbf{B}_1$ power in a sample coil (single coil).

$$\mathbf{B}_1 \simeq 3\left(\frac{PQ}{\nu_0 V}\right)^{1/2} \approx 3.7\left(\frac{PT/r}{V}\right)^{1/2},$$

where $\mathbf{B}_1$ is the power in the sample coil; $P$ is the transmitter power in watts; $Q$ is the quality factor of the sample circuit; $\nu_0$ is the resonance frequency in megahertz; $V$ is the volume of the sample coil in cubic centimeters; $T/r$ is the rise and fall time of the envelope of the rf pulse in microseconds; $2\mathbf{B}_1$ is the peak-to-peak amplitude of the rf field in gauss; and $Q = R/2\pi\nu_0 L = R/\omega_0 L$ for a parallel LCR-tuned circuit (where $R$ is the resistance in ohms; $\nu_0$ is the frequency in hertz; $L$ is the inductance in henrys; and $Q \sim 1.5\,\nu_0 T/r$).

5. The signal-to-noise ratio in a pulsed NMR experiment.

$$S/N \propto \zeta\gamma I\;(I+1)\;C\left(\frac{QV\nu_0 T_2}{\beta T_1}\right)^{1/2},$$

where $\zeta$ is the filling factor of the receiver coil; $I$ is the nuclear spin quantum number; $C$ is a constant dependent on many other constants; $\gamma$, $Q$, $V$, and $\nu_0$ have the same definitions as above; $T_1$ and $T_2$ are the spin–lattice and spin–spin relaxation times; and $\beta$ is the band width of the receiver-detector system.

6. Arrhenius relation.

$$\tau_c = \tau_c^0\, e^{E_a/kT},$$

where $\tau_c$ is the correlation time for the molecular motion; $\tau_c^0$ is the preexponential factor, the reciprocal of which gives jump frequency; $k$ is the Boltzmann constant; $T$ is the absolute temperature; and $E_a$ is the activation energy for the molecular motion under consideration.

7. Solomon-Bloembergen equations.

The relaxation rate of water molecules in the coordination sphere of a paramagnetic metal ion in an aqueous solution is given by the following equations, known as the Solomon-Bloembergen equations:

$$\frac{1}{T_{1_M}} = \frac{C}{r^6}\left(\frac{3\tau_{c_1}}{1+\omega_I^2\tau_{c_1}^2} + \frac{7\tau_{c_2}}{1+\omega_S^2\tau_{c_2}^2}\right) + \frac{2}{3}\left(\frac{A}{h}\right)^2 S(S+1)\left(\frac{\tau_{e_2}}{1+\omega_s^2\tau_{e_2}^2}\right), \text{ and}$$

$$\frac{1}{T_{2_M}} = \frac{C}{2r^6}\left(4\tau_{c_1} + \frac{3\tau_{c_1}}{1+\omega_I^2\tau_{c_1}^2} + \frac{13\tau_{c_2}}{1+\omega_S^2\tau_{c_2}^2}\right) + \frac{1}{3}\left(\frac{A}{h}\right)^2 S(S+1)\left(\tau_{e_1} + \frac{\tau_{e_2}}{1+\omega_s^2\tau_{e_2}^2}\right),$$

where $1/T_{1_M}$ is the spin–lattice relaxation rate of water protons in the coordination sphere of the metal ion, and $1/T_{2_M}$ is the spin–spin relaxation rate of the water molecules in the coordination sphere of the metal ion.

In the above equations, the first term is the dipolar interaction term and the second term is the scalar interaction term. The terms in the above equation are defined as follows:

$$C = \frac{2}{15}\gamma_I^2 g^2 S(S+1)\beta^2,$$

where $\gamma$ is the magnetogyric ratio of the protons; $g$ is the electronic $g$-factor; $S$ is the total electron spin of the metal ion; $\beta$ is the Bohr magneton; $r$ is the ion–proton distance; $\tau_{c_1}$ and $\tau_{c_2}$ are the effective correlation times for the dipolar interactions; $\omega_I$ is the proton Larmor frequency; $\omega_s$ is the electron Larmor frequency ($658\,\omega_I$); $A/h$ is the scalar coupling constant; and $\tau_{e_1}$ and $\tau_{e_2}$ are the correlation times for scalar coupling.

$$\frac{1}{\tau_{ci}} = \frac{1}{\tau_R} + \frac{1}{\tau_{ei}}, \text{ and}$$

$$\frac{1}{\tau_{ei}} = \frac{1}{\tau_M} + \frac{1}{T_{1_{ei}}},$$

where $\tau_R$ is the rotational correlation time of the ion–water complex; $\tau_M$ is the residence life time of water molecules in the metal ion hydration sphere; and $T_{1_e}$ is the electron spin relaxation time of the metal ion.

In the case of water protons in paramagnetic ion–macromolecule solutions, these equations can be simplified, with some assumptions, to the following:

$$\frac{1}{T_{1_M}} = \frac{C}{r^6}\left(\frac{3\tau_c}{1+\omega_I^2\tau_c^2}\right), \text{ and}$$

$$\frac{1}{T_{2_M}} = \frac{C}{2r^6}\left(4\tau_c + \frac{3\tau_c}{1+\omega_I^2\tau_c^2}\right),$$

where $\dfrac{1}{\tau_c} = \dfrac{1}{\tau_R} + \dfrac{1}{\tau_M} + \dfrac{1}{T_{1_e}}.$

8. Fast exchange.

If there is fast exchange between two sites A and B with resonance frequencies $\omega_A$ and $\omega_B$, and if $\Delta\omega_{0_A}$ and $\Delta\omega_{0_B}$ are the corresponding line widths in the absence of exchange, then a single resonance is observed as an intermediate frequency $\omega$ given by

$$\omega = x\,\omega_A + (1-x)\,\omega_B,$$

where $x$ and $(1-x)$ are the fractional populations of the exchanging species in the two environments $A$ and $B$. The line width of this line at the intermediate frequency is

$$\Delta\omega = x\,\Delta\omega_{0_A} + (1-x)\,\Delta\omega_{0_B}.$$

Similarly, if there is fast exchange of the molecules between two environments $A$ and $B$ with fractional population $x$ of the exchanging species in the $A$ environment, then the average relaxation rate of the nucleus is given by

$$\frac{1}{T_1} = x\left(\frac{1}{T_{1_A}}\right) + (1-x)\frac{1}{T_{1_B}},$$

where $1/T_{1_A}$ and $1/T_{1_B}$ are the spin–lattice relaxation rates in $A$ and $B$ environments, respectively. This is referred to as the two-fraction fast-exchange model.

(*Continued*)

**Table 8.29. Formulae used in NMR** (*Continued*)

9. Correlation for bulk susceptibility effect in high resolution NMR experiments using cylindrical samples.

$$\delta_{correct} = \delta_{obs} + \frac{2\pi}{3}(\chi_v^{ref} - \chi_v)$$

for the case where the static magnetic field $B_0$ is perpendicular to the sample spinning axis (i.e., electromagnet or permanent magnet).

$$\delta_{correct} = \delta_{obs} - \frac{4\pi}{3}(\chi_v^{ref} - \chi_v)$$

for the case where the static magnet field $B_0$ is parallel to the sample spinning axis (i.e., superconducting magnet). In the above equations $\delta_{correct}$ = chemical shift corrected for bulk susceptibility and $\delta_{obs}$ = observed chemical shift. $\chi^{ref}$ and $\chi_v$ are the volume magnetic susceptibilities of the reference and sample solutions, respectively.

CHAPTER 9

# *BIOLOGICAL DATA AND BIBLIOGRAPHY*

## 9.1. BIOLOGICAL DATA TABLES

### Table 9.1. Physiological data for humans

- **Total body water** (TBW)

| Fetus weight (g) | % $H_2O$ |
|---|---|
| 0.5 | 93–95 |
| 10 | 92–94 |
| 100 | 90 |
| 200 | 88 |
| 1,000 | 82–84 |
| 3,450 | 70–72 |

Postnatal

Male:

$$\frac{\text{TBW (L)}}{\text{Wt (kg)}} \times 100 = 79.45 - 0.24\ \text{Wt} - 0.15\ \text{age (yr)}$$

Female:

$$\frac{\text{TBW (L)}}{\text{Wt (kg)}} \times 100 = 69.81 - 0.26\ \text{Wt} - 0.12\ \text{age (yr)}$$

- **Extracellular water** (ECW)

| | % $H_2O$ |
|---|---|
| Prenatal | 75–96 |
| Postnatal | ECW = 0.239 Wt (kg) + 0.325 kg |
| Male children | ECW = 0.277 Wt (kg) + 0.916 kg |
| Female children | ECW = 0.211 Wt (kg) + 0.989 kg |
| Male adult | ECW = 7.35 + 0.135 Wt (kg) |
| Female adult | ECW = 5.27 + 0.135 Wt (kg) |

- **Intracellular water** (ICW = TBW − ECW

| | % $H_2O$ |
|---|---|
| Prenatal | 17–38 |
| Boys | ICW = 0.3236 Wt (kg) + 1.773 |
| Girls | ICW = 0.3288 Wt (kg) + 0.930 |

Men $$\frac{\text{ICW} \times 100}{\text{TBW}} = 55.3 - 0.07\ \text{(yr)}$$

Women $$\frac{\text{ICW} \times 100}{\text{TBW}} = 62.3 - 0.16\ \text{(yr)}$$

i.e., for an average male, 340 ml/kg; average female, 300 ml/kg

- **Intracellular water in body organs**

| | |
|---|---|
| Muscle | 235 ml/kg (Wt) |
| Bone | 20 ml/kg (Wt) |
| Connective tissue | 20 ml/kg (Wt) |
| Erythrocytes | 30 ml/kg (Wt) |
| Other tissues | 35 ml/kg (Wt) |
| Total ICW | 340 ml/kg (Wt) |

- **Total body volume**

| | |
|---|---|
| Adult male | 5,200 ml |
| Adult female | 3,900 ml |

- **Volume of red cells**—Hematocrit

| | |
|---|---|
| Male | 2,200 ml |
| Female | 1,350 ml |

- **Water content of blood** 78.9–80.8%

- **Hemoglobin content of blood** 11–17 g/100 ml

- **Changes in body fat with age**

| Male: Age | % Fat | Female: Age | % Fat |
|---|---|---|---|
| 20 | 11 | 20 | 29 |
| — | — | 35 | 29 |
| — | | | |
| 47 | 21 | 45 | 35 |
| 51 | 26 | 56 | 41 |
| 70 | 31 | 65 | 45 |

*(continued)*

**Table 9.1 Physiological data for humans** (*Continued*)

• **Water content of body organs** (adult)

| | Weight (g) | % $H_2O$ | % Fat |
|---|---|---|---|
| Skin | 2,600 | 62 | 19 |
| Hair | 20 | 4–13 | 2.3 |
| Nails | 10 | 0.07–13 | — |
| Skeleton | 10,000 | 28–40 | 18–25 |
| Cartilage | 1,100 | 78 | 1.3 |
| Bone marrow | | | |
| Red | 1,500 | 40 | 40 |
| Yellow | 1,500 | 15 | 35–80 |
| Lymphocytes | 1,500 | 80 | — |
| Spleen | 180 | 72–79 | 0.85–3 |
| Thymus | 20 | 82 | 2.8 |
| Skeletal muscle | 28,000 | 79 | 2.2–3 |
| Heart | 570 | 63–83 | 2.7–17 |
| Tongue | 70 | 60–72 | 15–24 |
| Esophagus | 40 | 76 | — |
| Stomach | 150 | 60–78 | — |
| Intestine | 1,200 | 77–82 | 1–9 |
| Liver | 1,800 | 63–74 | 1–12 |
| Gallbladder | 10 | 6 | — |
| Pancreas | 100 | 66–73 | 3–20 |
| Larynx | 28 | 68 | — |
| Trachea | 10 | 60 | — |
| Lung | 1,000 | 72–84 | 1–1.5 |
| Kidneys | 310 | 71–81 | 2–7 |
| Bladder | 45 | 65 | — |
| Filled with urine | 545 | 90 | — |
| Testes | 35 | 81 | 3 |
| Prostate | 16 | 83 | 1.2 |
| Ovaries | 30 | 78 | 1–2 |
| Uterus | — | 79 | 1–2.2 |
| Breast | 26 (male) | 50 | 3 |
| | 100 (female) | 39 | 40 |
| Thyroid | 20 | 72–78 | — |
| Adrenals | 14 | 64 | 6–26 |
| Brain | | | |
| Total | 1,400 | 76–79 | 9–17 |
| Gray matter | | 84–86 | 5.3 |
| White matter | | 68–77 | 18 |
| Spinal cord | 30 | 63–75 | 2–19 |
| Cerebrospinal fluid | | 99 | |
| Cornea stroma (ox) | | 78 | |
| Aqueous humor | 15 | 98.1 | — |
| Vitreous humor | 15 | 99 | — |
| Lens | 0.5 | 68 | 1–3 |
| Placenta | — | 84.6 | 0.11 |
| Fat | 13,500 | — | — |

• **Metabolic variables**

| | | |
|---|---|---|
| Energy expenditure | 3,000 | kcal/day |
| Metabolic rate | 17 | cal/min – kg (Wt) |
| Oxygen inhaled | 920 | g/day |
| Carbon dioxide exhaled | 1,000 | g/day |
| Nutrient intake | 605 | g/day |
| Weight of feces | 135 | g/day |
| Lung capacity | 7.5 | l/min |
| Urine volume | 1,400 | ml/day |

• **Water balance** (gain)

| | | |
|---|---|---|
| Fluid intake | 1,950 | ml/day |
| In food | 700 | ml/day |
| Metabolic product | 350 | ml/day |
| Total | 3,000 | ml/day |

• **Water balance** (lost)

| | | |
|---|---|---|
| In urine | 1,400 | ml/day |
| In feces | 100 | ml/day |
| In perspiration | 650 | ml/day |
| Insensible loss | 850 | ml/day |
| Total | 3,000 | ml/day |

All values in this table are for an average adult male unless otherwise stated. All can change with age, sex, and physiological state, and should not be taken as constants. (Data from *Report of the Task Group on Reference Man,* Pergamon Press, 1981.)

**Table 9.2. NMR proton relaxation times for water in normal and pathological nonmammalian tissues**

| Tissue | Frequency (MHz) | $T_1$ (ms) (mean ± SD) | $T_2$ (ms) (mean ± SD) | Reference |
|---|---|---|---|---|
| **Bacteria** | | | | |
| *E. coli* | 100 | 557 | 30 | Damadian et al., 1974 |
| *Halobacterium* | 100 | 40 | 21 | " |
| *E. coli* | — | | Line broadening | Cerbon, 1965 |
| *Nocardia asteroides* | — | | Line broadening | " |
| **Yeast** | 30 | 2 components | | Koga et al., 1966 |
| **Slime mold** | | | | |
| *Physarum* | 20.7 | 131 ± 13 | 30 ± 2 | Walter and Hope, 1971 |
| **Bluegreen algae** | | | | |
| *Aphamothece halophytica* | 60 | 200 | 20 | Venable et al., 1978 |
| **Eggs** | | | | |
| Frog | 30 | 216 | 36 | Mild et al., 1972 |
| Frog | 30 | 330 ± 10 | | Cameron et al., 1983 |
| Frog (stimulated) | 30 | 500 ± 100 | | " |
| **Fish** | | | | |
| Cod muscle | 60 | | Line broadening | Sussman and Chin, 1966 |
| **Embryonic cysts of brine shrimp** | | | | |
| Dry cysts | 30 | 178 | 67.9 | Seitz et al., 1981 |
| Hydrated | 30 | 255 | 54.2 | " |
| Hydrated | 60; 200 | (H, D, O study) | | Hazlewood et al., 1983 |
| **Squid** | | | | Chang and Hazlewood, 1980 |
| Axon | 30 | 1,490 ± 50 | 348 ± 36 | |
| Axoplasm | 30 | 1,500 ± 40 | 326 ± 25 | |
| Membrane | 30 | 178 ± 26 | 14 ± 5 | |
| **Frog** | | | | |
| Nerve | | | | |
| Polarized | 60 | 260 | 34 | Swift and Fritz, 1969 |
| Depolarized | | 150 | 24 | " |
| Liver | 20 | 270 | 46 | Abetsedarskaya et al., 1968 |
| Unspecified muscle | 4.3 | 335 ± 10 | | Shporer and Civan, 1975 |
| | 8.1 | 423 ± 13 | | " |
| | 24 | 400 | 40 | Bratton et al., 1965 |
| Peronius | 60 | 1,100 ± 100 | 50 ± 10 | Tanner, 1979 |
| Semitendinous | 60 | 1,100 ± 200 | 55 ± 15 | " |
| Sartorius | 60 | 1,400 ± 100 | 44 ± 3 | " |
| Gastrocnemius | 8.1 | 423 ± 13 | | Civan and Shporer, 1975 |
| | 8.1 ($^2$D) | 143 ± 6 | | Civan and Shporer, 1975 |
| | 10 | 520 | 30 | Held et al., 1973 |
| | | 450 | 45 | " |
| | 20 | 658 | 47 | Abetsedarskaya et al., 1968 |
| | 27.5 | 700 | 44 | Finch and Homer, 1974 |
| Relaxed | 30 | | 48 | Finch et al., 1971 |
| Contracted | 30 | | 50 | " |
| | 60 | 730 | | Belton et al., 1972 |
| **Toad** | | | | |
| Muscle | 21 | 230 ± 20 | 37 ± 3 | Walter and Hope, 1971 |
| | 31 | 730 | | Outhred and George, 1973b |

*(continued)*

**Table 9.2. NMR proton relaxation times for water in normal and pathological nonmammalian tissues *(Continued)***

| Tissue | Frequency (MHz) | $T_1$ (ms) (mean ± SD) | $T_2$ (ms) (mean ± SD) | Reference |
|---|---|---|---|---|
| **Newt** | | | | Sandhu and Friedman, 1978 |
| Tail (alive) | 20 | 676 | | |
| Tail (dead 5 h) | 20 | 910 | | |
| **Chicken** | | | | |
| Kidney | 32 | 403 ± 37 | | Ratkovic and Rusov, 1974* |
| Kidney (with virus) | | 481 ± 36 | | Ratkovic and Rusov, 1974* |
| Liver | 32 | 208 ± 16 | | |
| Liver (with virus) | | 370 ± 36 | | |
| Lung (with virus) | 32 | 735 ± 50 | | |
| Spleen (with virus) | | 641 ± 19 | | |
| Bone marrow | | 408 ± 53 | | |
| Bone marrow (with virus) | | 717 ± 42 | | |
| Blood | 32 | 920 ± 20 | | |
| Blood (with virus) | | 1,320 ± 130 | | |
| Plasma | 32 | 1,580 ± 110 | | |
| Plasma (with virus) | | 1,810 ± 300 | | |
| Plasma (normal) | 20 | 2,246 ± 53 | | Misra et al., 1983a† |
| Plasma (dystrophic) | | 2,222 ± 42 | | |
| Muscle | | | | Ratkovic and Rusov, 1974* |
| Red muscle | 32 | 642 ± 22 | | |
| Red muscle (with virus) | | 665 ± 48 | | |
| White muscle | 32 | 598 ± 20 | | |
| White muscle (with virus) | | 1,320 ± 130 | | |
| Breast | 30 | 654 ± 23 | 39 ± 2 | Beall et al., 1977a |
| Breast (dystrophic) | | 694 ± 46 | 53 ± 6 | |
| Pectoralis (chick) | 20 | | | |
| 1 day old | | 1,246 ± 40 | | Misra et al., 1983b |
| 16 days old | | 647 ± 10 | | |
| Pectoralis (adult) | 30 | 654 ± 22 | 39 ± 4 | Chang et al., 1981 |
| Pectoralis | 20 | 616 ± 7 | | Misra, 1983† |
| Pectoralis | 1.69 | 110 ± 8 | | |
| Pectoralis (dystrophic) | 30 | 692 ± 41 | 52 ± 7 | Chang et al., 1981 |
| Pectoralis | 20 | 693 ± 12 | | Misra, 1983† |
| Pectoralis | 1.69 | 217 ± 17 | | |
| Gastrocnemius (normal) | 20 | 667 ± 20 | | |
| Gastrocnemius (dystrophic) | 20 | 693 ± 18 | | |
| Posterior LD (normal) | 20 | 666 ± 19 | | |
| Posterior LD (dystrophic) | | 714 ± 13 | | |
| Anterior LD (normal) | 20 | 650 ± 20 | | |
| Anterior LD (dystrophic) | | 654 ± 19 | | |

*Chickens treated with erythroleukosis.
†Normal chickens (strain 412) compared with strain 413 birds with progressive muscular dystrophy.

**Table 9.3. NMR proton relaxation times for water in plants**

| Plant tissue | Frequency (MHz) | $T_1$ (ms) (mean ± SD) | $T_2$ (ms) (mean ± SD) | Reference |
|---|---|---|---|---|
| Maize leaves | 20 mc/s | 250 | 96 | Abetsedarskaya et al., 1968 |
| Bean leaves | | 300 | 110 | |
| Maize roots | | 885 | 133 | |
| Maize leaves | 20 mc/s | 540 | 210 | Fedotov et al., 1969 |
| Pea leaves | | 300 | 71 | |
| Bean leaves | | 384 | 180 | |
| Maize roots | | 867 | 285 | |
| Apple pulp | | 1,600 | 970 | |
| Onion bulb | | 910 | 620 | |
| Potato tuber | | 435 | 190 | |
| Crinum leaves | | 374 | | |
| Pea roots | | | | |
| Meristem | 20.8 | 180 ± 10 | 47 ± 4 | Walter and Hope, 1971 |
| Mature | | 300 ± 50 | 56 ± 4 | |
| Blue-green algae | 60 | 200 | 20 | Venable et al., 1978 |
| *Zea mays* | | | | |
| Leaf | 32 | 770 ± 20 | | Ratkovic et al., 1982 |
| Root | | 833 | | |
| Birch leaf | 32 | 1,000 | | |
| Maize seed | 32 | 4–5 | | |

**Table 9.4. NMR proton relaxation times in normal and pathological tissues from mammals**

| Tissue | Frequency (MHz) | $T_1$ (ms) (mean ± SD) | $T_2$ (ms) (mean ± SD) | Reference |
|---|---|---|---|---|
| **Blood and body fluids** | | | | |
| Whole blood | | | | |
| Rat | 10.7 | 472 | | Chandra et al., 1983 |
| Rabbit | | | | |
| Clotted | 2.5 | 404 ± 30 | — | Ling and Foster, 1981 |
| | 24 | 867 ± 88 | — | |
| Heparinized | 2.5 | 372 ± 34 | — | |
| | 24 | 872 ± 43 | — | |
| Mouse | | | | |
| Normal cells | 20 | 902 | — | McLachlan and Hamilton, 1979 |
| Killed cells | 20 | 908 | — | |
| **Serum** | | | | |
| Mouse | 8 | 1,370 | — | Floyd et al., 1974 |
| Mouse | 30 | 1,554 ± 19 | | Beall et al., 1977b, 1981a |
| Rat | 24 | 1,304 ± 24 | — | Hollis et al., 1974 |
| Rabbit | 2.5 | 820 ± 12 | — | Ling and Foster, 1981 |
| | 24 | 1,590 ± 113 | | |
| **Bile** | | | | |
| Rabbit | 2.5 | 888 ± 388 | — | Ling and Foster, 1981 |
| | 24 | 1,078 ± 44 | — | |
| **Bone marrow** | | | | |
| Rabbit (red femur) | 2.5 | 175 ± 50 | — | Ling and Foster, 1981 |
| | 24 | 202 ± 14 | — | |
| **Brain** | | | | |
| Normal | | | | |
| Rat | 10.7 | 405 | | Chandra et al., 1983 |
| Rat | 13.5 | 474 ± 58 | — | Block and Maxwell, 1974 |
| Rat (cortex) | 14 | 405 ± 11 | 83.4 ± 1.9 | Gwan Go and Edzes, 1975 |
| Rat (white matter) | 14 | 347 ± 10 | 79.5 ± 1.5 | " |
| Rat | 24 | 595 ± 7 | — | Damadian, 1971 |
| Rat | 25 | 472 | | Chaughule et al., 1974a |
| Rat | 60 | 866 ± 42 | 71 ± 7 | Kiricuta and Simplacenu, 1975 |
| Rat | 60 | — | 29 | Villey and Martin, 1974 |
| Mouse | 2.7 | 284 ± 1 | 75.2 ± 0.4 | Coles, 1976 |
| Mouse (ddYF) | 10.7 | 240 | 24 | Ujeno, 1980 |
| Mouse | 15 | 488 ± 4 | 77 ± 0.4 | Coles, 1976 |
| Mouse | 20 | 584 | | McLachlan and Hamilton, 1979 |
| Mouse (BUF) | 24 | 590 | — | Hollis et al., 1975 |
| Mouse (BUF) | 24 | 526 | — | Hollis et al., 1973 |
| Mouse | 30 | 646 ± 4 | 45 | Frey et al., 1972 |
| Mouse | 30 | 720 | — | Fung, 1977c |
| Mouse | 100 | 840 | 88 | Iijima et al., 1973 |
| Mouse | 90 | 913 | 75 | de Certaines et al., 1979 |
| Rabbit (gray matter) | 2.5 | 332 ± 22 | — | Ling and Foster, 1981 |
| | 24 | 644 ± 40 | — | " |
| Rabbit (white matter) | 2.5 | 264 ± 11 | — | " |
| Rabbit (cerebellum) | 2.5 | 326 ± 11 | — | " |
| | 24 | 570 ± 24 | — | " |
| Rabbit (medulla) | 2.5 | 308 ± 2 | — | " |
| | 24 | 493 ± 47 | — | " |
| Rabbit (spinal cord) | 2.5 | 325 ± 19 | — | " |
| | 24 | 464 ± 22 | — | " |

*(continued)*

**Table 9.4. NMR proton relaxation times in normal and pathological tissues from mammals *(Continued)***

| Tissue | Frequency (MHz) | $T_1$ (ms) (mean ± SD) | $T_2$ (ms) (mean ± SD) | Reference |
|---|---|---|---|---|
| Pathological | | | | |
| Vasogenic edema | | | | |
| Rat (cortex) | 14 | 651 ± 47 | 115 ± 3.9 | Gwan Go and Edzes, 1975 |
| Rat (white matter) | 14 | 591 ± 22 | 131.5 ± 7.4 | " |
| Osmotic edema | | | | |
| Rat (cortex) | 14 | 463 ± 23 | 91.6 ± 3.1 | " |
| Rat (white matter) | 14 | 392 ± 24 | 91.2 ± 3.1 | " |
| Triethyltin edema | | | | |
| Rat (cortex) | 14 | 428 ± 35 | 82.5 ± 2.2 | " |
| Rat (white matter) | 14 | 455 ± 35 | 90.3 ± 6.8 | " |
| Vasopressin-treated | | | | |
| Rat | 60 | — | 29 | Villey and Martin, 1974 |
| Fetal brain | | | | |
| Rat | 60 | 1,361 | 149 | Kiricuta and Simplacenu, 1975 |
| **Breast** | | | | |
| Normal | | | | |
| Mouse (virgin) | 30 | 275 ± 46 | 84 ± 8 | Beall et al., 1981b |
| Mouse (pregnant) | 30 | 357 ± 47 | 47.2 ± 0.9 | Hazlewood et al., 1974a |
| | 30 | 380 ± 41 | — | Beall et al., 1981b |
| Rat (lactating) | 13.5 | 337 ± 26 | — | Block and Maxwell, 1974 |
| **Benign cancer** | | | | |
| Rat (fribroadenoma) | 24 | 492 | — | Damadian, 1971 |
| Preneoplastic | | | | |
| Mouse (HAN)* | | | | |
| $D_1$ | 30 | 657 ± 68 | 54.7 ± 1.8 | Hazlewood et al., 1972 |
| $D_2$ | 30 | 709 ± 29 | 54.1 ± 1.2 | " |
| $D_2$ | 30 | 451 ± 21 | — | " |
| CDH3 | 30 | 350 ± 75 | 85 ± 18 | " |
| Cancers | | | | |
| Mouse | 24 | 798 | — | Hollis et al., 1973 |
| Mouse tumors | | | | |
| $D_2$ | 30 | 920 ± 47 | 91 ± 8 | Beall et al., 1981b |
| C3H | 30 | 887 ± 40 | 78.8 ± 8.2 | " |
| C3Hf | 30 | 906 ± 43 | 80.3 ± 5.9 | " |
| CDH3 | 30 | 721 ± 98 | 98 ± 20 | " |
| Rat (carcinoma) | 13.5 | 560 ± 32 | — | Block and Maxwell, 1974 |
| Rat (adenocarcinoma) | 60 | 1,113 | — | Bovée et al., 1974 |
| **Arteries and veins** | | | | |
| Rabbit (aortic arch) | 2.5 | 211 ± 41 | — | Ling and Foster, 1981 |
| | 24 | 451 ± 15 | — | " |
| Rabbit (vena cava) | 2.5 | 201 ± 5 | — | " |
| | 24 | 316 ± 56 | — | " |
| Rabbit (carotid artery) | 2.5 | 167 ± 17 | — | " |
| | 24 | 364 ± 30 | — | " |
| **Eye lens** | | | | |
| Rabbit (whole lens) | 25 | 490 | 29 | Neville et al., 1974 |
| Rabbit (cortex) | 25 | 551 | 40 | " |
| Rabbit (nucleus) | 25 | 268 | 7.7 | " |
| **Eye rod outer segment** | | | | |
| Cow | 90 | 8.6 | | Gaggelli et al., 1982 |

*(continued)*

**Table 9.4. NMR proton relaxation times in normal and pathological tissues from mammals *(Continued)***

| Tissue | Frequency (MHz) | $T_1$ (ms) (mean ± SD) | $T_2$ (ms) (mean ± SD) | Reference |
|---|---|---|---|---|
| **Glandular tissue** | | | | |
| Rabbit adrenal gland | 2.5 | 227 ± 13 | — | Ling and Foster, 1981 |
| | 24 | 448 ± 13 | — | " |
| Mouse (lymphatic node) | 24 | 867 | — | Cottam et al., 1972 |
| Rabbit (salivary gland) | 2.5 | 234 ± 38 | — | Ling and Foster, 1981 |
| | 24 | 340 ± 51 | — | " |
| Rat (normal thyroid) | 32 | 552 ± 57 | 37 ± 6 | Sinadinovic et al., 1977 |
| Rat (PTU-treated thyroid) | 32 | 698 ± 54 | 61 ± 4.5 | " |
| Rat ($NaClO_4$-treated thyroid) | 32 | 717 ± 42 | 62 ± 11 | " |
| **Heart** | | | | |
| Rat | 10 | 568 to 391 | — | Polak et al., 1983 |
| Rat | 10.7 | 434 | — | Chandra et al., 1983 |
| Rat | 20 | 590 ± 50 | 43 ± 4 | Boicelli et al., 1983 |
| Rat | 24 | 550 | — | Hollis et al., 1975 |
| Rat | 25 | 518 | | Chaughule et al., 1974b |
| Rat (12 months old) | 30 | 660 ± 17 | | Beall et al., 1983i |
| Rat (fetal) | 30 | 800 | 53 | " |
| Rat | 30 | 740 | 55 | Chaughule et al., 1974b |
| Rat | 60 | 873 ± 27 | 46 ± 11 | Kiricuta and Simplacenu, 1975 |
| Rat (fetal) | 60 | 936 | 69 | " |
| Mouse | 20 | 576 | — | McLachlan and Hamilton, 1979 |
| Mouse | 24 | 490 | — | Hollis et al., 1973 |
| Mouse | 30 | 650 ± 6 | 45 | Frey et al., 1972 |
| Mouse (infected with *Salmonella*) | 30 | 643 ± 7 | — | " |
| Mouse | 90 | 942 | 45 | de Certaines et al., 1979 |
| Rabbit (ventricle) | 2.5 | 243 ± 4 | — | Ling and Foster, 1981 |
| | 24 | 637 ± 28 | — | " |
| Rabbit (auricle) | 2.5 | 295 ± 9 | — | " |
| | 24 | 706 ± 42 | — | " |
| Dog (control) | — | 509 ± 6 | — | Williams et al., 1980 |
| Dog (ischemic for 120 min) | | 551 ± 10 | — | " |
| **Large intestine** | | | | |
| Rabbit (colon) | 2.5 | 227 ± 20 | — | Ling and Foster, 1981 |
| | 24 | 466 ± 29 | — | " |
| Rabbit (rectum) | 2.5 | 245 ± 6 | — | " |
| | 24 | 492 ± 56 | — | " |
| **Small intestine** | | | | |
| Rat | 24 | 257 ± 30 | — | Damadian, 1971 |
| Rat | 30 | 504 ± 3 | 58 ± 3 | Udall et al., 1977 |
| Mouse | 24 | 255 | — | Damadian, 1971 |
| Mouse | 30 | 366 ± 19 | 35 | Frey et al., 1972 |
| Mouse (*Salmonella*-infected) | 30 | 329 ± 19 | — | " |
| Rabbit | 2.5 | 244 ± 36 | — | Ling and Foster, 1981 |
| Rabbit | 24 | 481 ± 98 | — | " |
| Cholera-toxin–treated intestine | | | | |
| Rat (closed loop) | | | | |
| Control | 30 | 521 ± 69 | 62 ± 9 | Udall et al., 1975 |
| Treated | 30 | 668 ± 119 | 80 ± 21 | " |
| Rat (open loop) | | | | |
| Control | 30 | 504 ± 32 | 58 ± 3 | Udall et al., 1977 |
| Treated | 30 | 598 ± 28 | 74 ± 5 | " |
| Rat (villi scrapings) | | | | |
| Control | 30 | 366 ± 59 | 49 ± 3 | " |
| Treated | 30 | 867 ± 25 | 114 ± 9 | " |

*(continued)*

Table 9.4. **NMR proton relaxation times in normal and pathological tissues from mammals** *(Continued)*

| Tissue | Frequency (MHz) | $T_1$ (ms) (mean ± SD) | $T_2$ (ms) (mean ± SD) | Reference |
|---|---|---|---|---|
| **Kidney** | | | | |
| Rat | 10.7 | 342 | — | Chandra et al., 1983 |
| Rat | 13.5 | 410 ± 41 | — | Block and Maxwell, 1974 |
| Rat | 17.1 | 634 | — | Ling and Tucker, 1980 |
| Rat | 24 | 395 | — | Hollis et al., 1975 |
| Rat | 24 | 480 ± 26 | — | Damadian, 1971 |
| Rat | 25 | 577 | — | Chaughule et al., 1974b |
| Rat | 60 | 668 ± 31 | | Bovée et al., 1974 |
| Rat | 60 | 685 | 56 | Kiricuta and Simplacenu, 1975 |
| Rat | 60 | — | 38 | Villey and Martin, 1974 |
| Rat (treated with vasopressin) | | — | 36 | " |
| Mouse | 2.7 | 214 ± 2 | 56 ± 1 | Coles, 1976 |
| Mouse | 9 | 329 ± 7 | — | Modak et al., 1982 |
| Mouse | 10 | 370 | — | Fung, 1977c |
| Mouse | 10.7 | 190 | 22 | Ujeno, 1980 |
| Mouse | 15 | 390 ± 4 | 59 ± 5 | Coles, 1976 |
| Mouse | 17.1 | 524 | | Ling and Tucker, 1980 |
| Mouse | 20 | 391 | — | McLachlan and Hamilton, 1979 |
| Mouse | 24 | 272 | | Economou et al., 1973 |
| Mouse | 24 | 322 ± 11 | — | Hollis et al., 1974 |
| Mouse | 24 | 370 | — | Coles, 1976 |
| Mouse | 24.3 | 396 ± 43 | 52 | Cottam et al., 1972 |
| Mouse | 25 | 581 ± 25 | — | Modak et al., 1982 |
| Mouse | 30 | 480 ± 20 | — | Peemoeller et al., 1979 |
| Mouse (*Salmonella*-infected) | 30 | 484 ± 12 | | Frey et al., 1972 |
| Mouse | 30 | 470 ± 6 | | " |
| Mouse | 30 | 452 ± 5 | | Inch et al., 1974b |
| Mouse (pregnant) | 30 | 488 ± 6 | — | " |
| Mouse (regenerating liver) | 30 | 497 ± 5 | — | " |
| Mouse | 30 | 503 ± 11 | 47 ± 2.4 | Hazlewood et al., 1974a |
| Mouse | 30 | 551 ± 9 | 50.0 ± 0.3 | " |
| Mouse | 33.8 | 700 | 60 | Rustgi et al., 1978 |
| Mouse | 38 | 490 ± 50 | 50 ± 8 | Peemoeller et al., 1982 |
| Mouse | 90 | 508 | | Escanye et al., 1983 |
| Mouse | 90 | 735 | 55 | de Certaines et al., 1979 |
| Mouse | 100 | 612 ± 15 | — | Shah et al., 1982 |
| Rabbit | 24.3 | 363 | — | Cottam et al., 1972 |
| Hamster | 24.3 | 398 ± 36 | — | " |
| Dog | 24.3 | 534 | — | " |
| Rat | 5 | 165 ± 3 | — | Jacobs et al., 1983 |
| Rat | 5 | 186 | 49 | Thickman et al., 1983 |
| Rat | 8 | 163 ± 9 | — | Floyd et al., 1975 |
| Rat | 10.7 | 225 | | Chandra et al., 1983 |
| Rat | 13.5 | 238 ± 32 | — | Block and Maxwell, 1974 |
| Rat | 17.1 | 285 | — | Ling and Foster, 1981 |
| Rat | | | | |
| Normal | 20 | 281 ± 15 | 32 ± 2 | Ratner et al., 1983 |
| Ethanol | 20 | 278 ± 16 | 40 ± 4 | " |
| Azaserine | 20 | 290 ± 29 | 43 ± 10 | " |
| *l*-Ethionine | 20 | 286 ± 21 | 39 ± 2 | " |
| $CCl_4$ | 20 | 307 ± 12 | 48 ± 2 | " |
| *d*-Galactosamine | 20 | 301 ± 22 | 39 ± 3 | " |
| Rat | 24 | 275 ± 15 | | Chilton et al., 1982 |
| Rat | 24 | 270 | — | Hollis et al., 1975 |
| Rat | 24 | 293 ± 10 | — | Damadian, 1971 |
| Rat | 25 | 273 | — | Chaughule et al., 1974b |
| Rat | 25 | 180 ± 10 | — | Tanaka et al., 1974 |
| Rat | 30 | 440 | — | Raaphorst et al., 1975 |
| Rat (12 months old) | 30 | 240 ± 35 | — | Beall et al., 1983f |
| Rat (19 months old) | 30 | 253 ± 5 | — | " |
| Rat (3.5 months old) | 30 | 285 ± 22 | — | " |



*(continued)*

**Table 9.4. NMR proton relaxation times in normal and pathological tissues from mammals *(Continued)***

| Tissue | Frequency (MHz) | $T_1$ (ms) (mean ± SD) | $T_2$ (ms) (mean ± SD) | Reference |
|---|---|---|---|---|
| Rat (normal) | 32 | 250 | 40 | de Certaines et al., 1982b |
| Rat (regenerating) | 32 | 280 | 50 | " |
| Rat | 32 | 310 | | Sentjurc et al., 1974 |
| Rat (fetal) | 32 | 490 | | " |
| Rat | 60 | 467 ± 26 | — | Bovée et al., 1974 |
| Rat | 60 | — | 54 | Villey and Martin, 1974 |
| Rat (vasopressin) | 60 | — | 54 | " |
| Rat | 60 | 340 | | Kiricuta and Simplacenu, 1975 |
| Rat (fetal) | 60 | 527 | 43 | " |
| Rat | 100 | 420 | 37 | Iijima et al., 1973 |
| Mouse | 2.7 | 138 ± 2 | 36.7 ± 0.5 | Coles, 1976 |
| Mouse | 6 | 179 | | Escanye et al., 1983 |
| Mouse | 8 | 233 ± 8 | — | Barroilhet and Moran, 1976 |
| Mouse | 9 | 235 ± 5 | — | Modak et al., 1982 |
| Mouse | 10.7 | 166 | 19 | Ujeno, 1980 |
| Mouse | 15 | 293 ± 4 | 37 | Coles, 1976 |
| Mouse | 17.1 | 396 | — | Ling and Tucker, 1980 |
| Mouse | 20 | 411 | — | McLachlan and Hamilton, 1979 |
| Mouse | 24 | 243 | — | Economou et al., 1973 |
| Mouse | 24 | 225 ± 5 | — | Hollis et al., 1974 |
| Mouse | 24 | 263 | — | Hollis et al., 1973 |
| Mouse | 24 | 290 | 52 | Damadian, 1971 |
| Mouse | 24.3 | 350 ± 47 | 51 | Cottam et al., 1972 |
| Mouse | 25 | 412 ± 13 | — | Modak et al., 1982 |
| Mouse | 30 | 486 ± 13 | 25 | Inch et al., 1974b |
| Mouse | 30 | 341 ± 5 | 25 | " |
| Mouse (*Salmonella*-infected) | 30 | 371 ± 7 | — | Frey et al., 1972 |
| Mouse | 30 | 333 ± 10 | — | Peemoeller et al., 1979 |
| Mouse | 30 | 303 | — | " |
| Mouse | 30 | 438 ± 36 | 32.5 ± 1.7 | Hazlewood et al., 1974a |
| Mouse | 30 | 412 ± 12 | 32.7 ± 0.9 | " |
| Mouse (adult) (CWFD) | 30 | 361 | — | Inch et al., 1974b |
| Mouse (pregnant) | 30 | 348 ± 20 | — | " |
| Mouse (fetal liver) | 30 | 541 | — | " |
| Mouse (adult) (C3H/HeJ) | 30 | 341 ± 5 | — | " |
| Mouse (regenerating liver) | 30 | 414 ± 9 | — | " |
| Mouse | 38 | 380 ± 30 | 39 ± 6 | Peemoeller et al., 1982 |
| Mouse | 90 | 436 | | Escanye et al., 1983 |
| Mouse | 90 | 635 | 37 | de Certaines et al., 1979 |
| Mouse (ICRC) | 100 | 375 ± 29 | — | Shah et al., 1982 |
| Mouse (Swiss) | 100 | 426 ± 12 | — | " |
| Mouse | 100 | 420 | 37 | Iijima et al., 1973 |
| Mouse (liver homogenate) | 24 | 700 | — | Floyd et al., 1974 |
| Rabbit | 2.5 | 141 ± 16 | — | Ling and Foster, 1981 |
| | 24 | 311 ± 15 | — | " |
| | 24.3 | 353 ± 45 | — | Cottam et al., 1972 |
| Hamster | 24.3 | 303 ± 45 | — | " |
| | 25.6 | — | 53 ± 3 | Lewa and Zbytnlewski, 1976 |
| Dog | 24.3 | 301 | — | Cottam et al., 1972 |
| Rabbit (cortex) | 2.5 | 206 ± 27 | — | Ling and Foster, 1981 |
| | 24 | 406 ± 41 | — | " |
| Rabbit (medulla) | 2.5 | 426 ± 61 | — | " |
| | 24 | 801 ± 35 | — | " |
| Rabbit (ureter) | 2.5 | 172 ± 10 | — | " |
| | 24 | 252 ± 13 | — | " |
| **Lung** | | | | |
| Rat | 10.7 | 576 | — | Chandra et al., 1983 |
| Mouse | 90 | 942 | 45 | de Certaines et al., 1979 |
| Rabbit (trachea) | 2.5 | 199 ± 7 | — | Ling and Foster, 1981 |
| | 24.0 | 276 ± 11 | — | " |
| Rabbit (upper lung) | 2.5 | 283 ± 19 | — | " |
| | 24.0 | 624 ± 37 | — | " |
| Rabbit (lower lung) | 2.5 | 303 ± 43 | — | " |
| | 24.0 | 686 ± 57 | — | " |

*(continued)*

**Table 9.4. NMR proton relaxation times in normal and pathological tissues from mammals *(Continued)***

| Tissue | Frequency (MHz) | $T_1$ (ms) (mean ± SD) | $T_2$ (ms) (mean ± SD) | Reference |
|---|---|---|---|---|
| Rabbit (esophagus wall) | 2.5 | 250 ± 14 | — | " |
| | 24.0 | 534 ± 42 | — | " |
| Pathological lung | | | | |
| Hamster (edema) | 25.6 | — | 106 ± 3 | Lewa and Zbytnlewski, 1976 |
| Mouse (infected with *Salmonella*) | 30 | 644 ± 1 | — | Frey et al., 1972 |
| **Muscle** | | | | |
| Gastrocnemius | | | | |
| Rat | 30 | 608 | — | Chang et al., 1976 |
| Rat | 30 | — | 52 ± 4 | Finch and Homer, 1974 |
| Rat | 30 | 720 ± 50 | 45 ± 2 | Chang et al., 1972 |
| Rat (19 months old) | 30 | 663 ± 22 | — | Beall et al., 1983f |
| Rat (3.5 months old) | 30 | 671 ± 20 | — | " |
| Rat | 60 | 850 | — | Fung and McGaughy, 1974 |
| Thigh | | | | |
| Rat (mature) | 30 | 723 ± 49 | 47 ± 4 | Hazlewood et al., 1971 |
| Rat (immature) | 30 | 1,206 ± 55 | 127 ± 9 | " |
| Rat | 32 | 830 | — | Sentjurc et al., 1974 |
| Rat (turpentine inflammation) | 2.5 | 310 | 90 | Ling and Foster, 1982 |
| Rat (normal) | 24 | 600 | — | " |
| Rat (inflamed) | 24 | 640 | — | " |
| Rabbit (thigh) | 2.5 | 182 ± 12 | | " |
| | 24 | 554 ± 32 | | " |
| Rabbit (thigh) | 2.5 | 163 ± 3 | | " |
| Rectus | | | | |
| Rat | 13.5 | 404 ± 48 | — | Chaughule et al., 1974 |
| Rat | 24 | 538 ± 15 | 55 ± 5 | Damadian, 1971 |
| Rabbit (rectus ab.) | 2.5 | 183 ± 9 | — | Ling and Foster, 1981 |
| | 24 | 534 ± 23 | — | " |
| Rabbit (thigh with inflammation *in vitro*) | 2.5 | 285 ± 5 | | Ling and Foster, 1982 |
| Rabbit (thigh with inflammation *in vivo*) | 2.5 | 329 ± 5 | | " |
| Tibialis | | | | |
| Rat (parallel) | 30 | 668 ± 1 | 42.6 ± 0.4 | Kasturi et al., 1980 |
| Rat (perpendicular to field) | 30 | 674 ± 1 | 44.0 ± 0.3 | |
| Unspecified | | | | |
| Rat | 10.7 | 376 | — | Chandra et al., 1983 |
| Rat | 30 | 720 ± 40 | 44.7 ± 0.8 | Hazlewood et al., 1974a |
| Mouse | 38 | 610 ± 60 | 46 ± 4 | Peemoeller et al., 1982 |
| Rat | 60 | 850 ± 29 | — | Bovée et al., 1974 |
| Rat | 2.7 | 184.7 ± 1.9 | 48.5 ± 0.4 | Coles, 1976 |
| Mouse | 6 | 215 | | Escanye et al., 1983 |
| Mouse | 10.7 | 219 | 20 | Ujeno, 1980 |
| Mouse | 15 | 476 ± 5 | 51 ± 16 | " |
| Mouse | 20 | 525 | | McLachlan and Hamilton, 1979 |
| Mouse | 24 | 540 | 55 | Damadian, 1971 |
| Mouse | 24 | 491 | — | Hollis et al., 1973 |
| Mouse | 24 | 465 ± 47 | — | Hollis et al., 1974 |
| Mouse | 24 | 530 | — | Hollis et al., 1975 |
| Mouse | 30 | 563 | — | Inch et al., 1974b |
| Mouse | 30 | 590 ± 8 | — | " |
| Mouse | 30 | 615 ± 10 | 37 | Frey et al., 1972 |
| Mouse (infected with *Salmonella*) | 30 | 625 ± 10 | | " |
| Mouse | 33.8 | 555 ± 18 | — | Peemoeller et al., 1979 |
| Mouse | 33.8 | — | 143 ± 6 | Peemoeller et al., 1980 |
| Mouse | 33.8 | 800 | 70 | Rustgi et al., 1978 |
| Mouse | 60 | 700 | — | Fung, 1977b |
| Mouse | 90 | 547 | — | Escanye et al., 1983 |
| Mouse | 90 | 891 | 46 | de Certaines et al., 1979 |
| Mouse | 100 | 850 | 57 | Iijima et al., 1973 |

*(continued)*

**Table 9.4. NMR proton relaxation times in normal and pathological tissues from mammals *(Continued)***

| Tissue | Frequency (MHz) | $T_1$ (ms) (mean ± SD) | $T_2$ (ms) (mean ± SD) | Reference |
|---|---|---|---|---|
| Rabbit (psoas) | 100 | 756 | 66 | Cooke and Wein, 1971 |
| Rabbit (diaphragm) | 2.5 | 202 ± 32 | — | Ling and Foster, 1981 |
| | 24 | 458 ± 62 | — | " |
| Rabbit (intercostal) | 2.5 | 191 ± 14 | — | " |
| | 24 | 519 ± 43 | — | " |
| Pig | 10.7 | 405 ± 10 | 53 ± 2 | Pearson et al., 1974 |
| Cow | 25.6 | 500 | 50 | Lewa and Baczkowski, 1976 |
| Immature | | | | |
| Rat | 30 | 1,206 ± 55 | 127 ± 9 | Hazlewood et al., 1971 |
| **Pancreas** | | | | |
| Rat | 10.7 | 218 | — | Chandra et al., 1983 |
| **Reproductive organs** | | | | |
| Ovary | | | | |
| Rabbit | 2.5 | 310 | — | Ling and Foster, 1981 |
| | 24 | 573 | — | " |
| Uterus | | | | |
| Rat (3.5 months old) | 30 | 766 ± 83 | — | Beall et al., 1983f |
| Rat (12 months old) | 30 | 640 ± 35 | — | " |
| Rabbit | 2.5 | 296 | — | Ling and Foster, 1981 |
| Rabbit | 24 | 653 | — | " |
| Myometrium | | | | |
| Mouse (castrated) | | 600 ± 10 | 60 ± 2 | Odeblad and Ingelman-Sundberg, 1966 |
| Mouse (with estrogens) | | 800 ± 12 | 200 ± 5 | " |
| Testis | | | | |
| Rabbit | 2.5 | 463 ± 47 | — | Ling and Foster, 1981 |
| Rat | 10.7 | 674 | — | Chandra et al., 1983 |
| Mouse (yydF) | 10.7 | 375 | 37 | Ujeno, 1980 |
| Epididymis | | | | |
| Rabbit | 2.5 | 309 ± 6 | — | Ling and Foster, 1981 |
| | 24 | 625 ± 5 | — | " |
| **Whole fetus** | | | | |
| Rat | 30 | 634 ± 19 | 170 ± 7 | Hazlewood et al., 1982 |
| | 60 | 1,422 ± 7 | 200 ± 4 | Kiricuta and Simplacenu, 1975 |
| Mouse | 30 | 990 ± 34 | — | Inch et al., 1974b |
| **Skin** | | | | |
| Rat | 25 | 272 | — | Iijima et al., 1973 |
| Mouse | 24 | 199 | — | Hollis et al., 1975 |
| Mouse | 30 | 390 ± 39 | 35 | Frey et al., 1972 |
| Mouse | 30 | 505 | — | Inch et al., 1974b |
| Hamster | 25.6 | — | 72 ± 3 | Lewa and Zbytnlewski, 1976 |
| Rabbit | 2.5 | 198 ± 13 | — | Ling and Foster, 1981 |
| | 24 | 328 ± 28 | — | " |
| **Spleen** | | | | |
| Rat | 5 | 261 | 63 | Thickman et al., 1983 |
| Rat | 8 | 273 | — | Floyd et al., 1975 |
| Rat | 10.7 | 443 | — | Chandra et al., 1983 |
| Rat | 17.1 | 461 | | Ling and Tucker, 1980 |
| Rat | 60 | 582 ± 18 | — | Boveé et al., 1974 |
| Mouse | 2.7 | 262 ± 2 | 65.7 ± 1 | Coles, 1976 |
| Mouse | 6 | 230 | — | Escanye et al., 1983 |
| Mouse | 9 | 416 ± 10 | — | Modak et al., 1982 |
| Mouse | 10.7 | 250 | 36 | Ujeno, 1980 |
| Mouse | 15 | 493 ± 3 | 69.6 ± 0.6 | Coles, 1976 |
| Mouse | 17.1 | 641 | — | Ling and Tucker, 1980 |
| Mouse | 25 | 811 ± 112 | — | Modak et al., 1982 |

*(continued)*

**Table 9.4. NMR proton relaxation times in normal and pathological tissues from mammals *(Continued)***

| Tissue | Frequency (MHz) | $T_1$ (ms) (mean ± SD) | $T_2$ (ms) (mean ± SD) | Reference |
|---|---|---|---|---|
| Mouse | 30 | 523 ± 7 | — | Inch et al., 1974b |
| Mouse | 30 | 671 ± 9 | 44.8 ± 2.7 | Hazlewood et al., 1974a |
| Mouse | 30 | 500 ± 12 | — | Peemoeller et al., 1979 |
| Mouse | 30 | 737 ± 6 | 55.0 ± 1.7 | Hazlewood et al., 1974a |
| Mouse | 33.8 | 600 | 40 | Rustgi et al., 1978 |
| Mouse | 38 | 530 ± 20 | 60 ± 4 | Peemoeller et al., 1982 |
| Mouse | 90 | 891 | 55 | de Certaines et al., 1979 |
| Mouse | 100 | 712 ± 26 | — | Shah et al., 1982 |
| Mouse (adult) | 30 | 512 | — | Inch et al., 1974b |
| Mouse (pregnant) | 30 | 521 ± 18 | — | " |
| Mouse (adult) | 30 | 523 ± 7 | — | " |
| Mouse (regenerated liver) | 30 | 556 ± 5 | — | " |
| Mouse (*Salmonella*-infected) | 30 | 589 ± 13 | — | Frey et al., 1972 |
| Mouse | 90 | 549 | — | Escanye et al., 1983 |
| Rabbit | 2.5 | 258 ± 4 | — | Ling and Foster, 1981 |
| | 24 | 509 ± 11 | — | " |
| **Stomach—normal** | | | | |
| Rat | 24 | 270 ± 16 | — | Damadian, 1971 |
| Mouse | 30 | 294 ± 23 | 35 | Frey et al., 1972 |
| Mouse (*Salmonella*-infected) | 30 | 399 ± 26 | — | " |
| Dog (gastric mucosa) | 30 | 684 | 74 | Seitz et al., 1977 |
| Rabbit | 2.5 | 227 ± 20 | — | Ling and Foster, 1981 |
| | 24 | 468 ± 94 | — | " |
| **Tail** | | | | |
| Rat | 25 | 258 | 0 | Chaughule et al., 1974b |
| Mouse (*in vitro*) | 18 | 300 | — | Weisman et al., 1973 |
| Mouse (*in vivo*) | 18 | 700 | — | " |
| Mouse | 24 | 218 | — | Hollis et al., 1973 |
| **Various Cancers** | | | | |
| Hepatoma | | | | |
| Rat | | | | |
| Dunning hepatoma | 13.5 | 422 ± 18 | — | Block and Maxwell, 1974 |
| Hepatoma | 24 | 826 ± 13 | — | Damadian, 1971 |
| 3924A | 24 | 710 | — | Hollis et al., 1975 |
| 9633F | 24 | 520 | — | " |
| 7800 | 24 | 460 | — | " |
| R7 | 24 | 440 | — | " |
| 21 | 24 | 385 | — | " |
| Lymphosarcoma | | | | |
| Mouse | | | | |
| 6C3HED | 24 | 674 ± 10 | — | Hollis et al., 1975 |
| EL4 | 24 | 670 | — | " |
| Lymphosarcoma | 24.3 | 669 ± 63 | — | Cottam et al., 1972 |
| Fibrosarcoma | | | | |
| Mouse | | | | |
| MFS | 9 | 593 ± 55 | — | Modak et al., 1982 |
| MC3 | 24 | 705 | — | Hollis et al., 1975 |
| Fibrosarcoma | 24 | 662 | — | Hollis et al., 1973 |
| | 25 | 749 | — | Chaughule et al., 1974b |
| MFS | 25 | 1,086 ± 56 | — | Modak et al., 1982 |
| MC-1 | 30 | 853 ± 19 | 47 ± 13 | Frey et al., 1972 |
| 180 | 17.1 | 802 ± 15 | 86 ± 3 | Ling and Tucker, 1980 |
| Meth A | 17.1 | 805 ± 10 | 68 ± 1 | " |
| Ascites sarcoma | | | | |
| Mouse | 2.7 | 326 ± 5 | 82 ± 1.6 | Coles, 1976 |
| | 25 | 635 ± 10 | 84 ± 2 | " |

*(continued)*

**Table 9.4. NMR proton relaxation times in normal and pathological tissues from mammals *(Continued)***

| Tissue | Frequency (MHz) | $T_1$ (ms) (mean ± SD) | $T_2$ (ms) (mean ± SD) | Reference |
|---|---|---|---|---|
| Lewis lung carcinoma | | | | |
| Mouse | 90 | 1,140 ± 52 | 82 ± 5 | de Certaines et al., 1979 |
| Recticulum sarcoma | | | | |
| Rat | 60 | 1,128 | — | Boveé et al., 1974 |
| Rhabdomyosarcoma | | | | |
| Mouse | 6 | 372 | — | Escanye et al., 1983 |
| Rat | 60 | 1,150 | — | Boveé et al., 1974 |
| Mouse | 90 | 656 | — | Escanye et al., 1983 |
| Walker sarcoma | | | | |
| Rat | 24 | 736 ± 22 | 100 | Damadian, 1971 |
| | 60 | 1,093 | 77 | Kiricuta and Simplacenu, 1975 |
| Ehrlich solid tumor | | | | |
| Rat | 60 | 1,025 ± 7 | 38.2 ± 0.6 | " |
| Ehrlich ascites carcinoma | | | | |
| Mouse | 17.1 | 815 ± 7 | 62 ± 2 | Ling and Tucker, 1980 |
| Melanoma | | | | |
| Hamster (melanotic) | 25.6 | — | 67 ± 2 | Lewa and Baczkowski, 1976 |
| Hamster (amelanotic) | 25.6 | — | 128 ± 3 | " |
| Mouse | 24 | 708 | | Hollis et al., 1973 |
| Mouse | 25.6 | 725 | 82 | Lewa and Baczkowski, 1976 |
| Yoshida sarcoma | | | | |
| Rat | 24 | 739 ± 42 | — | Ling and Tucker, 1980 |
| Mammary adenocarcinoma | 30 | | | Beall, 1983e |
| Mouse—$D_1B$ | | 887 ± 84 | 84 | |
| Mouse—$C_4$ | | 712 ± 41 | 78 | " |
| Mouse—$D_2$ | | 903 ± 41 | 90 | " |
| EMT6 fibrosarcoma | | | | Peemoeller et al., 1982 |
| Mouse | 38 | 760 ± 50 | 170 ± 50 | |
| | | | 67 ± 15 | " |

*HAN, hyperplastic alveolar nodules.

**Table 9.5. NMR proton relaxation times (means ± SD) in normal and pathological human tissues**

| Tissue | Frequency (MHz) | $T_1$ (ms) | | $T_2$ (ms) | | Reference |
|---|---|---|---|---|---|---|
| | | Normal | Cancerous | Normal | Cancerous | |
| **Blood and body fluids** | | | | | | |
| Whole blood | | | | | | |
| Normal | 6 | 559 | | | | Brooks et al., 1975 |
| Normal | 10.7 | 830 | | 210 | | Fullerton et al., 1982 |
| Normal | 19.8 | 900 ± 90 | | | | Koivula et al., 1982 |
| Lysed | 20 | 550 | | | | Eisenstadt, 1981 |
| Blood cancers | 19.8 | | 1,100 ± 160 | | | Koivula et al., 1982 |
| | | | 1,150 ± 170 | | | |
| Remission | | | 970 ± 70 | | | |
| Normal | 44.4 | 560 | | 88 | | Zipp et al., 1974 |
| Sickle cell | 44.4 | | 530 | | 84 | " |
| Red blood cells | | | | | | |
| Normal | 10.7 | 540 | | 154 | | Fullerton et al., 1982 |
| Normal | 19.8 | 590 ± 60 | | | | Koivula et al., 1982 |
| Blood cancers | 19.8 | | 720 ± 120 | | | |
| Leukocytes | | | | | | |
| Normal | 24 | 715 ± 23 | | | | Ekstrand et al., 1977 |
| Leukemia (active) | 24 | | 855 ± 42 | | | |
| Leukemia (remission) | 24 | | 739 ± 54 | | | |
| Plasma | | | | | | |
| Normal | 6 | 1,098 | | | | Brooks et al., 1975 |
| Normal | 10.7 | 1,450 | | 282 | | Fullerton et al., 1982 |
| Normal | 20 | 1,357 ± 77 | | 582 ± 47 | | de Certaines et al., 1981 |
| Normal | 20 | 1,480 ± 250 | | 565 ± 50 | | McLachlan, 1980 |
| Tumor-bearing | 20 | | 1,566 ± 200 | | 598 ± 50 | |
| Other illnesses | 20 | 1,460 ± 333 | | 525 ± 50 | | |
| Normal | 19.8 | 1,410 ± 80 | | | | Koivula et al., 1982 |
| Normal | 24 | 1,260 ± 60 | | | | Ekstrand et al., 1977 |
| Blood cancers | 19.8 | | 1,560 ± 170 | | | Koivula et al., 1982 |
| Leukemia | 24 | | 1,105 ± 100 | | | Ekstrand et al., 1977 |
| Serum* | | | | | | |
| Normal | 20 | 1,371 ± 60 | 627 ± 47 | | | de Certaines et al., 1981 |
| Normal | 24 | 1,230 ± 170 | | | | Ekstrand et al., 1977 |
| Leukemia (remission) | 24 | 1,170 ± 100 | | | | |
| Cervical mucus | 21-157 | (changes during cycle due to changing proportions of various mucus types) | | | | Odeblad et al., 1984 |
| Human milk | 21.0 | 700 | | | | Odeblad and Westin, 1958 |
| **Bladder** | | | | | | |
| Normal | 100 | 891 ± 61 | | | | Damadian et al., 1974 |
| Tumor | 100 | 1,241 ± 165 | | | | |
| **Bone** | | | | | | |
| Normal | — | 103 ± 2 | | 28 ± 0.5 | | Takhavieva et al., 1974 |
| Embalmed | 8 | 165 | | | | Barroilhet and Moran, 1976 |
| Hydrated | 24 | 350–620 | | 10–130 | | Hopkins et al., 1983 |

*(continued)*

**Table 9.5. NMR proton relaxation times (means ± SD) in normal and pathological human tissues *(Continued)***

| Tissue | Frequency (MHz) | $T_1$ (ms) | | $T_2$ (ms) | | Reference |
|---|---|---|---|---|---|---|
| | | Normal | Cancerous | Normal | Cancerous | |
| **Bone tissue** | | | | | | |
| Marrow | | | | | | |
| Normal | 25 | 803 ± 100 | | | | Ranade et al., 1977 |
| Acute myeloid leukemia | 25 | 1,111 ± 70 | | | | |
| Chronic myeloid leukemia | 25 | | 1,112 ± 70 | | | |
| Acute lymphoid leukemia (remission) | 25 | 734 | | | | |
| Osteogenic sarcoma | 25 | | 754 (626–1,086) | | | Kasturi et al., 1976 |
| Osteogenic sarcoma | 24.3 | | 1,140 | | | Cottam et al., 1972 |
| Normal | 100 | 554 ± 27 | | | | Damadian et al., 1974 |
| Tumor | 100 | | 1,027 ± 152 | | | |
| **Brain** | | | | | | |
| White matter | 20 | 687 ± 95 | | 107 ± 26 | | Ngo et al., 1983 |
| Gray matter | | 825 ± 59 | | 110 ± 19 | | |
| Glioma | | | 980 ± 56 | | 139 ± 11 | |
| Gray matter | 60 | 435 | | 105 | | Parrish et al., 1974 |
| White matter | 60 | | | 65 | | |
| Malignant tumor | 60 | | 588 | | 108 | |
| Benign tumor | 60 | | 526 | | 87 | |
| Normal | 100 | 998 ± 16 | | | | Damadian et al., 1974 |
| Tumors | 100 | | | | | Naruse et al., 1983 |
| Glioblastoma | | | 1,100–1,700 | | 90–200 | |
| Astrocytoma | | | 1,300–1,400 | | 90–130 | |
| Meningioma | | | 1,000–1,600 | | 70–170 | |
| Metastasis | | | 1,300–1,500 | | 100–150 | |
| **Breast** | | | | | | |
| Normal | 22.5 | 449 ± 136 | | | | Koutcher et al., 1978 |
| Tumor | 22.5 | | 438 ± 173 | | | |
| Normal | 24 | 500 | | | | Mallard et al., 1979 |
| Tumor | 24 | | 550 | | | |
| Normal | 24.3 | 191 | | | | Eggleston et al., 1975 |
| Tumor | 24.3 | | 575 | | | |
| Fibrocystic | 24.3 | 450 | | | | |
| Tumor (lobular carcinoma) | 24.3 | | 658 | | | Cottam et al., 1972 |
| Normal | 25 | 208–524 | | | | Kasturi et al., 1976 |
| Tumor | 25 | | 493–1,152 | | | |
| Chronic mastitis | 25 | 636 to 937 | | | | |
| Normal | 30 | 682 ± 32 | | 35. ± 3.5 | | Medina et al., 1975 |
| Tumor (adenocarcinoma) | 30 | | 874 ± 28 | | 68.6 ± 2.3 | |
| Fibrocystic | 30 | 655 ± 21 | | 37.0 ± 3 | | |
| Fibroadenoma (benign) | 30 | | 980 ± 51 | | 62.5 ± 7.3 | |
| Normal | — | 585 | | | | Hardter, 1976 |
| Tumor | — | | 941 | | | |
| Normal | 60 | 850 ± 20 | | | | Tanaka et al., 1974 |
| Tumor | 60 | | 980 | | | |
| Normal | 60 | 900 | | | | Boveé et al., 1974 |
| Fatty breast | 60 | 200 | | | | |
| Tumor | 60 | | 901 ± 20 | | | " |
| Normal | 100 | 367 ± 79 | | | | Damadian et al., 1974 |
| Tumor | 100 | | 1,080 ± 80 | | | |

*(continued)*

**Table 9.5. NMR proton relaxation times (means ± SD) in normal and pathological human tissues *(Continued)***

| Tissue | Frequency (MHz) | $T_1$ (ms) Normal | $T_1$ (ms) Cancerous | $T_2$ (ms) Normal | $T_2$ (ms) Cancerous | Reference |
|---|---|---|---|---|---|---|
| **Esophagus** | | | | | | |
| Normal | 25 | 460–984 | | | | Kasturi et al., 1976 |
| Tumor | 25 | | 708–1,095 | | | |
| Normal | 100 | 790 ± 39 | | | | Shah et al., 1982 |
| Tumor | 100 | | 959 ± 39 | | | |
| Normal | 100 | 804 ± 108 | | | | Damadian et al., 1974 |
| Tumor | 100 | | 1,040 | | | |
| **Heart** | | | | | | |
| Normal | 24.3 | 873 ± 118 | | | | Cottam et al., 1972 |
| Normal | 100 | 906 ± 46 | | | | Damadian et al., 1974 |
| **Eye** | | | | | | |
| Aqueous humor | 21 mc/s | 3,000 | | | | Huggert and Odeblad, 1959 |
| Corpus vitrium | | 1,000 | | | | |
| Retina | | 1,000 | | | | |
| Cornea | | 300–400 | | | | |
| Cortex lentis | | 200–400 | | | | |
| Optic nerve | | 100 | | | | |
| Muscle | | 300–600 | | | | |
| Lacrimal gland | | 300–600 | | | | |
| Sclera | | 100 | | | | |
| Nucleus lentis | | 100 | | | | |
| **Glands** | | | | | | |
| Adrenal | | | | | | |
| Normal | 24.3 | 585 | | | | Cottam et al., 1972 |
| Normal | 100 | 608 ± 20 | | | | Damadian et al., 1974 |
| Tumor | 100 | | 683 | | | |
| Pancreas | | | | | | |
| Normal | 24.3 | 320 ± 15 | | | | Cottam et al., 1972 |
| Normal | 100 | 605 ± 36 | | | | Damadian et al., 1974 |
| Prostate | | | | | | |
| Normal | 24.3 | 767 | | | | Cottam et al., 1972 |
| Normal | 100 | 803 ± 14 | | | | Damadian et al., 1974 |
| Tumor | 100 | | 1,110 | | | |
| Thyroid | | | | | | |
| Papillary carcinoma | 10.7 | | 617–643 | | 70–82 | Tennvall et al., 1983 |
| Benign lesions | 10.7 | 433–721 | | 60–96 | | |
| Goiter | 20 | 513 ± 55 | 541 ± 76 | 74 ± 13 | 88 ± 25 | de Certaines et al., 1982a |
| Benign tumor | 20 | 448 ± 23 | 590 ± 116 | 59 ± 7 | 118 ± 32 | |
| Active tumor | 20 | 490 ± 69 | 529 ± 79 | 69 ± 13 | 108 ± 38 | |
| Normal | 24.3 | 586 | | | | Cottam et al., 1972 |
| Normal | 24.3 | 410 | | | | Eggleston et al., 1975 |
| Tumor | 24.3 | | 679 | | | |
| Tumor | 25 | | 669 | | | Kasturi et al., 1976 |
| Tumor | — | | 520–590 | | | de Certaines et al., 1982a |
| Normal | 32 | 700 | | | | Sentjurc et al., 1974 |
| Tumor | 32 | | 700 | | | |
| Normal | 100 | 882 ± 45 | | | | Damadian et al., 1974 |
| Tumor | 100 | | 1,072 | | | |
| Thymus | | | | | | |
| Normal | 24.3 | 809 | | | | Cottam et al., 1972 |
| **Kidney** | | | | | | |
| Normal | 24 | 459 | | | | Hollis et al., 1973 |
| Tumor | 24 | | 638 | | | |
| Normal | 24.3 | 770 | | | | Cottam et al., 1972 |
| Normal | 25 | 773 | | | | Kasturi et al., 1976 |
| Tumor | 25 | | 853 | | | |
| Normal | 100 | 862 ± 33 | | | | Damadian et al., 1974 |

*(continued)*

**Table 9.5. NMR proton relaxation times (means ± SD) in normal and pathological human tissues *(Continued)***

| Tissue | Frequency (MHz) | $T_1$ (ms) | | $T_2$ (ms) | | Reference |
|---|---|---|---|---|---|---|
| | | Normal | Cancerous | Normal | Cancerous | |
| **Intestine** | | | | | | |
| Small | | | | | | |
| Normal | 22.5 | 374 ± 108 | | 58 ± 10 | | Goldsmith et al., 1978 |
| Tumor | 22.5 | | 594 ± 122 | | 106 ± 24 | |
| Normal | 100 | 641 ± 80 | | | | Damadian et al., 1974 |
| Tumor | 100 | | 1,122 ± 40 | | | |
| Large | | | | | | |
| Normal | 22.5 | 330 ± 129 | | | | Koutcher et al., 1978 |
| Tumor | 22.5 | | 626 ± 58 | | | |
| Normal | 24 | 475 | | | | Hollis et al., 1973 |
| Tumor | 24 | | 553 | | | |
| Normal | 24.3 | 552 | | | | Eggleston et al., 1975 |
| Tumor | 24.3 | | 559 | | | |
| Normal | 25 | 803 | | | | Kasturi et al., 1976 |
| Tumor | 25 | | 783–1,270 | | | |
| Normal | 100 | 641 ± 43 | | | | Damadian et al., 1974 |
| Normal | 100 | 857 ± 57 | | | | Shah et al., 1982 |
| Tumor | | | 1,088 ± 67 | | | |
| Rectum | | | | | | |
| Normal | 25 | 563 to 979 | | | | Kasturi et al., 1976 |
| Tumor | 25 | | 703–1,782 | | | |
| **Liver** | | | | | | |
| Embalmed normal | 8 | 100 | | | | Barroilhet and Moran, 1976 |
| Normal | 24.3 | 383 ± 48 | | | | Cottam et al., 1972 |
| Normal | | 400 to 600 | | | | Schmidt et al., 1973 |
| Tumor | | | 720–1,210 | | | |
| Normal | 24 | 339 ± 42 | | | | Mallard et al., 1979 |
| Tumor | 24 | | 650–800 | | | |
| Normal | 100 | 570 ± 29 | | | | Damadian et al., 1974 |
| Tumor | 100 | | 832 ± 12 | | | |
| **Lung** | | | | | | |
| Normal | 22.5 | 505 ± 63 | | | | Koutcher et al., 1978 |
| Tumor | 22.5 | | 711 ± 207 | | | |
| Normal | | 425 | | | | Schmidt et al., 1973 |
| Tumor | | | 790 to 970 | | | |
| Normal | 24.5 | 555 | | | | Eggleston et al., 1975 |
| Tumor | 24.5 | | 555 | | | |
| Tumor | 25 | | 1,003 | | | Kasturi et al., 1976 |
| Tumor | 100 | | 1,110 ± 57 | | | Damadian et al., 1974 |
| Pleural fluid | 10.7 | 839–2,650 | | 253–493 | | Kamath et al., 1983 |
| **Lymph nodes** | | | | | | |
| Normal | 32 | 500 | | | | Sentjurc et al., 1974 |
| Tumor (melanoma) | 32 | | 850 | | | |
| Normal | 100 | 720 ± 76 | | | | Damadian et al., 1974 |
| Tumor (melanoma) | 100 | | 724 ± 147 | | | |
| **Muscle** | | | | | | |
| Embalmed normal | 8 | 233 | | | | Barroilhet and Moran, 1976 |
| Normal | 24 | 600 | | | | Mallard et al., 1979 |
| Normal | 24.3 | 807 | | | | Cottam et al., 1972 |
| Normal | 24.3 | 404 | | | | Hollis et al., 1973 |
| Normal | 25 | 984 | | | | Kasturi et al., 1976 |
| Normal | 100 | 788 ± 63 | | | | Damadian et al., 1974 |
| Normal | 100 | 1,023 ± 29 | | | | |
| Tumor | 100 | | 1,413 ± 82 | | | |
| Tumor (benign) | 100 | | 1,037 ± 154 | | | |

*(continued)*

**Table 9.5. NMR proton relaxation times (means ± SD) in normal and pathological human tissues *(Continued)***

| Tissue | Frequency (MHz) | $T_1$ (ms) | | $T_2$ (ms) | | Reference |
|---|---|---|---|---|---|---|
| | | Normal | Cancerous | Normal | Cancerous | |
| **Nerve** | | | | | | |
| Normal | 100 | 557 ± 158 | | | | Damadian et al., 1974 |
| Tumor | 100 | | 1,204 | | | |
| **Oral tissue (mouth)** | | | | | | |
| Normal | 16.7 | — | — | 6.8 | | Forsslund et al., 1962 |
| Tumor | 16.7 | | | 9.6 | | |
| Normal | 25 | 600–1,000 | | | | Ranade et al., 1976 |
| Tumor | 25 | | 500–1,300 | | | |
| Normal | 25 | 302–700 | | | | Kasturi et al., 1976 |
| Tumor | 25 | | 710–1,244 | | | |
| Tumor | 100 | | 1,288 | | | Damadian et al., 1974 |
| **Peritoneum** | | | | | | |
| Normal | 100 | 476 | | | | Damadian et al., 1974 |
| **Reproductive tissues** | | | | | | |
| Cervix | | | | | | |
| Normal | 22.5 | 510 | | 41 | | Fruchter et al., 1978 |
| Dysplasia | 22.5 | 880 | | 82 | | |
| Tumor | 22.5 | | 684 | | 84 | |
| Normal | 25 | 825 | | | | Kasturi et al., 1976 |
| Tumor | 25 | | 1,089,–1,376 | | | |
| Normal | 100 | 827 ± 26 | | | | Damadian et al., 1974 |
| Tumor | 100 | | 1,101 | | | |
| Endometrium | | | | | | |
| Normal | 22.5 | 801 | | 95 | | Fruchter et al., 1978 |
| Adenocarcinoma | 22.5 | | 673 | | 98 | |
| Hyperplasia | 22.5 | 757 | | 83 | | |
| Gestational | 22.5 | 1,406 | | 161 | | |
| Secretory | 22.5 | 774 | | 107 | | |
| Proliferative | 22.5 | 846 | | 119 | | |
| Fallopian tubes | | | | | | |
| Normal | 22.5 | 546 | | 73 | | Fruchter et al., 1978 |
| Myometrium | | | | | | |
| Normal | 22.5 | 553 | | 66 | | Fruchter et al., 1978 |
| Benign tumor | 22.5 | | 634 | | 58 | " |
| Leiomyoma | 22.5 | | 476 | | 66 | |
| Malignant sarcoma | 22.5 | | 888 | | 155 | " |
| Ovary | | | | | | |
| Normal | 10 | 210 ± 5 | | | | Tanaka et al., 1974 |
| Tumor | 10 | | 400 ± 10 | | | |
| Normal | | 480 | | | | Schmidt et al., 1973 |
| Tumor | | | 960 | | | |
| Normal | 22.5 | 553 | | 66 | | Fruchter et al., 1978 |
| Tumor | 22.5 | | 691 | | 101 | |
| Brenner tumor | 22.5 | | 507 | | 85 | |
| Normal | 24 | 650 | | | | Gordon et al., 1978 |
| Normal | 60 | 770 ± 10 | | | | Tanaka et al., 1974 |
| Tumor | 60 | | 820 ± 10 | | | |
| Normal | 100 | 989 ± 47 | | | | Damadian et al., 1974 |
| Tumor | 100 | | 1,282 ± 118 | | | |
| Placenta | | | | | | |
| Normal | 22.5 | 668 | | 114 | | Fruchter et al., 1978 |
| Normal | 24.3 | 1,000 | | | | Cottam et al., 1972 |
| Testis | | | | | | |
| Normal | 24.3 | 1,010 | | | | Cottam et al., 1972 |
| Normal | 100 | 1,200 ± 48 | | | | Damadian et al., 1974 |
| Tumor | 100 | | 1,223 | | | |

*(continued)*

**Table 9.5. NMR proton relaxation times (means ± SD) in normal and pathological human tissues *(Continued)***

| Tissue | Frequency (MHz) | $T_1$ (ms) | | $T_2$ (ms) | | Reference |
|---|---|---|---|---|---|---|
| | | Normal | Cancerous | Normal | Cancerous | |
| Uterus | | | | | | |
| Normal | 25 | 842 | | | | Kasturi et al., 1976 |
| Tumor | 25 | | 1,013 | | | |
| Normal | | 880 | | | | Schmidt et al., 1973 |
| Tumor | | | 990 | | | |
| Normal | 100 | 924 ± 38 | | | | Damadian et al., 1974 |
| Tumor | 100 | | 1,392–973 | | | |
| Vagina | | | | | | |
| Normal | 22.5 | 609 | | 60 | | Fruchter et al., 1978 |
| Radiation-treated | 22.5 | 678 | | 97 | | |
| Vulva | | | | | | |
| Normal | 22.5 | 451 | | 34 | | Fruchter et al., 1978 |
| Tumor | 22.5 | | 718 | | 80 | |
| **Skin** | | | | | | |
| Normal | 24.3 | 423 | | | | Eggleston et al., 1975 |
| Normal | 24 | 535 ± 116 | | | | Mallard et al., 1979 |
| Normal | 25 | 360 | | | | Kasturi et al., 1976 |
| Tumor | 25 | | 1,037 | | | |
| Normal | 100 | 616 ± 47 | | | | Damadian et al., 1974 |
| Tumor | 100 | | 1,047 ± 108 | | | |
| **Spleen** | | | | | | |
| Normal | 24 | 471 ± 40 | | | | Mallard et al., 1979 |
| Normal | 60 | 688 ± 47 | | | | Shah et al., 1982 |
| Tumor | 100 | | 1,113 ± 6 | | | Damadian et al., 1974 |
| **Stomach** | | | | | | |
| Normal | 24.3 | 541 | | | | Cottam et al., 1972 |
| Normal | 24.3 | 535 | | | | Eggleston et al., 1975 |
| Tumor | 24.3 | | 535 | | | |
| Tumor | 24 | | 800 | | | Gordon et al., 1978 |
| Normal | 25 | 687–1,178 | | | | Kasturi et al., 1976 |
| Tumor | 25 | | 826–1,020 | | | |
| Normal | 100 | 765 ± 75 | | | | Damadian et al., 1974 |
| Tumor | 100 | | 1,238 ± 109 | | | |
| **Vein** | | | | | | |
| Normal | 24 | 520 | | | | Gordon et al., 1978 |

*See Tables 9.14 and 9.15.

**Table 9.6. NMR determination of nonfrozen water in muscle**

| Sample | Temperature (°C) | % $H_2O$ | Reference |
|---|---|---|---|
| Tropomyosin (5% solution) | −20 | 10 | Cyr et al., 1971 |
| Frog gastrocnemius | −30 | 20 | Belton et al., 1972 |
| Porcine skeletal | −20 | 10 | Derbyshire and Parsons, 1972 |
| Rat gastrocnemius | −8 to −70 | 7–12 | Fung and McGaughy, 1974 |
| Mouse gastrocnemius | −20 | 9 | Fung, 1974 |
| Mouse gastrocnemius with tumor | −20 | 12 | Fung, 1974 |
| Mouse muscle | −40 | 10 | Rustgi et al., 1978 |
| Mouse muscle | −10 | 9.8 | Peemoeller et al., 1980 |
| Mouse muscle | −10 | 15.8 | Escanye et al., 1983 |
| Rat tibialis anterior | −5 | 25 | Kasturi et al., 1980 |
| | −10 | 20 | |

**Table 9.7. Self-diffusion coefficients of water in biological systems**

| System | Method | $D_{H_2O}$ ($\times 10^{-5}$ cm$^2$ s$^{-1}$) | Reference |
|---|---|---|---|
| **Pure water** | NMR | 2.5 | Chang et al., 1973 |
| **5% Bovine serum albumin** | NMR | 1.93 | Abetsedarskaya et al., 1968 |
| **6% Agar rose gel** | NMR | 2.26 | Walter and Hope, 1971 |
| 18% | Quasi-elastic neutron scattering | 1.70 | Trantham, 1980 |
| **20% Polyox polymer** | NMR | 1.10 | Bearden, 1983 |
| **2% Collagen gel** | NMR | 2.24 | Westover and Dresden, 1974 |
| **Eggs** | | | |
| Frog | NMR | 0.67 | Mild et al., 1972 |
| Frog | Isotope | 0.996 | Ling et al., 1967 |
| Chicken (yolk) | NMR | 0.60 | James and Gillen, 1972 |
| Chicken (white) | NMR | 1.71 | James and Gillen, 1972 |
| **Skeletal muscle** | | | |
| Barnacle | NMR | 1.38 | Clark et al., 1982 |
| Barnacle | Isotope | 2.42 | Bunch and Kallsen, 1969 |
| Barnacle | Isotope | 1.35 | Ernst and Hazlewood, 1978 |
| Barnacle | Isotope | 1.34 | Caille and Hinke, 1974 |
| Frog | NMR | 1.56 | Abetsedarskaya et al., 1968 |
| Frog | NMR | 1.13 | Finch et al., 1971 |
| Frog | NMR | 1.12 | Finch et al., 1971 |
| Frog | NMR (pulsed) | 1.74 | Tanner, 1979 |
| Frog | NMR (pulsed) | 1.41 | Yoshizaki et al., 1982 |
| Frog | Isotope | 1.18 | Ling et al., 1967 |
| Frog | Isotope | 1.43 | Ling et al., 1967 |
| Rat | NMR | 1.43 | Chang et al., 1973 |
| Rat | NMR | 1.50 | Chang et al., 1973 |
| Rat | NMR | 1.09 | Finch et al., 1971 |
| Rat | NMR | 0.9 | Cleveland et al., 1976; Rorschach et al., 1973 |
| Toad | NMR | 1.2 | Rustgi et al., 1978 |
| Toad | NMR | 1.5 | " |

*(continued)*

**Table 9.7. Self-diffusion coefficients of water in biological systems *(Continued)***

| System | Method | $D_{H_2O}$ ($\times 10^{-5}$ cm$^2$ s$^{-1}$) | Reference |
|---|---|---|---|
| **Cardiac muscle** | | | |
| Rat | NMR | 0.9 | Cooper et al., 1974 |
| Rabbit | NMR | 0.65 | Cooper et al., 1974 |
| **Liver** | | | |
| Rat | NMR | 0.625–0.75 | Cooper et al., 1974 |
| Rabbit | NMR | 0.625 | Cooper et al., 1974 |
| Rabbit | NMR | 0.49 | Cooper et al., 1974 |
| **Red blood cells** | | | |
| Human | NMR | 0.2–0.55 | Cooper et al., 1974 |
| Human | NMR | 0.25–0.625 | Tanner, 1976 |
| Human | NMR | 1.16 | Andrasko, 1976 |
| Human | NMR | 0.63 | Tanner, 1983 |
| **Cultured cells** | | | |
| HeLa | NMR | 0.48 | Beall et al., 1982c |
| CHO | NMR | 0.40 | Beall et al., 1982c |
| BHK | NMR | 0.50 | Beall et al., 1982c |
| **Lens** | | | |
| Rabbit (whole) | NMR | 0.9 | Neville et al., 1974 |
| Rabbit (cortex) | NMR | 1.06 | Neville et al., 1974 |
| Rabbit (nucleus) | NMR | 0.97 | Neville et al., 1974 |
| **Pea roots** | | | |
| Meristematic | NMR | 0.59 | Abetsedarskaya et al., 1968 |
| Mature cells | NMR | 0.68 | Abetsedarskaya et al., 1968 |
| **Wheat endosperm** | NMR | 1.2 | Callaghan et al., 1979 |
| **Slime mold** | NMR | 1.26 | Walter and Hope, 1971 |
| **Brine shrimp cysts** | | | |
| 50% $H_2O$ | NMR | 0.24 | Seitz et al., 1981 |
| 30–50% $H_2O$ | NMR | 0.4–1.0 | Tanner, 1983 |
| **Yeast** | NMR | 0.35–0.68 | Tanner, 1983 |
| | NMR | 0.4–1.0 | Valikanov and Volkov, 1979 |

**Table 9.8. Effect of age and hormone status on $T_1$ of rat tissues**

| Tissue | 3-month-old breeders with estrogen (5) $T_1$ | 12-month-old retired breeders—low estrogen (10) $T_1$ (ms) | 12-month-old ovariectomized retired breeders—no estrogen (8) $T_1$ (ms) |
|---|---|---|---|
| Heart | 703 ± 15 | 660 ± 17 | 716 ± 64 |
| Skeletal muscle | 671 ± 20 | 693 ± 21 | 651 ± 34 |
| Uterus | 766 ± 83 | 640 ± 35 | 575 ± 69 |
| Liver | 285 ± 22 | 240 ± 35 | 228 ± 18 |

Data from Beall, 1983f. Mean ± SD, 30 MHz, 25°C.

*Note:* Alterations in $T_1$ values may be associated with old age and the loss of estrogen stimulation of organs. $T_1$ values for organs generally decrease with age, as does the water content of the organ. When deprived of estrogen input for 6 months, changes are seen in reproductive tract organs and possibly in the ventricular muscle of the heart. These findings have significance for postmenopausal patients and aged individuals.

**Table 9.9. Effect of development and aging on $T_1$ values of rat tissues**

| Rat age | Cardiac muscle $T_1$ (ms) | Skeletal muscle $T_1$ (ms) | Skeletal muscle $T_2$ (ms) | Uterus $T_1$ (ms) | Liver $T_1$ (ms) |
|---|---|---|---|---|---|
| 1 day | | 1,495 | 181 | | |
| 3 days | | 1,235 | 167 | | |
| 7 days | | 1,058 | 109 | | |
| 14 days | | 890 | 87 | | |
| 24 days | | 750 | 56 | | |
| 31 days | | 722 | 56 | | |
| 49 days | | 710 | 43 | | |
| 105 days | 703 ± 15 | 671 ± 20 | | 766 ± 83 | 285 ± 22 |
| 12 months | 660 ± 7 | 693 ± 21 | | 640 ± 35 | 240 ± 35 |
| 18 months | 627 ± 3 | 679 ± 12 | | — | 319 ± 16 |

*Note:* Fetal tissues are wet and have high $T_1$ values. Immediately after birth, developmental changes begin which result in a rapid decrease in $T_1$ and $T_2$ and a reduction in the water content of tissues with maturation. Mature adult animals maintain reproducible plateaus of water content and relaxation times over long periods. Variations from these numbers suggest physiological abnormalities. In old age, the data suggest a continued and slight decline in water content, which may be accompanied by reduced $T_1$ values, or, if aging of the tissue results in macromolecular disorganization there may be a higher $T_1$ value, such as is seen in the liver of older rats. Reference: Beall, 1983f.

**Table 9.10. NMR proton relaxation times of cell suspensions and cultured cells**

| Cell type | Frequency (MHz) | $T_1$ (ms) | $T_2$ (ms) | % $H_2O$ | Reference |
|---|---|---|---|---|---|
| **Animal cells** | | | | | |
| Rat cancer cells | | | | | |
| Suspensions | 17.1 | 815 | 61 ± 2 | 80.8 ± 0.34 | Ling and Tucker, 1980 |
| Ehrlich ascites | 60 | 1,150 | | | Kiricuta and Simplacenu, 1975 |
| Walker 256 | 60 | 1,093 | | | " |
| Ehrlich solid | 60 | 1,025 ± 7 | 38.2 ± 6 | | " |
| Novikoff hepatoma ascites | 17.1 | 855 ± 8.6 | 97 ± 1.2 | 82.7 ± 0.12 | Ling and Tucker, 1980 |
| AS-30 D hepatoma ascites | 17.1 | 844 ± 16.4 | 81 ± 2.1 | 81.3 ± 0.3 | " |
| Normal suckling rat | | | | | |
| Kidney cells | | | | | |
| NRK (normal) | 40 | 1,575 ± 40 | | | Schmidt et al., 1973 |
| NRK (virus-infected) | 40 | 2,550 ± 70 | | | " |
| Mouse mammary cells | | | | | |
| Primary cultures | | | | | |
| Normal (pregnant) | 30 | 916 ± 24 | 158 ± 6 | 90.8 ± 0.4 | Beall et al., 1980a |
| Preneoplastic (D2-HAN) | 30 | 1,029 ± 24 | 187 ± 7 | 90.0 ± 0.5 | " |
| Neoplastic (adenocarcinoma) | 30 | 1,155 ± 42 | 206 ± 8 | 91.4 ± 0.2 | " |
| Established lines | | | | | |
| ESD/BALB CL3 | 30 | 632 ± 8 | 113 ± 8 | 85.1 ± 0.7 | Beall et al., 1980a |
| MTV-L/BALB CL2 | 30 | 762 ± 26 | 105 ± 5 | 83.9 ± 0.7 | |
| DMBA/BALB CL2 | 30 | 739 ± 33 | 109 ± 8 | 83.8 ± 0.2 | " |
| Mouse lung fibroblasts | | | | | |
| 3T3 | 30 | 1,250 | 160 | 85.4 | Beall et al., 1980b |
| SV3T3 (transformed) | 30 | 757 ± 58 | 84 ± 5 | 83.2 | " |
| Mouse embryonic cells | | | | | |
| L929 | 90 | 300 | | | Vucelic et al., 1978 |
| Trypsinized embryo | 90 | 420 | | | " |
| Chinese hamster cells | | | | | |
| CHO ovary | | | | | |
| Random | 30 | 630 | 97 | 85.0 | Beall 1979 |
| Mitotic | 30 | 889 ± 27 | 113 ± 8 | 89.8 ± 1.0 | " |
| Interphase | 30 | 681 ± 35 | 98 ± 3 | 85.0 ± 0.4 | " |
| CHO lung fibroblasts | 30 | 800 | | | Raaphorst et al., 1975 |
| V79-5171 | 30 | 900 | | 75.5 | Raaphorst et al., 1978 |
| Mouse and rat cells | | | | | |
| Normal rat kidney NRK (density-dependent) | 30 | 1,360 | 270 | 91.4 | Ader and Cohen, 1979 |
| | | 910 | 100 | 84.2 | |
| | | 1,170 | 210 | | |
| Normal rat kidney NRK | 30 | 1,100 | | 88 | " |
| | | 1,370 | | 88 | " |
| | | 1,320 | | 88 | |
| | | 990 | 130 | 86.5 | |
| | | 970 | 110 | 84.2 | |

*(continued)*

**Table 9.10. NMR proton relaxation times of cell suspensions and cultured cells *(Continued)***

| Cell type | Frequency (MHz) | $T_1$ (ms) | $T_2$ (ms) | % $H_2O$ | Reference |
|---|---|---|---|---|---|
| BALB 3T3 mouse lung (density-dependent) | 30 | 1,440 | 270 | | " |
| | | 1,220 | 230 | | |
| BALB R4 mouse lung (density-independent) | 30 | 1,490 | 270 | | |
| E4 (SV40-transformed) | 30 | 1,560 | 250 | 91.7 | " |
| M3 (transformed) | 30 | 1,480 | 310 | 93.8 | " |
| CEM (human leukemia) | 30 | 890 | | 86.5 | " |
| | | 820 | | | |
| | | 1,120 | | | |
| FU55 (rat hepatoma ) | 30 | 990 | | 85.2 | " |
| | | 890 | | 87.5 | |
| NEO (embryo-transformed fibroblasts) | 30 | 1,550 | 330 | 97.0 | " |
| NOR | 30 | 1,200 | 240 | 91.4 | " |
| Chicken cells | | | | | |
| Embryonic | 40 | 1,840 ± 45 | | | Schmidt et al., 1973 |
| With virus 37°C | 40 | 2,270 ± 60 | | | |
| With virus 35°C | 40 | 2,610 ± 80 | | | |
| With virus 35°C | 40 | 2,030 ± 65 | | | |
| CEC embryo | 25 | 784 | | | Wagh et al., 1977 |
| With virus | | 523 | | | |
| Chick red cells | 25 | 470 | | | " |
| With virus | | 560 | | | |
| Monkey cells | | | | | |
| Vero monkey kidney | 25 | 724 | | | Wagh et al., 1977 |
| **Human cell lines** | | | | | |
| HeLa cells | | | | | |
| $M_0$ (0 min) | 30 | 1,020 ± 84 | 130 ± 13 | 88.2 ± 0.3 | Beall et al., 1976a, 1978 |
| $M_{30}$ (30 min) | 30 | 817 ± 76 | 127 ± 18 | 87.5 ± 0.1 | |
| $G_1$ (4 h) | 30 | 638 ± 110 | 110 ± 11 | 85.8 ± 0.6 | |
| $G_1$ (8 h) | 30 | 570 ± 56 | 117 ± 11 | 85.5 ± 0.7 | |
| S (12 h) | 30 | 534 ± 43 | 100 ± 9 | 84.4 ± 0.7 | |
| $G_2$ (18 h) | 30 | 621 ± 25 | 96 ± 8 | 84.5 ± 0.2 | |
| $G_2$ (19 h) | 30 | 690 ± 4 | | 84.3 ± 0.6 | |
| $G_2$ (20 h) | 30 | 739 ± 59 | 116 ± 7 | 84.3 ± 1.1 | |
| Random | 30 | 990 | | | Ranade et al., 1975 |
| | | 1,038 ± 75 | | | |
| Breast cancer | | | | | |
| MDA-MB 231 | 30 | 934 ± 78 | 123 ± 31 | 86.5 ± 1.3 | Beall et al., 1982a |
| 157 | 30 | 907 ± 10 | 135 ± 4 | 87.4 ± 0.1 | |
| 361 | 30 | 849 ± 25 | | 88.4 ± 0.2 | |
| 134 | 30 | 717 ± 64 | 145 ± 39 | 88.9 ± 0.4 | |
| 453 | 30 | 770 ± 15 | 113 ± 18 | 87.4 ± 2.1 | |
| 330 | 30 | 752 ± 39 | | 88.5 ± 1.5 | |
| 435 | 30 | 607 ± 9 | 112 ± 13 | 87.3 ± 0.9 | |
| 331 | 30 | 549 ± 136 | 75 ± 20 | 88.5 ± 0.1 | |
| 431 | 30 | 521 | 126 | | |
| 436 | 30 | 499 ± 49 | 100 ± 20 | 84.3 ± 0.4 | |
| 231 (slow) | 30 | 622 ± 70 | 104 ± 18 | | |
| 157 (slow) | 30 | 669 ± 52 | 102 ± 18 | 85.2 ± 1.0 | |
| Breast | | | | | |
| SW 527 | 30 | 727 | 110 | 86.2 | Beall et al. (unpublished) |
| SW 613 | 30 | 606 | 104 | 85.0 | |
| Leukocytes | | | | | |
| Normal | 24 | 715 ± 23 | | | Ekstrand et al., 1977 |
| Leukemic (active) | 24 | 855 ± 42 | | | |
| Leukemic (remission) | 24 | 739 ± 54 | | | |

*(continued)*

**Table 9.10. NMR proton relaxation times of cell suspensions and cultured cells *(Continued)***

| Cell type | Frequency (MHz) | $T_1$ (ms) | $T_2$ (ms) | % $H_2O$ | Reference |
|---|---|---|---|---|---|
| Colon cancer | | | | | |
| LS 180 | 30 | 643 ± 60 | | 85.5 ± 2.4 | Beall et al., 1982b |
| LS 174T | 30 | 744 ± 16 | 121 ± 3 | 86.6 ± 0.5 | |
| LS 174T 3–5 clone | 30 | 663 ± 65 | 127 ± 11 | 89.7 ± 1.9 | |
| LS 174T 6–6 clone | 30 | 716 ± 39 | 114 ± 16 | 86.4 ± 1.2 | |
| HT 29 | 30 | 686 ± 21 | 108 ± 7 | | |
| SW 480 | 30 | 982 ± 9 | 176 ± 6 | 90.1 ± 1.4 | |
| SW 1345 | 30 | 460 ± 45 | 83 ± 6 | 83.6 ± 1.8 | |
| Noninvolved colon | | | | | |
| NBV | 30 | 1,214 ± 46 | 207 ± 7 | 92.3 ± 0.4 | Beall et al., 1982b |
| NBM | 30 | 1,009 ± 7 | 191 ± 3 | 91.4 ± 0.4 | |
| Fetal colon | | | | | |
| Hs 0677 | 30 | 1,058 ± 13 | 221 ± 3 | 90.6 ± 0.4 | |
| Hs 0074 | 30 | 1,106 ± 24 | 163 ± 23 | | |
| Other human cells | | | | | |
| VTC-4 (human amnion) | 25 | 749 | | | Wagh et al., 1977 |
| HLS-2 (human liposarcoma) | 25 | 1,093–914 | | | |
| HEp-2 (laryngeal carcinoma) | 90 | 940 | | | Valensin et al., 1982 |
| HEp-2 (with polio virus) | 90 | 710 | | | |
| HEp-2 (with coxsackie virus) | 90 | 730 | | | |
| HEp-2 (with adenovirus) | 90 | 800 | | | |
| HEp-2 (with measles virus) | 90 | 1,060 | | | |
| HEp-2 (with syncytial virus) | 90 | 1,230 | | | |
| WI38 cells | 30 | 1,154 | 165 | | Beall et al., 1978 |
| Red blood cells | | | | | |
| Normal oxy | 60 | 382 | | | Zipp et al., 1974 |
| Normal deoxy | 60 | 260 | | | |
| Sickle cell oxy | 60 | 879 | | | |
| Sickle cell deoxy | 60 | 864 | | | |
| Normal | 90 | 800 | | | Valensin et al., 1981 |
| Ghosts | 90 | 950 | | | |
| Normal | 90 | 730 | | | |
| With echovirus | 90 | 510 | | | |
| With concanavalin A | 90 | 610 | | | |
| With cytochalasin D | 90 | 650 | | | |
| With colchicine | 90 | 700 | | | |

**Table 9.11 NMR proton relaxation times of isolated cell organelles**

| Organelle | Frequency (MHz) | $T_1$ (ms) | $T_2$ (ms) | Reference |
|---|---|---|---|---|
| **Nuclei** | | | | |
| Human spleen cells | | | | |
| (whole cells $T_1$ = 658) | 60 | 999 ± 76 | | Shah et al. 1982 |
| HeLa—human carcinoma | | | | |
| (whole cells $T_1$ = 534) | 30 | | | Beall et al., 1981a |
| S phase nuclei | 30 | 457 ± 29 | 53 | |
| S nuclei with spermine | 30 | 617 ± 43 | | |
| (whole cells $T_1$ = 690) | | | | |
| $G_2$ phase nuclei | 30 | 692 ± 20 | 114 ± 15 | |
| Mouse—liver nuclei | | | | |
| (whole cells $T_1$ = 426) | 60 | 695 ± 20 | | Shah et al., 1982 |
| Mouse—liver nuclei from tumor bearer | | | | |
| (whole cells $T_1$ = 615) | 60 | 804 ± 37 | | |
| Mouse—spleen nuclei | | | | |
| (whole cells $T_1$ = 712) | 60 | 1,044 ± 59 | | |
| Mouse—kidney | | | | |
| (whole cells $T_1$ = 612) | 60 | 873 ± 52 | | Shah et al., 1982 |
| Mouse—liver nuclei | 9 | 826 ± 41 | | Modak et al., 1982 |
| Mouse—kidney nuclei | 9 | 851 ± 63 | | |
| Mouse—spleen nuclei | 9 | 1,124 ± 106 | | |
| Mouse—fibrosarcoma | 9 | 1,056 ± 85 | | |
| **Mitochondria** | | | | |
| Human spleen | 60 | 548 ± 19 | | Shah et al., 1982 |
| Mouse liver | 60 | 688 ± 35 | | |
| Mouse liver with tumor | 60 | 702 ± 23 | | |
| Mouse fibrosarcoma | 60 | 998 ± 64 | | |
| Dog cardiac mitochondria | | | | |
| Subsarcolemmal | 30 | 1,135 ± 20 | 110 ± 4 | Michael et al., 1980 |
| Ischemic | 30 | 915 ± 11 | 88 ± 3 | |
| Intermyofibrillar | 30 | 930 ± 6 | 71 ± 5 | |
| Ischemic | 30 | 684 ± 8 | 48 ± 2 | |
| **Ribosomes** | | | | |
| *E. coli* | 100 | measured hydration water, in fragments 70s,50s,30s | | White et al., 1971 |

**Table 9.12. Systemic effect of cancers on uninvolved tissues in tumor-bearing animals**

| Animal and type of cancer | Frequency (MHz) | Tissue | $T_1$ (ms) in nontumor-bearing animals | $T_1$ (ms) in tumor-bearing animals | Reference |
|---|---|---|---|---|---|
| Rat—carcinogen diet for 4 |  | Serum | 700 ± 40 | 810 ± 20 | Floyd et al., 1975 |
| weeks | 8 | Liver | 165 ± 10 | 195 ± 12 |  |
|  |  | Spleen | 287 ± 16 | 181 ± 03 |  |
| Mouse—fibrosarcoma |  | Liver | 235 ± 5 | 287 ± 9 | Modak et al., 1982 |
| ($T_1$ = 593 ± 55) | 9 | Kidney | 329 ± 7 | 365 ± 17 |  |
|  |  | Spleen | 416 ± 10 | 474 ± 10 |  |
| Mouse (A.W.S.) |  | Liver | 293 ± 4 | 342 ± 6 | Coles, 1976 |
| (MSWBS ascites sarcoma) | 15 | Kidney | 390 ± 4 | 415 ± 3 |  |
| ($T_1$ = 635 ± 10) |  | Muscle | 476 ± 5 | 486 ± 5 |  |
|  |  | Spleen | 493 ± 3 | 531 ± 3 |  |
|  |  | Brain | 488 ± 4 | 491 ± 5 |  |
| Mouse—sarcoma 180 | 20 | Blood | 902 | 905 | McLachlan and |
|  |  | Leg muscle | 528 | 607 | Hamilton, 1979 |
|  |  | Liver | 411 | 421 |  |
|  |  | Spleen | 566 | 582 |  |
|  |  | Kidney | 391 | 420 |  |
|  |  | Heart | 576 | 583 |  |
|  |  | Brain | 584 | 583 |  |
| Mouse—injected ascites |  | Liver homogenate |  |  | Floyd et al., 1974 |
| (Ehrlich) | 24.3 |  |  |  |  |
|  |  | Day 1 | 700 (normal) | 820 (treated) |  |
|  |  | Day 2 | 800 (normal) | 830 (treated) |  |
|  |  | Day 3 | 750 (normal) | 790 (treated) |  |
|  |  | Day 4 | — | — |  |
|  |  | Day 5 | 700 (normal) | 720 (treated) |  |
| Mouse (lymphosarcoma |  | Liver | 225 ± 5 | 283 ± 8 | Hollis et al., 1974 |
| 6C3HED) |  | Kidney | 322 ± 11,365 ± 23 |  |  |
| ($T_1$ = 674 ± 10) | 24 | Muscle | 465 ± 49 | 458 ± 51 |  |
|  |  |  | $\Delta T_1$ between normal and tumor-bearing after 46 days |  |  |
| Rat—Morris hepatoma (leg) | 24 | Liver | Up 120 ms |  |  |
|  |  | Kidney | Up 75 ms |  |  |
|  |  | Spleen | Up 140 ms |  |  |
|  |  | Heart | No change |  |  |
|  |  | Skeletal muscle | No change |  |  |
| Mouse—fibrosarcoma | 24 | Liver | 243–262 | 293–357 | Economou et al., 1973 |
|  |  | Kidney | 272–307 | 337–422 |  |
|  |  | Liver mitochondria | 354 | 701 |  |
| Mouse—fibrosarcoma |  | Liver | 412 ± 13 | 624 ± 25 | Modak et al., 1982 |
|  |  | Kidney | 581 ± 25 | 661 ± 25 |  |
| ($T_1$ = 1086 ± 56) | 25 | Spleen | 811 ± 112 | 831 ± 34 |  |
| Mouse—MCI tumor |  | Heart | 650 ± 6 | 705 ± 5 | Frey et al., 1972 |
| ($T_1$ = 853 ± 19) | 30 | Brain | 646 ± 4 | 618 ± 41 |  |
|  |  | Lung | 641 ± 9 | 665 ± 47 |  |
|  |  | Muscle | 615 ± 10 | 643 ± 10 |  |
|  |  | Spleen | 571 ± 8 | 731 ± 8 |  |
|  |  | Kidney | 470 ± 6 | 601 ± 19 |  |
|  |  | Liver | 386 ± 13 | 461 ± 9 |  |
|  |  | Skin | 390 ± 29 | 442 ± 37 |  |
|  |  | Intestine | 366 ± 19 | 399 ± 24 |  |
|  |  | Stomach | 294 ± 23 | 317 ± 29 |  |
| Mouse—mammary |  | Liver | 341 ± 5 | 335 ± 28 | Inch et al., 1974b |
| carcinoma C3H | 30 | Spleen | 523 ± 7 | 492 ± 27 |  |
|  |  | Kidney | 452 ± 5 | 463 ± 19 |  |
|  |  | Muscle | 590 ± 8 | 599 ± 25 |  |

*(continued)*

**Table 9.12. Systemic effect of cancers on uninvolved tissues in tumor-bearing animals *(Continued)***

| Animal and type of cancer | Frequency (MHz) | Tissue | $T_1$ (ms) in nontumor-bearing animals | $T_1$ (ms) in tumor-bearing animals | Reference |
|---|---|---|---|---|---|
| Mouse—isoimplants of C3H | 30 | Liver | 341 ± 5 | 347 ± 7 | Inch et al., 1974b |
| | | Spleen | 523 ± 7 | 551 ± 24 | |
| | | Kidney | 452 ± 5 | 448 ± 10 | |
| | | Muscle | 590 ± 8 | 586 ± 7 | |
| Mouse (C3HBA) | 30 | Liver | 341 ± 5 | 392 ± 15 | Inch et al., 1974b |
| | | Spleen | 523 ± 7 | 628 ± 19 | |
| | | Kidney | 452 ± 5 | 527 ± 13 | |
| | | Muscle | 590 ± 8 | 619 ± 14 | |
| Mouse (MCA) | 30 | Liver | 341 ± 5 | 383 ± 21 | Inch et al., 1974b |
| | | Spleen | 523 ± 7 | 609 ± 19 | |
| | | Kidney | 452 ± 5 | 519 ± 12 | |
| | | Muscle | 590 ± 8 | 592 ± 11 | |
| Mouse (CFWD strain) | 30 | Liver | 361 | 348 ± 20 (pregnant) | Inch et al., 1974b |
| | | Spleen | 512 | 521 ± 18 | |
| | | Kidney | 449 | 488 ± 6 | |
| | | Muscle | 563 | 586 ±18 | |
| Mouse (CEH/HeJ strain) | 30 | Liver | 341 ± 5 | 414 ± 9 (partial hepatectomy) | |
| | | Spleen | 523 ± 7 | 556 ± 5 | |
| | | Kidney | 452 ± 5 | 497 ± 5 | |
| Mouse BALB/C (EMT6 fibrosarcoma) | 30 | Spleen | 500 | 625 | Peemoeller et al., 1979 |
| | | Kidney | 476 | 555 | |
| | | Liver | 333 ± 10 | 384 | |
| | | Muscle | 555 | 588 | |
| Mouse C3H/HeJ (BA carcinoma) | 30 | Spleen | 500 | 588 | Peemoeller et al., 1979 |
| | | Kidney | 454 | 500 | |
| | | Liver | 303 | 344 | |
| | | Muscle | 625 | 666 | |
| Mouse C3H (mammary carcinoma) ($T_1$ = 887 ± 4) | 30 | Liver | 438 ± 36 | 468 ± 35 | Hazlewood et al., 1974a |
| | | Spleen | 737 ± 6 | 734 ± 12 | |
| | | Kidney | 503 ± 11 | 577 ± 14 | |
| Mouse C3HF ($T_1$ = 906 ± 43) | 30 | Liver | 412 ± 12 | 716 ± 7 | Hazlewood et al., 1974 |
| | | Spleen | 671 ± 9 | 716 ± 7 | |
| | | Kidney | 551 ± 9 | 537 ± 17 | |
| Mouse | | | | | |
| Mammary tumor $D_1B$ | 30 | Breast | 373 ± 9 | 733 ± 12 | Beall, and Hazlewood 1983 |
| Mammary tumor $C_4$ | | | 373 ± 9 | 594 ± 51 | |
| Mammary tumor $D_2$ | | | 373 ± 9 | 680 ± 55 | |

*(continued)*

**Table 9.12. Systemic effect of cancers on uninvolved tissues in tumor-bearing animals *(Continued)***

| Animal and type of cancer | Frequency (MHz) | Tissue | $T_1$ (ms) in nontumor-bearing animals | $T_1$ (ms) in tumor-bearing animals | Reference |
|---|---|---|---|---|---|
| Rat—rhabdomyosarcoma | 60 | Liver | 467 ± 26 | 556 | Boveé et al., 1974 |
| ($T_1$ = 1,150) | | Muscle | 850 ± 29 | 882 | |
| | | Kidney | 668 ± 31 | 881 | |
| | | Spleen | 582 ± 18 | 686 | |
| Rat—adenocarcinoma 4658 | 60 | Liver | 467 ± 26 | 628 | Boveé et al., 1974 |
| ($T_1$ = 1,113) | | Muscle | 850 ± 29 | 894 | |
| Rat—reticulum sarcoma 2880 | 60 | Liver | 467 ± 26 | 652 | Boveé et al., 1974 |
| ($T_1$ = 1,138) | | Muscle | 850 ± 29 | 921 | |
| Rat—adenocarcinoma 3207 | 60 | Liver | 467 ± 26 | 528 | Boveé et al., 1974 |
| ($T_1$ = 1,279) | | Muscle | 850 ± 29 | 903 | |
| Mouse—Lewis lung carcinoma, in thigh | 90 | Brain | 982 ± 53 | 972 ± 59 | de Certaines et al., 1979 |
| ($T_1$ = 1,140 ± 52) | | Heart | 924 ± 34 | 976 ± 62 | |
| | | Kidney | 752 ± 23 | 860 ± 46 | |
| | | Lung | 909 ± 75 | 950 ± 62 | |
| | | Liver | 616 ± 21 | 707 ± 63 | |
| | | Muscle | 940 ± 47 | 1,012 ± 50 | |
| | | Spleen | 778 ± 72 | 1,018 ± 49 | |
| Mouse—rhabdomyosarcoma | | Muscle | 547 | (no difference | Escanye et al., 1983 |
| ($T_1$ = 622–656) | 90 | Liver | 436 | detected in | |
| | | Kidney | 508 | tumor-bearing | |
| | | Spleen | 549 | animals) | |
| Mouse—MHI 134 tumor | | Muscle | 850 | 880 | Iijima et al., 1973 |
| ($T_1$ = 1,000) | 100 | Liver | 420 | 480 | |
| | | Brain | 840 | 930 | |
| Mouse—Swiss hepatoma | 100 | Liver | 426 ± 12 | 615 ± 17 | McLachlan and Hamilton, 1979 |
| | | Spleen | 712 ± 26 | 782 ± 19 | |
| | | Kidney | 612 ± 15 | 669 ± 24 | |
| Mouse—ICRC hepatoma | 100 | Liver | 375 ± 29 | 366 ± 17 | |
| Mouse—Swiss hepatoma | 100 | Liver | 426 ± 12 | 615 ± 17 | Shah et al., 1982 |
| | | Spleen | 712 ± 26 | 782 ± 19 | |
| | | Kidney | 612 ± 15 | 669 ± 24 | |
| Mouse—ICRC hepatoma | 100 | Liver | 375 ± 29 | 366 ± 17 | Shah et al., 1982 |

**Table 9.13. Systemic effect of cancers on uninvolved human tissues**

| Tissue | Frequency (MHz) | $T_1$ (ms) | Reference |
|---|---|---|---|
| **Gastrointestinal tissue** | | | |
| Autopsy (normal) | 22 | 416 ± 103 | Goldsmith et al., 1978 |
| Uninvolved host tissue | | 574 ± 110 | " |
| Cancer tissue | | 644 ± 136 | " |
| **Colon** | | | |
| Accident fatality 22.5 | normals | 330 ± 129 | Koutcher et al., 1978 |
| Uninvolved host tissue | | 590 ± 70 | " |
| Cancer tissue | | 620 ± 58 | " |
| **Breast** | | | |
| Accident fatality normals | 22.5 | 449 ± 136 | Koutcher et al., 1978 |
| Uninvolved host tissue | 25 | (400–650) | Ranade et al., 1975, 1976 |
| Cancer tissue | 22.5 | 438 ± 173 | Koutcher et al., 1978 |
| Cancer tissue | 25 | >550 | Mallard et al., 1979 |
| **Lung** | | | |
| Accident fatality normals | 22.5 | 505 ± 63 | Koutcher et al., 1978 |
| Uninvolved host tissue | 24.5 | <555 | Eggleston et al., 1975 |
| Cancer tissue | 22.5 | 711 ± 207 | Koutcher et al., 1978 |
| **Esophagus** | | | |
| Autopsy (normal) | 60 | 804 ± 108 | Damadian et al., 1974 |
| Uninvolved host tissue | 25 | 797 ± 39 | Ranade et al., 1975 |
| Cancer tissue | 25 | 959 ± 39 | " |
| **Liver** | 25 | | |
| Normal | | 339 ± 42 | Mallard et al., 1979 |
| Uninvolved host tissue | | <300 | " |
| Metastases in liver | | 650–800 | " |

**Table 9.14. Systemic effect of cancers on the relaxation times of sera in tumor-bearing animals**

| Animal | Time | $T_1$ (ms) | | Reference |
|---|---|---|---|---|
| | | Normal animals | Tumor-bearing animals | |
| Mouse (Ehrlich ascites) | Day 1 | 1,370 | 1,580 | Floyd et al., 1974 |
| | Day 2 | 1,370 | 1,380 | |
| | Day 3 | 1,370 | 1,580 | |
| | Day 4 | 1,370 | 1,580 | |
| | Day 5 | 1,370 | 1,610 | |
| | | | $\Delta T_1 = (T_1\ \text{can}_s - T_1\ \text{nor}_s)$ | Hollis et al., 1974 |
| Rat (hepatoma) | Day 5 | | +45 | |
| | Day 8 | | +0 | |
| | Day 12 | | +55 | |
| | Day 15 | | +60 | |
| | Day 20 | | +55 | |
| | Day 21 | | +75 | |
| | Day 28 | | +100 | |
| | Day 40 | | +160 | |
| | Day 46 | | +210 | |
| Rat (carcinogen-fed) | 1 week | 750 Control | 770 Treated | Floyd et al., 1975 |
| | 2 weeks | 740 Control | 780 Treated | |
| | 4 weeks | 700 Control | 810 Treated | |
| | 4 weeks | 770 Control | 780 Treated | |
| Mouse mammary tumors | [n] | | | Beall et al., 1981a |
| Normal virgin | [24] | 1,554 ± 19 | | |
| Normal virgin | [22] | 1,504 ± 9 | | |
| Protein malnourished virgins | [5] | 1,565 ± 6 | | |
| Duct hyperplasia (CD-1) | [4] | 1,585 ± 22 | | |
| HAN C4 | | 1,576 ± 13 | | |
| HAN C3 | | 1,463 ± 12 | | |
| Ductal papilloma (benign) | [4] | | 1,719 ± 22 | |
| Mammary carcinoma (C4) | [17] | | 1,801 ± 46 | |
| Mammary carcinoma (C3) | [19] | | 1,625 ± 23 | |

**Table 9.15. Systemic effect of cancers on the relaxation times of human sera**

| Type of cancer | *n* | $T_1$ (ms) | $T_2$ (ms) | Reference |
|---|---|---|---|---|
| **Breast cancer** (30 MHz) | | | | Beall et al., 1981a |
| Normal females (fresh) | 10 | 1,535 ± 59 | | |
| Normal males (frozen) | 30 | 1,503 ± 146 | 488 ± 63 | |
| Normal females (frozen) | 30 | 1,543 ± 178 | 475 ± 68 | |
| Biopsy females (small tumors, preoperative) | 10 | 1,612 ± 45 | | |
| Patients in remission (chemotherapy and radiation) | 20 | 1,460 ± 91 | 474 ± 46 | |
| Patients with cancer (on drugs, not in remission) | 22 | 1,618 ± 112 | 491 ± 61 | |
| Patients with metastases (at time serum was taken) | 5 | 1,734 ± 71 | 551 ± 46 | |
| **Assorted cancers** (30 MHz) | | | | Beall et al., 1981a |
| Normal | 10 | 1,535 ± 59 | | |
| Meningioma of brain | 1 | 1,615 | | |
| Carcinoma of lung (A) | 1 | 1,648 | | |
| (B) | 1 | 1,642 | | |
| (C) | 1 | 1,658 | | |
| Carcinoma of chest wall | 1 | 1,513 | | |
| Carcinoma of kidney (A) | 1 | 1,669 | | |
| (B) | 1 | 1,656 | | |
| **Assorted cancers** (20 MHz) | | | | de Certaines et al., 1981 |
| Normal males | 224 | 1,387 ± 57 | 652 ± 40 | |
| Normal females | | 1,394 ± 66 | 646 ± 55 | |
| Patients in remission | 155 | 1,354 ± 77 | 627 ± 51 | |
| Active GI, breast, and other cancers | 156 | 1,428 ± 108 | 652 ± 70 | |
| Carcinoma of the lymph node (A) | 1 | 1,583 | | |
| Metastatic (B) | 1 | 1,690 | | |
| (C) | 1 | 1,852 | | |
| Squamous carcinoma of the lymph node | 1 | 1,632 | | |
| Carcinoma of omentum | 1 | 1,906 | | |
| Carcinoma of prostate | 1 | 1,798 | | |

*(continued)*

**Table 9.15. Systemic effect of cancers on the relaxation times of human sera** ***(Continued)***

| Type of cancer | $n$ | $T_1$ (ms) | $T_2$ (ms) | Reference |
|---|---|---|---|---|
| Ovarian adenocarcinoma | 1 | 1,743 | | |
| Carcinoma of colon | 1 | 1,802 | | |
| Adenocarcinoma of small bowel | 1 | 1,605 | | |
| | Mean | 1,688 ± 105 | | |
| **Human** (20 MHz) | | | | Beall et al., 1983 |
| Normal | 23 | 1,628 ± 110 | | |
| Chronic myeloid leukemia | 4 | 1,680 ± 66 | | |
| Chronic lymphatic leukemia | 4 | 1,437 ± 109 | | |
| Acute myeloid leukemia | 5 | 1,409 ± 112 | | |
| Plasma cell myeloma | 7 | 1,584 ± 116 | | |
| Monoclonal gammopathies | 13 | 1,426 ± 186 | | |
| Stomach cancer | 2 | 1,984 ± 25 | | |
| Rectal and colon cancers | 12 | 1,711 ± 192 | | |
| **Melanoma** (30 MHz) | | | | Beall (unpublished), 1977 |
| Normal | | 1,430 ± 48 | 444 ± 5 | |
| Melanotic | | 1,585 ± 60 | 520 ± 41 | |
| **Blood cancers** (19.8 MHz) | | | | Koivula et al., 1982 |
| Normal plasma | | 1,410 ± 60 | | |
| Cancerous plasma (no cancer cells) | | 1,560 ± 170 | | |
| Plasma (in remission) | | 1,500 | | |
| **Assorted cancers** | | | | Ekstrand et al., 1977 |
| Normal | 7 | 1,230 ± 170 | | |
| Leukemia (remission) | 7 | 1,170 ± 100 | | |
| Carcinoma of cervix | 15 | 1,230 ± 90 | | |
| Carcinoma of breast | 5 | 1,260 ± 80 | | |
| **Other Conditions** | | | | |
| Cirrhosis of liver | 5 | 1,500 ± 198 | | |
| Chronic active hepatitis | 12 | 1,667 ± 101 | | |

**Table 9.16. Expected $T_1$ values for human and rabbit tissues at various frequencies**

| Tissue | 100 MHz $T_1$ (ms) (human) | 24 MHz $T_1$ (ms) (rabbit) | 2.5 MHz $T_1$ (ms) (rabbit) |
|---|---|---|---|
| Blood | | | |
| Whole blood, clotted | | 867 ± 88 | 404 ± 30 |
| Whole blood, heparinized | | 872 ± 43 | 372 ± 34 |
| Blood serum | | 1,590 ± 113 | 820 ± 12 |
| Brain | | | |
| Gray cerebral tissue | 998 ± 16 (8) | 644 ± 40 | 332 ± 22 |
| White cerebral tissue | | 469 ± 13 | 264 ± 11 |
| Cerebellum | | 570 ± 24 | 326 ± 11 |
| Medulla oblongata | | 493 ± 47 | 308 ± 2 |
| Spinal cord | 557 ± 158 (2) | 464 ± 22 | 325 ± 19 |
| Gastrointestinal Tract | | | |
| Parotid salivary gland | | 340 ± 51 | 234 ± 38 |
| Oesophagus wall | 804 ± 108 (5) | 534 ± 42 | 250 ± 14 |
| Stomach wall | 765 ± 075 (8) | 468 ± 94 | 227 ± 20 |
| Small intestine wall | 641 ± 080 (8) | 481 ± 98 | 224 ± 26 |
| Colon wall | 641 ± 43 (12) | 466 ± 29 | 227 ± 20 |
| Rectum wall | | 492 ± 56 | 245 ± 6 |
| Heart | | | |
| Heart ventricle | 906 ± 46 (9) | 637 ± 28 | 243 ± 4 |
| Heart auricle | | 706 ± 42 | 295 ± 9 |
| Aortic arch | | 451 ± 15 | 211 ± 41 |
| Inferior vena cava | | 361 ± 56 | 201 ± 5 |
| Carotid artery | | 364 ± 30 | 167 ± 17 |
| Kidney | | | |
| Kidney cortex | 827 ± 26 (4) | 406 ± 41 | 206 ± 27 |
| Kidney medulla | | 801 ± 35 | 426 ± 61 |
| Ureter | | 252 ± 13 | 172 ± 10 |
| Adrenal gland | 476 (1) | 448 ± 13 | 227 ± 13 |
| Bladder | 891 ± 61 (4) | | |

(continued)

**Table 9.16. Expected $T_1$ values for human and rabbit tissues at various frequencies *(Continued)***

| Tissue | 100 MHz $T_1$ (ms) (human) | 24 MHz $T_1$ (ms) (rabbit) | 2.5 MHz $T_1$ (ms) (rabbit) |
|---|---|---|---|
| Liver | | | |
| Liver | 570 ± 29 (14) | 311 ± 15 | 141 ± 16 |
| Bile | | 1,078 ± 446 | 888 ± 388 |
| Spleen | 701 ± 45 (17) | 509 ± 11 | 258 ± 4 |
| Pancreas | 605 ± 36 (10) | | |
| Lung | | | |
| Trachea | | 276 ± 11 | 199 ± 17 |
| Upper lung | | 624 ± 37 | 283 ± 19 |
| Lower lung | 788 ± 63 (5) | 686 ± 37 | 303 ± 43 |
| Muscle | | | |
| Diaphragm | | 458 ± 62 | 202 ± 32 |
| Intercostal muscle | | 519 ± 43 | 191 ± 4 |
| Rectus abdominis | | 534 ± 23 | 183 ± 9 |
| Skeletal muscle (thigh) | 1,023 ± 29 (17) | 554 ± 32 | 182 ± 12 |
| Reproductive Tract | | | |
| Ovary | 989 ± 47 (5) | 573 | 310 |
| Uterus wall | 924 ± 38 (4) | 653 | 296 |
| Testis | | 855 ± 110 | 463 ± 47 |
| Epididymis | 803 ± 14 (2) | 625 ± 5 | 309 ± 6 |
| Prostate | 608 ± 20 (5) | | |
| Cervix | 1,200 ± 48 (4) | | |
| Breast | 367 ± 79 (5) | | |
| Other | | | |
| Skin | 616 ± 19 (9) | 328 ± 28 | 198 ± 13 |
| Bone | 554 ± 27 (10) | 202 ± 14 | 175 ± 50 |
| Lymphatic | 720 ± 76 (6) | | |
| Thyroid | 882 ± 45 (7) | | |
| Adipose | 279 ± 8 (5) | | |

*Notes:* 100 MHz data from Damadian et al., 1974. 24 MHz and 2.5 MHz data from Ling and Foster, 1981.

$T_1$ values are frequency dependent with higher values being measured at higher frequencies. Earlier *in vitro* data collected at 100, 60, 30, and 22 MHz are difficult to directly compare with *in vivo* imaging data taken at less than 10 MHz. However, Ling and Foster (1981) have determined that most tissues maintain the same relative differences at 24 and 2.5 MHz. The use of lower frequencies for patient safety and imaging advantages may reduce the absolute differences between tissues and with the error due to machine and sample handling problems, may reduce the probability of distinction of normal and cancerous tissues.

**Table 9.17. Comparison of $T_1$ values for *in vivo* and *in vitro* measurements**

| Type of Tissue | *In vitro* $T_1$ of rabbit tissue at 2.5 MHz and 25°C (ms) | *In vivo* $T_1$ of human tissue at 1.7 MHz and 37°C (ms) |
|---|---|---|
| **Rabbit and human tissues** | (Ling and Foster, 1981) | (Smith et al., 1982) |
| Blood | 372 ± 34 | 340 ± 370 |
| Bone | 175 ± 50 | 190–210 |
| Brain | | |
| Gray matter | 332 ± 22 | 275 |
| White matter | 264 ± 11 | 225 |
| Cerebellum | 326 ± 11 | 230–280 |
| Spinal cord | 325 ± 19 | 230–280 |
| Heart | 243 ± 4 | 240–260 |
| Kidney | | |
| Cortex | 206 ± 27 | 300–340 |
| Medulla | 426 ± 61 | |
| Liver | 141 ± 16 | 70–170 |
| Muscle skeletal | 182 ± 12 | 120–140 |
| Spleen | 258 ± 4 | 250–290 |
| **Rabbit thigh** | (Ling and Foster, 1982) | |
| (Turpentine inflammation) | | |
| Center of reaction | 328 ± 5 | 364 ± 5 |
| Surrounding surface | 285 ± 5 | 329 ± 5 |
| Underlying muscle | 220 ± 10 | 273 ± 10 |
| Control normal muscle | 163 ± 3 | 184 ± 2 |

Data given as Mean ± *SD*.

*Note:* Based on an evaluation of available data, there is no significant difference between $T_1$ values for living tissues versus excised fresh tissues using proton NMR. However, $^{31}P$ NMR is much more sensitive to the metabolic state of the tissue and can readily sense ischemia in tissues. If the physiological damage causes a disruption of water and electrolyte balance, such as in coronary infarcts or brain edema, proton NMR may be sensitive to *in vivo* and *in vitro* differences. Insufficient data on pathological states are available to make this determination currently.

**Table 9.18. Some publications giving $T_1$ values for *in vivo* NMR images of living tissues**

| Type of Tissue | $T_1$ (msec) | $T_2$ (msec) | Reference |
|---|---|---|---|
| Rat | | | |
| Hepatoma | 900 | 60 | Davis et al., 1981 |
| Rat brain (15 MHz) | | | |
| Normal | 730 | 80 | Richards et al., 1983 |
| Irradiated | 1,300 | 80 | " |
| Gerbil brain (6.25 MHz) | | | |
| Normal | 1,280 ± 40 | 56.6 ± 5.6 | |
| Asymptomatic (CH) | 1,270 ± 30 | 59.8 ± 2.9 | |
| Symptomatic (CH) | 1,280 ± 50 | 58.7 ± 3.9 | |
| Asymptomatic (litigated carotid) | 1,270 ± 200 | 61.0 ± 1.8 | |
| Symptomatic (litigated carotid) | 1,470 ± 120 | 76.0 ± 9 | |
| Rat (15 MHz) | | | Davis et al., 1981 |
| Normal liver | 450 ± 50 | 40 ± 8 | |
| Hepatomas | 800 ± 200 | 55 ± 10 | |
| Hepatomas | 970 | 70 | |
| Abscess | 620 | 60 | |
| Brain | 890 | 57 | |
| Muscle | 720 | 32 | |
| Fat | 30 | 48 | |
| Rat muscle (15 MHz) | | | Moon et al., 1983a |
| Fibrosarcoma | 1,544 ± 387 | 68 ± 4.7 | |
| Muscle | 744 ± 98 | 30 ± 1.6 | |
| Fat | 305 ± 38 | 57 ± 3.4 | |
| Bladder contents | 3,740 ± 4,537 | 159 ± 149 | |
| Rat heart (15 MHz) | | | Herfkins et al., 1983b |
| Normal (24 h) | 620 ± 74 | 29.2 ± 1.7 | |
| Infarcted (24 h) | 1,079 ± 134 | 45.8 ± 4.2 | |
| Dog (15 MHz) | | | Moon et al., 1983d |
| Bile (fasting) | 770–840 | | |
| Bile (nonfasting) | 2,400 | | |
| Human abdomen | | | Ross et al., 1982b |
| Normal liver | 107 | | |
| Affected liver | 305 | | |
| Gastric mass (leiomyosarcoma) | 328–288 | | |
| Human abdomen (100 MHz) | | | Gore et al., 1981 |
| Muscle | 1,100 | | |
| Kidney | 850 | | |
| Liver | 650 | | |
| Fat | 250 | | |
| Human abdomen (1.7 MHz) | | | Smith et al., 1982 |
| | *Range* | | |
| Normal liver | 140–170 | | |
| Normal spleen | 250–290 | | |
| Normal pancreas | 180–200 | | |
| Bile | 250–300 | | |
| Blood | 340–370 | | |
| Cirrhosis | 180–300 | | |
| Hepatoma | 300–450 | | |
| Cholangiocarcinoma | 300–450 | | |
| Metastases | 300–450 | | |
| Hemangioma | 340–370 | | |
| Simple cyst | 800–1,000 | | |
| Pancreatitis | 200–275 | | |
| Carcinoma of pancreas | 275–400 | | |

*(continued)*

**Table 9.18. Some publications giving $T_1$ values for *in vivo* NMR images of living tissues *(Continued)***

| Type of Tissue | $T_1$ (msec) | $T_2$ (msec) | Reference |
|---|---|---|---|
| Human adrenal gland (15 MHz) | 394 | 73 | Moon et al., 1983a |
| Metastatic colon tumor | 1,266 | 45 | |
| Human brain (425 G) | | | Furuse et al., 1983 |
| Gray matter | 288 ± 42 | | |
| White matter | 227 ± 27 | | |
| Cerebellar cortex | 273 ± 29 | | |
| Medulla | 248 ± 29 | | |
| Pontine region | 250 ± 41 | | |
| Human brain (10.65 MHz) | | | Go et al., 1983 |
| White matter | 311 ± 25 | 80 ± 6 | |
| Fatty tissue | 158 ± 13 | 122 ± 4 | |
| Human brain (QED 80) | | | Rangel-Guerra et al., 1982 |
| Frontal lobes normal | 210 ± 10 | | |
| Affective disorders and schizophrenia | 264 ± 9 | | |
| After lithium therapy | 208 ± 8 | | |
| Human brain (QED 80) | | | Furuse et al., 1982 |
| Normal frontal cortex | 227 ± 27 | | |
| White matter | 288 ± 42 | | |
| Cerebellar cortex | 273 ± 29 | | |
| Medulla | 248 ± 29 | | |
| Pons | 250 ± 41 | | |
| Human bone (0.35 T) | | | Heller et al., 1983 |
| Femoral bone normal | 232–305 | 47–63 | |
| Femoral bone necrotic | 310–493 | 54–62 | |
| Human brain (1.69 MHz) | | | |
| Normal | 215 ± 42 | | Hazlewood et al., 1982 |
| Brain lesions | 397 | | |
| Brain diseases | wide range | | Bydder et al., 1983 |
| Human brain and tumors (1.5 kG) | | | Araki et al., 1983 |
| White matter | 290 ± 22 | | |
| Gray matter | 365 ± 40 | | |
| Astrocytomas | 546 ± 1,048 | | |
| Meningiomas | 360 ± 472 | | |
| Neurinoma | 690–785 | | |
| Lipoma | 235–290 | | |
| Ependymoma | 645 ± 70 | | |
| Pituitary adenocarcinoma | 764 ± 180 | | |
| Renal carcinoma metastases | 605–696 | | |
| Human nerve tissue (1.7 MHz) | | | Smith et al., 1982 |
| | *Range* | | |
| Brain gray matter | 275–325 | | |
| White matter | 255–250 | | |
| Cerebellum | 230–280 | | |
| C S fluid | 350–1,000 | | |
| Spinal cord | 230–280 | | |
| Blood | 340–370 | | |
| Edema | 360–420 | | |
| Infarct | 320–375 | | |
| Glioma | 250–360 | | |
| Metastases | 250–350 | | |
| Meningioma | 200–300 | | |
| Human brain (3.5 kG) | | | |
| Cerebral cortex | 1,100 | 80 | Crooks et al., 1982 |

*(continued)*

**Table 9.18. Some publications giving $T_1$ values for *in vivo* NMR images of living tissues *(Continued)***

| Type of Tissue | $T_1$ (msec) | $T_2$ (msec) | Reference |
|---|---|---|---|
| Human brain | | | |
| Normal | 241–309 | | Besson et al., 1983 |
| Senile dementia-Alzheimer's | 277–330 | | |
| Senile dementia-multi infarct | 283–340 | | |
| Human breast (4 MHz) | | | Mansfield et al., 1980 |
| Normal uninvolved | 164 ± 90 | | |
| Tumor | 750 ± 1,000 | | |
| Mixed | 420 ± 83 | | |
| Nipple | 750 ± 1,000 | | |
| Human breast | | | |
| Normal and fatty | 100–200 | | Ross et al., 1982a |
| Dysplasia | 100–300 | | |
| Cysts | 200–400 | | |
| Fibrosis | 300 | | |
| Carcinoma | 50–250 | | |
| Human gall bladder (15 MHz) | | | |
| Fasting | 302–914 | 57–119 | Hricak et al., 1983a |
| After meal | 3,000–3,250 | 115–158 | |
| Human heart (3.5 kG) | | | Higgins et al., 1983 |
| Septum | — | 39 ± 6 | |
| Wall | — | 37 ± 9 | |
| Skeletal muscle | — | 33 ± 4 | |
| Fat | — | 50 ± 6 | |
| Human kidney (15 MHz) | | | |
| Normal | 500–900 | 50–90 | London et al., 1983 |
| Diseased | 400–2,500 | 40–100 | |
| Normal (15 MHz) | 700–900 | 45–70 | Hricak et al., 1983b, 1983c |
| Cysts | 800–2,500 | 70–350 | |
| Masses | 1,000–1,600 | 60–90 | |
| Human kidney (15 MHz) | | | |
| Normal | 671 | 52 | Brasch et al., 1983 |
| With contrast | 266, 578 | 51–56 | |
| Human liver (6.5 MHz) | *Range* | | Doyle et al., 1982 |
| Normal | 210–270 | | |
| Hepatoma | 460–530 | | |
| Metastases | 560–810 | | |
| Bile ducts | 550–980 | | |
| Liver atrophy | 320 | | |
| Hepatic infarct | 410 | | |
| Infectious hepatitis | 310 | | |
| Chronic hepatitis | 290 | | |
| Hemochromatosis | 170–190 | | |
| Hemosiderosis | 180 | | |
| Steatosis | 210–240 | | |
| Cirrhosis | 280–450 | | |
| Primary biliary cirrhosis | 160–350 | | |
| Wilson disease | 230 | | |
| Human liver (3.5 kG) | — | 49 ± 12 | Wesbey et al., 1983 |
| Human musculoskeletal system (15 MHz) | | | |
| Muscle | 541 ± 141 | 35 ± 6 | Moon et al., 1983b |
| Subcutaneous fat | 218 ± 68 | 61 ± 18 | |
| Vertebral narrow | 420 ± 112 | 50 ± 12 | |
| Intervertebral disk | 2,623 ± 2,030 | 48 ± 7 | |
| Ligament/tendon | 864 ± 206 | 107 ± 16 | |
| Bone tumors | 2,800 | 57 | |

*(continued)*

**Table 9.18. Some publications giving $T_1$ values for *in vivo* NMR images of living tissues *(Continued)***

| Type of Tissue | $T_1$ (msec) | $T_2$ (msec) | Reference |
|---|---|---|---|
| Human spleen (3.5 kG) | — | 62 ± 10 | Wesbey et al., 1983 |
| Human thorax (15 MHz) | | | Gamsu et al., 1983 |
| Mediastinal fat | 250–600 | 29 | |
| Bronchogenic carcinoma | 750–1,650 | 107 | |
| Human tissues | *Range* | | Smith, 1983 |
| Skeletal muscle | 120–140 | | |
| Cardiac muscle | 240–260 | | |
| Fat | 160–180 | | |
| Bone | 190–210 | | |
| Blood | 340–370 | | |
| Liver | 140–170 | | |
| Spleen | 250–290 | | |
| Kidney | 300–340 | | |
| Pancreatitis | 200–275 | | |
| Carcinoma pancreas | 275–400 | | |
| Pseudocyst | 800–1,000 | | |
| Pus | 400–440 | | |
| Ascitic fluid | 1,000 | | |
| Human tissues (16 MHz) | | | Bené et al., 1980 |
| Blood | | 150 | |
| Soft tissues | | 30 | |
| Urine | | 2,170 | |
| Bladder | | 280–107 | |
| Human urinary tract (1.7 MHz) | *Range* | | Smith, 1983 |
| Kidney | 300–340 | | |
| Urine | 600–1,000 | | |
| Acute tubular necrosis | 400–420 | | |
| Simple cyst | 600–1,000 | | |
| Carcinoma | 400–450 | | |
| Metastases | 400–450 | | |
| Abscess | 490–420 | | |
| Bladder carcinoma | 200–240 | | |
| Prostate | 250–325 | | |
| Carcinoma | 350–450 | | |

*Note:* The majority of *in vivo* measurements have been made at less than 10 MHz and it is difficult to directly compare $T_1$ and $T_2$ values to earlier *in vitro* values at greater than 20 MHz. However, Ling and Foster (1981) have measured many types of rabbit tissues at 2.5 MHz *in vitro* and *in vivo*. Values obtained for excised organs such as brain, kidney, liver, spleen, and muscle are similar *in vivo* and *in vitro* considering the contribution of temperature differences. Proton NMR relaxation times of organs *in vivo* do not differ greatly from freshly excised organs. However $^{31}P$ NMR will be much more sensitive to alterations in temperature, pH, and oxygen levels.

**Table 9.19. $^{13}C$ NMR investigations in living systems**

| System | Properties measured | Reference |
|---|---|---|
| Human muscle, normal and diseased | Phospholipids, glycerol, lactic acid, glucose $^{13}C$ spectra as a function of temperature and field, $T_1 = 0.31$ to 0.52 sec. | Barany et al., 1982b |
| Frog gastrocnemius muscle | 118.2 MHz high resolution spectra, lactic acid, phospholipid mobility, effects of caffeine induced contraction | Barany et al., 1982a<br>Doyle and Barany, 1982 |
| Photosynthetic bacteria | Biosynthesis of amino acids, Calvin cycle, Krebs cycle glucose synthesis | Paalme et al. 1982a<br>Paalme et al., 1982b |
| *Rhodopseudomonas sphaeroides* | Metabolism of 2-$^{13}C$ acetate, dark and light cycles | Nicolay et al., 1982 |
| Chicken pectoralis muscle | 90.5 MHz high resolution spectra phospholipids, proteins, organics | Doyle et al., 1981 |
| Phosphatidylcholine vesicles | 90.5 MHz chemical shifts gel transitions | DeKruijff, 1976 |
| Mouse muscle | 25.2 MHz three peaks for aliphatic, aromatic, and carbonyl carbons | Fung, 1977b |
| Hemoglobin solutions (human, fetal, chick, bovine) | 15.18 MHz tryptophan assignment | Oldfield and Allerhand, 1975 |
| Biological proteins (ferricytochromes, lysozyme, myoglobin) | Chemical shift ranges for carbons of amino acids | Oldfield and Allerhand, 1975 |
| Collagen (rat skin) | Chemical shift and $T_1$ | Torchia et al., 1974 |

**Table 9.20. Deuterium ($D_2O$) NMR data on biological systems**

| System studied | Frequency (MHz) | Temp. | $T_1$ (ms) | $T_2$ (ms) | Reference |
|---|---|---|---|---|---|
| Pure $D_2O$ | 6 | 80°C | 480 | 450 | Hansen, 1971 |
| Pure $D_2O$ | 9.21 | −18 to 178°C | * | * | Hindman et al., 1971 |
| Frog sartorius and gastrocnemius muscle | 8.1 | 20 to 40°C | 134 ± 1.34<br>122 ± 1.44 | 13.7 ± 0.26<br>10.9 ± 0.25 | Civan et al., 1978 |
| Mouse muscle | 2 kHz | Different temperatures | Function of frequency | — | Fung, 1977a |
| Frog gastrocnemius muscle | 4.3<br>8.1 | 23°C<br>23°C | 116 ± 5.7<br>143 ± 5.5 | —<br>— | Civan and Shporer, 1975 |
| Mouse muscle—leg | 9.21 | 25°C | 130 | 15 | Fung and McGaughy, 1979 |
| Mouse liver | 4.5 | 37 to −20°C | n100 (at 4.5 MHz and 37°C) | — | Fung et al., 1975a |
| Pure $D_2O$ | 4 | 25°C | 420 | — | Cope, 1969 |
| Rat | | | | | |
| Brain | 4 | 25°C | 131 | 22 | " |
| Muscle | | | 92 | 9 | " |
| Rat | | | | | Hansen, 1971 |
| Brain | 6 | 30°C | 181 ± 5 | 33 ± 1.0 | |
| Muscle | | | 112 ± 2 | 18 ± 0.5 | |

*Refer to Hindman et al., 1971 for temperature dependence.

**Table 9.21. $^{39}K$ NMR relaxation times of normal and cancerous tissues**

| Tissues | Frequency (MHz) | Relaxation times (ms) | | Reference |
|---|---|---|---|---|
| | | Normal $T_1$ | Cancer $T_1$ | |
| **Various Tissues** | | | | |
| Spleen (rat) | 10 | 6.9 ± 0.9 | | Damadian and Cope, 1973, 1974 |
| Leukemia (rat) | | | 5.7 ± 0.8 | |
| Lymphoma (mouse) | | | 5.1 ± 0.5 | |
| Liver (rat) | | 5.9 ± 0.2 | | |
| Hepatoma (rat) | | | 6.6 ± 0.3 | |
| Brain (human) | | 11.0 ± 2.0 | | |
| Tumors (human) | | | 54.2 | |
| | | | 30.0 | |
| | | | 12.0 | |
| | | | 10.5 | |
| Infant cerebellum (human) | | | 20.2 | |
| Muscle (rat) | | 8.4 ± 0.7 | | |
| Spindle cell sarcoma | | | 6.7 ± 0.3 | |
| Kidney (rats) | | 7.4 ± 0.9 | | |
| Walker 256 | | | 5.4 ± 0.6 | |
| Small intestine (rats) | | 9.5 ± 0.6 | | |
| Testis (rat) | | 8.1 ± 1.2 | | |
| Mammary carcinoma (rat) | | | 7.3 ± 1.3 | |
| Muscle (rat) | | Chemical shift data | | Cope and Damadian, 1977, 1974 |
| **Mouse—Whole Embryos** | 10 | Alive $T_2$ = 0.240 ms<br>Dead $T_1$ = 0.330 ms | | Cope and Damadian, 1979 |
| *H. halobium* **Bacteria** | 10 | $T_2$ = 0.35 ms | | Cope and Damadian, 1970 |
| | | $T_1$ = 3.95 ms | $T_2$ = 0.18 ms | Shporer and Civan, 1977 |
| | | NMR invisible $K^+$ | | Magnuson and Magnuson, 1973 |
| *E. Coli* | | Line widths | | Damadian and Cope, 1973 |
| **Ion Exchange Resin** | | | | |
| Dowex—50 | | Chemical shift data | | Cope and Damadian, 1977 |
| IRC—50 | | | | Cope and Damadian, 1974 |
| **Frog Striated Muscle** | 10.2 | | | |
| KCl Solution 4–7°C | | 45.5 ± 6.2 | — | Civan et al., 1976 |
| 21–22°C | | 66.7 ± 8.9 | — | |
| Muscle 4–7°C | | 5.05 ± 0.79 | ($T_2$ = 3.06 ± 0.21) | |
| 21–22°C | | 12.7 ± 2.4 | ($T_2$ = 4.37 ± 0.16) | |

**Table 9.22. $^{23}Na$ NMR parameters on various tissues**

| Tissue (frequency) | Sodium parameters | | Reference |
|---|---|---|---|
| | $R_1$ (ms) | $T_2$ (ms) | |
| NaCl solution $^{23}Na$ | 18.4 ± 0.4 | 54.3 ± 1.18 | Edzes et al., 1977 |
| NaCl solution (26.45 MHz) | — | 59 | Chang and Woessner, 1978 |
| Frog muscle (15.9 MHz) | 63–72% invisible | | Cope, 1965, 1967 |
| | Line broadening | | Magnuson et al., 1973 |
| | 60–70% NMR invisible | | |
| | 58–65% NMR invisible | | Ling and Cope, 1969 |
| Frog muscle (7.8 MHz) | 36–65% NMR invisible | | Martinez et al., 1969 |
| Frog muscle (15.1 MHz) | 37% NMR visible | | Czeisler et al., 1970 |
| Frog muscle (15.8 MHz) | 53% NMR visible | | Yeh et al., 1973 |
| Frog muscle (15.7 MHz) | | | Shporer and Civan, 1974 |
| Fresh at | | | |
| 3°C | 12 | | |
| 22°C | 20 | | |
| 24°C | 22 | | |
| 39°C | 26 | | |
| 104 h old at | | | |
| 3°C | 17 | | |
| 22°C | 22 | | |
| 24°C | 24 | | |
| 39°C | 28 | | |
| Frog skin (60 MHz) | 30.8% NMR invisible | | Reisin et al., 1970 |
| Frog skin (15.8 MHz) 22°C | 42.6% NMR visible | | Rotunno et al., 1967 |
| Frog liver (7.8 MHz) | 66% NMR invisible | | Martinez et al., 1969 |
| Rat tissue (10 MHz) (25–30°C) | | | |
| 0.2 N NaCl | 55 | 56 | Cope, 1970b |
| Muscle | 12 | 14 | |
| Brain | 15 | 10 | |
| Kidney | — | 9 | |
| Rat tissue (10 MHz) (25 ± 2°C) | | | Goldsmith and Damadian, 1975 |
| Normal tissue | | | |
| Muscle | 12.1 ± 0.7 | | |
| Liver | 6.5 ± 0.5 | | |
| Kidney | 12.5 ± 0.3 | | |
| Intestine | 13.3 ± 1.8 | | |
| Testis | 17.8 ± 0.7 | | |

(continued)

**Table 9.22. $^{23}$Na NMR parameters on various tissues *(Continued)***

| Tissue (frequency) | Sodium parameters $R_1$ (ms) | $T_2$ (ms) | Reference |
|---|---|---|---|
| Tumor tissues | | | |
| Walker 256 carcinoma | 9.8 ± 1.1 | | |
| Novikoff hepatoma | 23.7 ± 2.7 | | |
| Solid Ehrlich ascites (mice) | 18.9 ± 0.9 | | |
| Sarcoma 180 | 15.1 ± 2.5 | | |
| Rat skeletal muscle (26.45 MHz) (mixture of tibialis anterior and gastrocnemius muscle) | 18.3 ± 0.5 | 16.1 ± 1.6<br>1.59 ± 0.16 | Chang and Woessner, 1978 |
| Rat heart | $^{23}$Na gated images of a perfused tissue | | DeLayre et al., 1981 |
| Rat liver (11.2 MHz) | 61% NMR invisible | | Monoi, 1974 |
| Rat liver homogenate (11.2 MHz) | 62% NMR invisible | | Monoi, 1974 |
| Rat testicle (60 MHz) | 14.6% NMR invisible | | Reisin et al., 1970 |
| Human red cell membranes (15.9 MHz) | Broadened sodium resonance | | Magnuson et al., 1970 |
| Human erythrocytes | 98% NMR visible | | Yeh et al., 1973 |
| Dog red cell ghosts | No visible Na | | Jardetzky and Wertz, 1956 |
| Pig red cell ghosts | 98% visible Na | | Monoi and Katsukura, 1976 |
| Chromatin calf thymus (24 MHz) | Binding to DNA | | Burton and Reimarsson, 1978 |
| Rabbit myelinated nerve | Binding of Na, 56% invisible | | Cope, 1970a |

| | Relaxation rates for several nuclei in bacteria | | | |
|---|---|---|---|---|
| | $^{23}$Na | $^{87}$Rb | $^{133}$Cs | |
| *Halobacterium sp.* | | | | |
| NMR visible signal | 80% | 40% | 15% | Edzes et al., 1977 |
| Spin–spin relaxation $T_2$ time (ms) | 12.35 ± 2.90 | 0.115 ± 0.013 | 28.6 ± 4.1 | |
| Relaxation rate of salt solution ($s^{-1}$) | 18.4 ± .4 | 400 ± 30 | 0.21 ± .02 | |
| Spin–spin relaxation $T_2$ time (ms) | 54.3 ± 1.18 | 2.5 ± 0.188 | 4762 ± 454 | |
| Yeast cells (132 MHz) | $^{23}$Na line broadening<br>$^{7}$Li chemical shift | | | Balschi et al., 1982 |

**Table 9.23. $^{17}O$ relaxation times in biological tissues**

| System studied | Frequency (MHz) | Temperature (°C) | $T_1$ (ms) | $T_2$ (ms) | Reference |
|---|---|---|---|---|---|
| $H_2{}^{17}O$ | | Relaxation in pure water | | | Glasel, 1967 |
| $H_2{}^{17}O$ (50% enriched) | 24.73 | 24 | 6.99 | 7.23 | Kasturi and Seitz (private communications, 1982) |
| $H_2{}^{17}O$ (16.94% enriched) | 8.13 | −14 to 180 | Temperature dependence | | Hindman et al., 1971 |
| Lysozyme | 4.3, 8.1 | Relaxation rates | | | Koenig et al., 1975 |
| Blood—Human | | | | | |
| Plasma | 25 | 25 | 3.9 | | Fabry and Eisenstadt, 1975 |
| Red cells | 25 | | 1.7 | | |
| Red cells | 8.13 | 24 | 1.0–1.05 | | Shporer and Civan, 1975 |
| Frog skeletal muscle | | 30 | | 1.22 | Swift and Barr, 1973 |
| | | 10 | | 1.18 | |
| Frog sartorius and gastrocnemius muscle | 8.1 | 20–24 | 2.08 | 1.29<br>1.152 | Civan et al., 1978 |
| | 7.55 | Multi component relaxation | | | Civan and Shporer, 1974 |
| Frog gastrocnemius muscle | 4.3 | 23 | 1.46 | — | Civan and Shporer, 1975 |
| | 8.1 | 23 | 1.86 | — | Civan and Shporer, 1975 |
| Mouse muscle, leg | 9.21 | 25 | 1.8 | 1.1 | Fung and McGaughy, 1979 |
| Rat lymphocytes (thymocytes) | 7.72 | 26.5 | 3.12* | 1.30* | Shporer et al., 1976 |
| " | 4.36 | 26.5 | 2.90* | 1.30* | " |
| Rat malignant thymocytes | | | | | |
| K127T | 7.72 | 26.5 | 3.75* | 2.08* | " |
| K127R | 7.72 | 26.5 | 3.62* | 2.04* | " |

*Initial relaxation times in a non-exponential decay.

**Table 9.24. $^{31}P$ NMR chemical shifts (ppm) of some common phosphate compounds in $H_2O$ (ppm from 85% $H_3PO_4$)**

| No. | Compound | pH | Chemical Shift α | β | γ | Coupling Constants J(α-β) | J(β-γ) | Reference |
|---|---|---|---|---|---|---|---|---|
| 1. | ATP = Adenosine 5′-triphosphate | 7.0 | −11.0 | −21.8 | −7.3 | — | — | Cohn and Rao, 1979 |
| 2. | Mg ATP = Adenosine 5′-triphosphate | 7.0 | −10.8 | −19.2 | −5.6 | — | — | " |
| 3. | ATP = Adenosine 5′-triphosphate | 7.4 | −11.45 | −22.66 | −7.33 | 19.75 | 19.75 | Labotka et al., 1976a |
| 4. | ADP = Adenosine 5′-diphosphate | 7.0 | −10.8 | −7.5 | — | — | — | O'Neill and Richards, 1980 |
| 5. | Mg ADP = Adenosine 5′-diphosphate | 7.0 | −9.9 | −5.9 | — | — | — | Cohn and Rao, 1979 |
| 6. | ADP = Adenosine 5′-diphosphate | 7.4 | −11.13 | −6.96 | — | 23.1 | — | Labotka et al., 1976 |
| 7. | Mg ADP = Adenosine 5′-diphosphate (Bound to arginine kinase) | 7.0 | −11.0 | −3.3 | — | — | — | Cohn and Rao, 1979 |
| 8. | Mg ADP = Adenosine 5′-diphosphate (Bound to creatine kinase) | 7.0 | −11.0 | −3.8 | — | — | — | " |
| 9. | Mg ADP = Adenosine 5′-diphosphate (Bound to pyrurate kinase) | 7.0 | −10.0 | −5.7 | — | — | — | " |
| 10. | Mg ATP = Adenosine 5′-triphosphate (Bound to arginine kinase) | 7.0 | −11.0 | −19.4 | −5.6 | — | — | " |
| 11. | Mg ATP = Adenosine 5′-triphosphate (Bound to creatine kinase) | 7.0 | −10.9 | −19.0 | −5.4 | — | — | " |
| 12. | Mg ATP = Adenosine 5′-triphosphate (Bound to pyruvate) | 7.0 | −10.9 | −19.2 | −5.5 | — | — | " |
| 13. | TDP= Thymidine 5′-diphosphate | 7.4 | −11.26 | −7.02 | — | 23.6 | — | Labotka et al., 1976 |
| 14. | UDP = Uridine 5′-diphosphate | 7.4 | −11.18 | −6.85 | — | 23.6 | — | O'Neill and Richards, 1980 |
| 15. | UTP = Uridine 5′-triphosphate | 7.4 | −11.50 | −22.92 | −7.52 | 20.53 | 20.53 | " |
| 16. | IDP = Inosine 5′-diphosphate | 7.4 | −11.03 | −7.11 | — | 22.1 | — | " |
| 17. | ITP = Inosine 5′-triphosphate | 7.4 | −11.49 | −22.84 | −7.63 | 20.75 | 20.75 | " |
| 18. | CDP = Cytidine 5′-diphosphate | 7.4 | −11.13 | −6.94 | — | 22.6 | — | " |
| 19. | CTP = Cytidine 5′-triphosphate | 7.4 | −11.51 | −22.77 | −7.25 | 20.25 | 20.25 | " |
| 20. | GDP = Guanosine 5′-diphosphate | 7.4 | −11.10 | −6.69 | — | 23.1 | — | " |
| 21. | GTP = Guanosine 5′-triphosphate | 7.4 | −11.49 | −22.85 | −7.70 | 19.25 | 19.25 | " |

*The numbers in parentheses represent the reference in the literature.

**High frequency values (down field) are taken to be positive.

**Table 9.25. $^{31}P$ NMR chemical shifts (ppm) of some common tissue metabolites**

| System | Temp °C | $P_i^e$ | S–P | Cr–P | γ ATP (Mg bound) | α ATP (Mg bound) | β ATP (Mg bound) | β ADP (Mg bound) | α ADP (Mg bound) | Others | Reference |
|---|---|---|---|---|---|---|---|---|---|---|---|
| A. Skeletal muscle—hind leg of mice (*in vitro*)[a] | 24 | 1.7 ± 0.1 | 3.9 ± 0.1 | −3.1 ± 0.1 | −5.6 ± 0.1 | −10.8 ± 0.1 | −19.6 ± 0.2 | — | — | Unknown phodiester −0.6 ± 0.4 | Koutcher and Damadian, 1977 |
| B. Rhabdomyosarcoma tumor (*in vitro*)[a] | 24 | 2.4 ± 0.2 | 4.3 ± 0.2 | — | — | — | — | — | — | −0.4 ± 0.5 | Koutcher and Damadian, 1977 |
| C. Normal liver—mouse (*in vitro*)[a] | 4 | 2.3 ± 0.2 | 4.2 ± 0.1 | −2.5 ± 0.1 | — | — | — | −5.9 ± 0.3 | −10.6 ± 0.5 | — | Koutcher et al., 1981 |
| D. Hepatoma—mouse (*in vitro*)[a] | 4 | 2.3 ± 0.1 | 4.2 ± 0.3 | −2.4 ± 0.4 | — | — | — | −5.5 ± 0.3 | −11.0 ± 0.5 | −0.4 ± 0.5 | Koutcher et al., 1981 |
| E. Human forearm muscle (*in vivo*) | 37 | (b) | Fructose -6-phosphate | | | | | | | Nicotinamide adenine dinucleotide | Edwards et al., 1982 |
| | | 5.65 to 3.1[c] 2.55 to 0.21 | 7.5 | 0 | −2.5 | −7.6 | −15.9 | — | — | −8.3 | |
| | | | 4.4 | −3.1[d] | −5.6 | −10.7 | −19.0 | — | — | −11.4 | |
| F. Chicken pectoralis muscle (*in vitro*) | 31 | — | — | — | −5.71 | −10.69 | −19.37 | | | 1/2 (line width) 24 Hz | Glonek et al., 1981 |

S–P = sugar phosphate
Cr–P = creatine phosphate
$P_i$ = inorganic phosphate
[a]Chemical shifts measured with reference to external 85% $H_3PO_4$ in the concentric tube coaxial with the sample. Numbers in parentheses below the chemical shift values represent *n* samples used by the authors to obtain the mean value reported.
[b]Chemical shifts taking creatine phosphate as 0 ppm.
[c]Chemical shifts calculated with reference to the 85% $H_3PO_4$ assuming Cr–P to be at 3.1 ppm with to 85% $H_3PO_4$.
[d]Chemical shift of Cr–P can vary between −2.3 ppm and −3.2 ppm with reference to 85% $H_3PO_4$ depending on sample geometry.
Muscle pH = 6.8 + $\log_{10}$(S − 8.31/5.845 − S) where S is the chemical shift of inorganic phosphate peak with reference to phosphocreatine.
[e]Chemical shift of inorganic phosphate depends on pH.

**Table 9.26. Studies of $^{31}P$ NMR in Biological Systems**

| System studied (frequency) | Function monitored | Reference |
|---|---|---|
| Bacteria | | |
| *E. coli* (145.7 MHz) | Metabolic phosphates pH 7.55 | Navon et al., 1977 |
| | ATPase activity | Ugurbil et al., 1978 |
| Yeast | | |
| Yeast (145.7 MHz) | Metabolic phosphates long chain polyphosphates pH 6.3 | Salhany et al., 1975 |
| | | Ogawa et al., 1978 |
| | Temp. sensitive cycle | Gillies and Benoit, 1983 |
| | ATP production | Kainosho et al., 1977 |
| Tetrahymena | Pyrophosphate, phosphonic acids, phospholipids | Deslauriers et al., 1982a |
| Liver flukes | Metabolism, ATP, ADP, phosphodiesters, sugar phosphates | Mansour et al., 1982 |
| Eggs and embryos | | |
| Artemia embryos (80.99 MHz) | Intracellular pH changes and metabolic status of dormant and developing embryos | Busa et al., 1982 |
| Frog eggs and developing tadpoles | Yolk proteins, nucleotide triphosphates | Colman and Gadian, 1976 |
| Snail eggs | Phosphates of invertebrates at fertilization | Miceli et al., 1978 |
| Sea Urchin eggs | | Winkler et al., 1982 |
| Cell organelles | | |
| Mitochondria—rat liver (145.7 MHz) | pH 7.4 | Navon et al., 1978 |
| | Oxidative phosphorylation | Ogawa et al., 1978, 1981 |
| Chromaffin granules (129 MHz) | Internal pH 5.6 ± 0.1 | Seeley et al., 1977 |
| | Glycerol-3-phosphate | Pollard et al., 1979 |
| | AMP, NADPH, PCr | |
| Cultured Cells | | |
| Ehrlich ascites tumor cells (145.7 MHz) | Phosphate metabolites ΔpH of $<0.2$ units | Navon et al., 1977 |
| Hela cells | Nucleotide phosphates | Evans and Kaplan, 1977 |
| | UTP levels | |
| Continuous flow culture | pH monitoring | Gonzalez-Mendez et al., 1982 |
| Lymphoid line, friend leukemia cells, astrocytoma | No ΔpH, high PCr | Navon et al., 1978 |
| Rat liver cells (145.7 MHz) | ΔpH, mitochondria sugar phosphates PCr, GPC, GPE | Cohen et al., 1978 |
| Mouse embryo fibroblasts | Difference in transformed cells | Ugurbil et al., 1981 |
| Animal tissues | | |
| Heart muscle | | |
| Rat (129 MHz) | Sugar phosphates, PCr | Gadian, 1982 |
| Rat (40 MHz) | ATP, pH change | Jacobus et al., 1977 |
| Rabbit (72.9 MHz) (beating) | Monitoring ischemia | Hollis, 1980 |
| | pH 6.6, infarct | Bailey et al., 1981 |
| | KCl arrest | Hollis, 1980 |
| | Calcium paradox | Bulkey et al., 1978 |
| Rat (perfused) | Creatine kinase | Brown et al., 1978 |
| | Phosphate metabolites | Nunnally and Hollis, 1978 |
| | | Dawson et al., 1978 |
| | | Fossel et al., 1980 |
| | | Ingwall, 1982 |
| | | Matthews et al., 1982 |
| Guinea pig | Free $Mg^{+2}$ 2.5 mM | Wu et al., 1981 |
| | Metabolites | Salhany et al., 1979 |
| Rat (*in vivo*) | ATP, PCr, Pi, change in ATP with beat | |

*(continued)*

**Table 9.26. Studies of $^{31}P$ NMR in Biological Systems *(Continued)***

| System studied (frequency) | Function monitored | Reference |
|---|---|---|
| Skeletal muscle | | |
| Rat (129 MHz) | First $^{31}P$ NMR of muscle | Hoult et al., 1974 |
| | AP-$Mg^{+2}$ comples, ΔpH | |
| Rat contracted and relaxed | Changes in ATP during muscle contraction | Dawson et al., 1976, 1977 |
| Frog | Creatine kinase kinetics | Gadian et al., 1981a |
| | ATP-$Mg^{+2}$ complexes | Barany et al., 1975 |
| | Measurement of free $Mg^{+2}$ | Hoult et al., 1974 |
| | Phosphate metabolites | Gupta and Moore, 1980 |
| | | Wu et al., 1981 |
| Cat (biceps) | Creatine kinase | Kushmerick et al., 1983 |
| Chicken—normal (36.43 MHz) | Phosphate metabolites | Chalovich et al., 1979 |
| —dystrophic | L-serine ethanolamine phosphodiesterase | Burt et al., 1976b |
| Rabbit-actin solution | ATP binding | Cozzone et al., 1974 |
| Rabbit-myosin solution | Calcium bridge phosphorylation | Mak et al., 1978 |
| Kidney | Monitoring of transplant | Sehr et al., 1977 |
| Rat (129 MHz) | PCr, ATP, Pi, GTP detection of ischemia | |
| Rat | *in vivo* spectra | Balaban et al., 1981 |
| Liver | | |
| Rat (60.7 MHz) | Phosphate metabolites | Salhany et al., 1979 |
| Rat—perfused (100 MHz) | KCN loss of ATP | Zaner and Damadian, 1975a |
| Mouse | ATP assay | McLaughlin et al., 1979 |
| Brain | | |
| Rat | ATP, PCr, Pi | Chance et al., 1980 |
| Rat (100 MHz) | Effect of low oxygen | Zaner and Damadian, 1975b |
| Intestine—Rat (100 MHz) | Phosphate metabolites | Zaner and Damadian, 1975a |
| Adrenal glands—Dog | Phosphate metabolites | Radda, 1975 |
| Nerve—Crayfish nerve cord | Effects of oxygen phosphorylarginine | Nishikawa et al., 1978 |
| Gastric mucosa—Dog | *in vitro* (Using chamber-membrane transport | Balaban, 1982 |
| Teeth (bovine) | Structure | Lee et al., 1977 |
| Lens | | |
| Human | Phosphate metabolites | Kopp et al., 1982 |
| Cat | " | |
| Dog | " | |
| Pig | " | |
| Cow | " | |
| Sheep | " | |
| Rat | " | |
| Skin | | |
| Frog | Ratios of PCr to ATP and ADP as to functions of time, oxygen, temperature and extracellular Pi | Lin-Er Lin et al., 1982 |

*(continued)*

**Table 9.26. Studies of $^{31}$P NMR in Biological Systems *(Continued)***

| System studied (frequency) | Function monitored | Reference |
|---|---|---|
| Blood and blood components | | |
| Erythrocytes | | |
| Rabbit | Hemoglobin pH, DPG | Moon and Richards, 1974 |
| Rabbit | ATP, DPG, Pi | Labotka et al., 1976a |
| Reticulocytes | | |
| Rabbit | CPG, CPE, ATP | Labotka et al., 1976a |
| Platelets—human | no DPG | Labotka et al., 1976b |
| Hemoglobin | Oxygen binding | Costello et al., 1976 |
| Hemoglobin | Cooperativity | Huestis and Raftery, 1975 |
| Hemoglobin | DPG shifts | Fossel and Solomon, 1976 |
| Hemoglobin | ATP binding, Mg | Gupta and Moore, 1978 |
| Whole blood | | |
| Mouse | Plasmodium infected, DPG | Deslauriers et al., 1982b, 1983 |
| Human | DPG, ATP, Pi | Henderson et al., 1974 |
| Dog (circulating) | Abnormal $^{31}$P NMR from 2,3 diphosphoglyceric acid (DPG) and phospholipids in circulating proteins | Burt et al., 1982 |
| Lipoproteins | Phospholipids | Assmann et al., 1974 |
| Lipoproteins | $Mn^{+2}$ EDTA quencher | Glonek et al., 1970 |
| Semen | (PCH), Pi, GPC | Burt et al., 1978 |
| Human tissues | | |
| Muscle | | |
| Normal | ATP, PCr, NADH | Burt et al., 1977 |
| Nemalin rod myopathy | Decrease in metabolites | Barany et al., 1977 |
| McArdle's syndrome | Metabolites | Ross et al., 1981 |
| Dystrophic | No PCr, no ATP | Burt et al., 1977 |
| Duchenne dystrophy | 5× normal NADH | Chalovich et al., 1979 |
| Aged | 1/2 Pi, decrease in PCr | Burt et al., 1977 |
| | ATP decline | Cresshull et al., 1981 |
| Normal | Pi, PCr | Edwards and Wiles, 1981 |
| Dystrophy | Pi, PCr | Edwards, 1978 |
| | Changes in Pi and PCr in patients with phosphofructokinase deficiency and Duchenne muscular dystrophy, energy exchange in human muscle | Edwards et al., 1982 |
| Rat (100 MHz) | | |
| Liver | 2.33 ± 0.14 | Zaner and Damadian, 1975 |
| Muscle | 2.03 ± 0.05 | |
| Intestine | 1.97 ± 0.12 | |
| Brain | 1.13 ± 0.05 | |
| Kidney | 1.43 ± 0.05 | |
| Brain | 1.20 | |
| Kidney | 1.03 | |
| Intestine | 1.60 | |
| Rat-tissue—cancerous | | |
| Walker sarcoma | 5.38 ± 0.68 | |
| Novikoff hepatoma | 5.98 ± 0.57 | |
| Crocker sarcoma | 5.19 ± 1.42 | |

## 9.2. BIBLIOGRAPHY


Abetsedarskaya, L.A., Miftahutdinova, F.G., and Fedotov, V.D. State of water in living tissues. *Biofizika* **13:** 750–758; 1968.

Abragam, A. *The Principles of Nuclear Magnetism.* London: Oxford University Press; 1961.

Ackerman, J.J., Grove, T.H., Wong, G.G., Gadian, D.G., and Radda, G.K. Mapping of metabolites in whole animals by $^{31}P$ NMR using surface coils. *Nature* **283:** 167–170; 1980a.

Ackerman, J.J., Bore, P.J., Gadian, D.G., Grove, T.H., and Radda, G.K. NMR studies of metabolism in perfused organs. *Phil. Trans.R. Soc. Lond. (Biol.)* **289:** 425–436; 1980b.

Ader, R., and Cohen, J.S. Relaxation time measurements of bulk water in packed mammalian density-dependent and independent cells grown in culture. *J. Mag. Res.* **34:** 349–356; 1979.

Andrasko, J. Water diffusion permeability of human erythrocytes studied by pulsed gradient NMR technique. *Biochim. Biophys. Acta.* **428:** 304–311; 1976.

Araki, T., Inouye, T., Matozaki, T., and Iio, M. *In vivo* $T_1$ measurement of brain tumors by NMR-CT. *Proc. SMRM,* San Francisco, p. 3; 1983.

Assmann, G., Sokoloski, E.A., and Brewer, H.B. NMR relaxation times in plants: phosporus-31 nuclear magnetic resonance spectroscopy of native and recombined lipoproteins, *PNAS. USA.* **71:** 549–553; 1974.

Bader, J.P., Ray, D.A., and Brown, N.R. Accumulation of water during transformation of cells by an avian sarcoma virus. *Cell* **3:** 307–312, 1974.

Bailey, I.A., Williams, S.R., Radda, G.K., and Gadian, D.G. Activity of phosphorylase in total global ischemia in the heart: A $^{31}P$-NMR study. *Biochem. J.* **196:** 171–178; 1981.

Balaban, R.S. Nuclear magnetic resonance studies of epithelial metabolism and function. *Fed. Proc.* **41:** 42–47; 1982.

Balaban, R.S., Gadian, D.G., and Radda, G.K. Phosphorous nuclear magnetic resonance study of the rat kidney in vivo. *Kidney International* **20:** 575–579; 1981.

Balschi, J.A., Cirillo, V.P., and Springer, C.S. Direct high-resolution nuclear magnetic resonance studies of cation transport *in vivo.* Na transport in yeast cells. *Biophys. J.* **38:** 323–326; 1982.

Barany, M., Doyle, D.D., Graff, G., Westler, W.M., and Markley, J.L. Changes in the natural abundance $^{13}C$ NMR spectra of intact frog muscle upon storage and caffeine contracture. *J. Biol. Chem.* **257**(6): 2741–2743; 1982a.

Barany, M., Doyle, D.D., Westler, W.M., and Markley, J.L. $^{13}C$ NMR of intact human muscle. Quantitation of lactic acid. Abstracts: First Annual Meeting of the Society of Magnetic Resonance in Medicine, Boston, MA; 1982b.

Barany, M, Burt, C.T., Labotka, R.J., Danon, M.J., Glonek, T., and Huncke, B.H. In: *Pathogenesis of Human Muscular Dystrophies,* L.P. Rowland, ed. Amsterdam: Excerpta Medica Foundation; p. 337–340; 1977.

Barany, M., Barany, K., Burt, C.T., Glonek, T., and Myers, T.C. Structural changes in myosin during contraction and the state of ATP in intact frog muscle. *J. Supramol. Struct.* **3:** 125–140; 1975.

Barroilhet, L.F., and Moran, P.R. NMR relaxation behavior in living and ischemically damaged tissue. *Med. Phys.* **3**(6): 410–414; 1976.

Beall, P.T., Amtey, S.R., and Kasturi, S.R. *NMR Data Handbook for Biomedical Applications.* New York: Pergamon Press; 1984.

Beall, P.T. Safe physiological methods for improved NMR contrast in mouse mammary cancer. *Proc. SMRM,* San Francisco; p. 28; 1983a.

Beall, P.T. States of water in biological systems. *Cryobiology* **20:** 324–334; 1983b.

Beall, P.T. A historical perspective on biomedical NMR. *Mag. Res. Imaging,* **1**(4): 189–190; 1983c.

Beall, P.T. Improved NMR contrast for mouse mammary cancer by safe physiological agents. *Physiol. Chem. Phys.* **14:** 399–403; 1983d.

Beall, P.T. Practical methods for biological NMR sample handling. *Mag. Res. Imaging* **1**(3): 165–180; 1983e.

Beall, P.T. NMR studies in vitro. In: *Proceedings of the International Conference on Pathologists and Surgeons.* S. Stipa, ed., New York: Academic Press (accepted for publication); 1983f.

Beall, P.T. NMR relaxation times of water protons in cultured cells under freezing and osmotic stress conditions. In: *Biophysics of Water,* F. Franks, ed. New York: John Wiley & Sons; pp. 323–326; 1983g.

Beall, P.T. The systemic effect in cancer—Its implications for NMR imaging. In: *NMR Imaging: Future Potential,* J. Jaklovsky, ed. Massachusetts: Addison-Wesley Publishing Co. (accepted for publication); 1983h.

Beall, P.T. *In vitro* distinction of cancer and other disease states by nuclear magnetic resonance. In: *NMR Imaging,* P. Conti, ed., Rome, Italy: Marrapese Editore (accepted for publication); 1983i.

Beall, P.T., and Hazlewood, C.F. Distinction of the normal, preneoplastic, and neoplastic states by water proton NMR relaxation times. In: *Nuclear Magnetic Resonance (NMR) Imaging,* L. Partain, E. James, F. Rollo, R. Price, eds. Philadelphia: W.B. Saunders & Co.; pp. 312–338; 1983.

Beall, P.T., Narayana, P.A., Amtey, S.R., Livio, S., Intra, E., Ridella, S., and Mela, G. The systemic effect of cancers on human sera. *Proc. SMRM,* San Francisco; p. 30, 1983.

Beall, P.T., Narayana, P.A., Amtey, S.R., and Mela, G. A systemic effect of several cancers and other diseases on the NMR relaxation times of human sera. *Mag. Res. Imaging* **1**(4): 235–236; 1982a.

Beall, P.T., Brinkley, B.R., Chang, D.C., and Hazlewood, C.F. Microtubule complexes correlated with growth rate and water proton relaxation times in human cancer cells. *Cancer Research* **42:** 4124–4130; 1982b.

Beall, P.T., Hazlewood, C.F., and Rutzky, L.P. NMR water relaxation times of human colon cancer cell lines and clones. *Cancer Biochem. Biophys.* **6:** 7–12; 1982c.

Beall, P.T., Izzat, N.N., and Cassidy, M.M. Effects of cytochalasin B and colchicine on in vivo intestinal absorption in rats. *FASEB;* 1982d.

Beall, P.T., Hazlewood, C.F., and Chang, D.C. Microtubule organization and the self-diffusion coefficient of water in baby hamster kidney cells as a function of temperature. *J. Cell Biol.* **44:** 556a; 1982e.

Beall, P.T. Contribution of cytoskeleton to cellular water properties in diverse functional states. *Fed. Proc.* **40**(2): 206–213; 1981.

Beall, P.T., Medina, D., and Hazlewood, C.F. The systemic effect of elevated tissue and serum relaxation times for water in animals and humans with cancers. In: *NMR Basic Principles and Progress,* Vol. 19, P. Diehl, E. Fluck,

R. Kosfield, eds. Berlin: Springer-Verlag: pp. 39–57; 1981a.

Beall, P.T., Asch, B.B., Medina, D., and Hazlewood, C.F. Distinction of normal, preneoplastic, and neoplastic mouse mammary cells and tissues by nuclear magnetic resonance techniques. In: *The Transformed Cell, I.* Cameron, ed. New York: Academic Press; pp. 293–325; 1981b.

Beall, P.T., Asch, B.B., Chang, D.C., Medina, D., and Hazlewood, C.F. Distinction of normal, preneoplastic, and neoplastic mouse mammary primary cell cultures by water nuclear magnetic resonance relaxation times. *J. Natl. Cancer Inst.* **64:** 335–338; 1980a.

Beall, P.T. Water-macromolecular interactions during the cell cycle. In: *Nuclear-Cytoplasmic Interactions,* G. Whitson, ed. New York: Academic Press; pp. 223–249; 1980b.

Beall, P.T. Application of cell biology to an understanding of biological water. In: *Cell Associated Water,* W.D. Hansen; J. Clegg, eds. New York: Academic Press; pp. 271–292; 1979.

Beall, P.T. Unpublished results (1977).

Beall, P.T., Chang, D.C., and Hazlewood, C.F. Structural changes in chromatin during the Hela cell cycle: Effect on water NMR relaxation times. In: *Biomolecular Structure and Function,* P. Argris, ed. New York: Academic Press; pp. 233–237; 1978.

Beall, P.T., Chang, D.C., Misra, L.K., Fanguy, R.C., and Hazlewood, C.F. Progressive muscular dystrophy in chickens: Electrolytes and NMR of cellular water. *The Physiologist* **20**(7): 41a; 1977a.

Beall, P.T., Medina, D., Chang, D.C., Seitz, P.K., and Hazlewood, C.F. A systemic effect of benign and malignant mammary cancer on the spin-lattice relaxation time, $T_1$, of water protons in mouse serum. *J. Natl. Cancer Inst.* **59:** 1431–1433; 1977b.

Beall, P.T., Hazlewood, C.F., and Rao, P.N. Nuclear magnetic resonance patterns of intracellular water as a function of HeLa cell cycle. *Science* **192:** 904–907; 1976a.

Beall, P.T., Cailleau, R.M., and Hazlewood, C.F. The relaxation times of water protons and division rate in human breast cancer cells: A possible relationship to survival. *Physiol. Chem. Phys.* **8:** 281–284; 1976b.

Beall, P.T., Hazlewood, C.F., and Rao, P.N. Non-linearity of relaxation time versus water content. *Science* **194:** 213–214, 1976c.

Bearden, D. Unpublished results (personal communication). (1983).

Belagyi, J. Water structure in striated muscle by spin labeling technique. *Acta Biochim. et Biophys. Acad. Sci. Hung* **10**(1–2): 63–70; 1975.

Belton, P.S., and Packer, K.J. Pulsed NMR studies of water in striated muscle. III. The effects of water content. *Biochim. Biophys. Acta* **354:** 305–314; 1974.

Belton, P.S., Packer, K.J., and Sellwood, T.C. Pulsed NMR studies of water in striated muscle. II. Spin–lattice relaxation times and dynamics of the nonfreezing fraction of water. *Biochim. Biophys. Acta* **304:** 45–52; 1973.

Belton, P.S., Jackson, R.R., and Packer, K.J. Pulsed NMR study of water in striated muscle. I. Transverse relaxation times and freezing effects. *Biochim. Biophys. Acta* **286:** 16–27; 1972.

Bené, B.C. Diagnosis of meconium in amniotic fluids by nuclear magnetic resonance spectroscopy. *Physiol. Chem. Phys.* **12:** 241–245; 1980.

Bené, G.J., Borcard, B., Hiltbrand, E., and Magnin, P. *In situ* identification of human physiological fluids by nuclear magnetism in the Earth's field. *Philos. Trans. R. Soc. Lond. (Biol.)* **289:** 501–502; 1980.

Bessman, S.P., and Geiger, P.J. Transport of energy in muscle: The phosphorylcreatine shuttle. *Science* **211:** 448–452; 1981.

Besson, J.A.O., Corrigan, F.M., Foreman, E.L., Eastwood, L.M., Smith, F.W., and Ashcroft, G.W. Proton NMR observations in dementia. *Proc. SMRM,* San Francisco; p. 43; 1983.

Block, R.E., Maxwell, G.P., Prudhomme, D.L., and Hudson, J.L. High resolution proton magnetic resonance spectral characteristic of water, lipid and protein signals from three mouse cell populations. *J. Natl. Cancer Inst.* **58:** 151–156; 1977.

Block, R.E., and Maxwell, G.P. Proton magnetic resonance studies of water in normal and tumor rat tissues. *Magn. Reson.* **14:** 329–334; 1974.

Block, R.E. Factors affecting proton magnetic resonance linewidths of water in several rat tissues. *FEBS Lett.* **34:** 109–112; 1973.

Boicelli, C.A., Guarnieri, C., and Toni, R. Induced cardiac hypertrophy in rats. *Proc. SMRM,* San Francisco; p. 47; 1983.

Borcard, B., Hiltbrand, E., Nagnin, P., Bene, G.J., Briguet, A., Duplan, J.C., Delman, J., Guiband, S., Bonnet, M., Dumont, M., and Fara, F. Estimating meconium (fetal feces) concentration in human amniotic fluid by nuclear magnetic resonance. *Physiol. Chem. Phys.* **14:** 189–192; 1982.

Boveé, W.M., Creyghton, J.H., Getreuer, K.W., Korbee, D., Lobregt, S., Smidt, J., Wind, R.A., Lindeman, J., Smid, L., and Posthuman, H. NMR relaxation and images of human breast tumors *in vitro. Phil. Trans. R. Soc. Lond.* **289:** 535–536; 1980.

Boveé, W., Huisman, P., and Smidt, J. Tumor detection and nuclear magnetic resonance. *J. Natl. Cancer Inst.* **52:** 595–597; 1974.

Brasch, R.C., London, D.A., Wesbey, G.E., Tozer, T.N., Nitecki, D.E., Williams, R.D., Doemeny, J., Tuck, L.D., and Lallemand, D.P. Work in progress: Nuclear magnetic resonance study of a paramagnetic nitroxide contrast agent for enhancement of renal structures in experimental animals. *Radiology* **147:** 773–779; 1983.

Bratton, C.B., Hopkins, A.L., and Weinberg, J.W. Nuclear magnetic resonance studies of living muscle. *Science* **147:** 738–741; 1965.

Brooks, R.A.; Battocletti, J.H., Sances, A., Larson, S.J., Bowman, R.L., and Kudravcev, V. Nuclear magnetic resonance in blood. *IEEE Trans. Biomed. Eng.* **22**(1): 12–18; 1975.

Brown, F.F., and Campbell, I.D. NMR studies of red cells. *Philos. Trans. R. Soc. Lond. (Biol.)* **289**(1037): 395–406; 1980.

Brown, T.R., Gadian, D.G., Garlick, P.B., Radda, G.K., Seeley, P.J., and Styles, P.J. Creatine kinase activities in skeletal and cardiac muscle measured by saturation transfer NMR. *Front. Biol. Energ.* **2:** 1341–1349; 1978.

Bulkey, B.H., Nunnally, R.L., and Hollis, D.P. "Calcium paradox" and the effect of varied temperature on its development: A phosphorous NMR and morphological study. *J. Lab. Invest.* **39:** 133–140; 1978.

Bunch, W.H., and Kallsen, G. Rate of intracellular diffusion as measured in barnacle muscle. *Science* **164:** 1178–1179; 1969.

Burt, C.T., Eisemann, A.E., Schofield, J.C., and Wyvwicz, A.M. Nuclear magnetic resonance studies on circulating blood. *J. Magnetic Res.* **46**(No.1): 176; 1982.

Burt, C.T., Cohen, S.M., and Barany, M. Analysis of intact tissue with $^{31}$P-NMR. *Ann. Rev. Biophys. Bioeng.* **8:** 1–25; 1979.

Burt, C.T., Arrata, S.M., and Corder, S. NMR investigations of sperm. *Fert. Ster.* **30:** 329–333; 1978.

Burt, C.T., Glonek, T., and Barany, M. Analysis of living tissue by phosphorus-31 magnetic resonance. *Science* **195:** 145–149; 1977.

Burt, C.T., Glonek, T., and Barany, M. Analysis of phosphate metabolites, the intracellular pH, and the state of adenosine triphosphate in intact muscle by phosphorous nuclear magnetic resonance. *J. Biol. Chem.* **251:** 2584–2591; 1976a.

Burt, C.T., Glonek, T., and Barany, M. Phosphorus-31 nuclear magnetic resonance detection of unexpected phosphodiesters in muscle. *Biochem.* **15**(22): 4850–4853; 1976b.

Burton, D.R., and Reimarsson, P. $^{23}$Na NMR as a probe of ion binding to chromatin. *FEBS Letters* **89:** 183–186; 1978.

Buorranno, F.S., Brady, T.J., and Pykett, I.L. Proton NMR imaging in experimental ischemic cerebral infarction. *Ann. Neurol.* **10:** 75–82; 1981.

Busa, W.B., Crowe, J.H., and Matson, G.B. Intracellular pH and metabolic status of dormant and developing artemia embryos. *Arch. Biochem. Biophys.* **216**(2); 1982.

Busby, S.J.W., Gadian, D.G., Radda, G.K., Richards, R.E., and Seely, P.J. Phosphorus nuclear magnetic resonance studies of compartmentation in muscle. *Biochem. J.* **170:** 103–114; 1978.

Bydder, G.M., Steiner, R.E., Thomas, D.J., Marshall, J., Gilderdale, D.J., and Young, I.R. Nuclear magnetic resonance imaging of the posterior fossa: 50 cases. *Clinical Radiology* **34:** 173–188; 1983.

Caille, J.P., and Hinke, J.A.M. The volume available to diffusion in the muscle fiber. *Can. J. Physiol. Pharmacol.* **52:** 814–828; 1974.

Callaghan, P.T., Jolley, K.W., and Leleivre, J. Diffusion of water in the endosperm tissue of wheat grains as studied by pulsed field gradient nuclear resonance. *Biophys. J.* **28:** 133–142; 1979.

Cameron, I.L., LaBadie, D.R.L., Hunter, K.E., and Hazlewood, C.F. Changes in water proton relaxation times and in nuclear to cytoplasmic element gradients during meiotic maturation of Xenopus Oocytes. *J. Cell Biol.* **116:** 87–92; 1983.

Cantor, C.R., and Schimmel, P.R. *Biophysical Chemistry Part II: Techniques for the Study of Biological Structure and Function.* San Francisco: Freeman; pp.481–538; 1980.

Cerbon, J. Variations of the lipid phase of living microorganisms during the transport process. *Biochim. Biophys. Acta* **102:** 449–458; 1965.

Chalovich, J.M., Burt, C.T., Danon, M.J., Glonek, I., and Barany, M. Phosphodiesters and muscular dystrophies. *Ann. NY Acad. Sci.* **317:** 649–668; 1979.

Chance, B., Eleff, S., and Leigh, J.S. Noninvasive, nondestructive approach to cell bioenergetics. *Proc. Natl. Acad. Sci. USA* **77:** 7430–7434; 1980.

Chandra, R., Pizzarello, D.J., Keegan, A.F., and Chase, N.E. A study of proton relaxation time ($T_1$) in rat tissues at various times after intracardiac manganese injection. *Proc. SMRM,* San Francisco; p. 90; 1983.

Chang, D.C. A voltage clamp study of the effects of colchicine on the squid giant axon. *J. Cell Physiol.* **115:** 260–264; 1983.

Chang, D.C., Misra, L.K., Beall, P.T., Fanguy, R.C., and Hazlewood, C.F. Nuclear magnetic resonance study of muscle water protons in muscular dystrophy of chickens. *J. Cell. Physiol.* **107:** 139–143; 1981.

Chang, D.C., and Hazlewood, C.F. Nuclear magnetic resonance study of squid giantaxon. *Biochim. Biophys. Acta* **630:** 131–136; 1980.

Chang, D.C., and Woessner, D.E. Spin–echo study of $^{23}$Na relaxation in skeletal muscle. Evidence of sodium ion binding inside a biological cell. *J. Mag. Res.* **30:** 185–191; 1978.

Chang, D.C., Hazlewood, C.F., and Woessner, D.E. The spin–lattice relaxation times of water associated with early post mortem changes in skeletal muscle. *Biochim. Biophys. Acta* **437:** 253–258; 1976.

Chang, D.C., Hazlewood, C.F., Nichols, B.L., and Rorschach, H.D. Implications of diffusion coefficient measurements for the structure of cellular water. *Ann. NY Acad. Sci.* **204:** 434–443; 1973.

Chang, D.C., Hazlewood, C.F., Nichols, B.L., and Rorschach, H.E. Spin echo studies on cellular water. *Nature,* (London) **235:** 170–171; 1972.

Chapman, G., and McLauchlan, K.A. Oriented water in the sciatic nerve of rabbit. *Nature,* (London) **215:** 391–392; 1967.

Chaughule, R.S., Kasturi, S.R., and Ranade, S.S. Proton spin–lattice relaxation times and water content in human tissues. *Proc. Nucl. Phys. Solid State Phys. Symp.* **17C**; 1974a.

Chaughule, R.S., Kasturi, S.R., Vijayaraghavan, R., and Ranade, S.S. Normal and malignant tissues. An investigation by pulsed nuclear magnetic resonance. *Indian J. Biochem. Biophys.* **11:** 256–258; 1974b.

Chilton, H.M., Ekstrand, K.E., Dixon, R.L., Karstaedt, N., and Jackels, S. Contrast enhancement of $T_1$ signal in rat liver tissues using manganese sulfide colloid. *Mag. Res. Imaging* **1**(4): 238; 1982.

Civan, M.M., Achlama, A.A., and Shporer, M. The relationship between the transverse and longitudinal nuclear magnetic resonance relaxation rates of muscle water. *Biophys. J.* **21:** 127–146; 1978.

Civan, M.M., McDonald, G.G., Pring, M., and Shporer, M. Pulsed nuclear magnetic resonance study of $^{39}$K in frog striated muscle. *Biophys. J.* **16:** 1385–1398; 1976.

Civan, M.M., and Shporer, M. Pulsed magnetic resonance study of $^{17}$O, $^{2}$D and $^{1}$H of water in frog striated muscle. *Biophys. J.* **15:** 299–306; 1975.

Civan, M.M., and Shporer, M. Pulsed NMR studies of $^{17}$O from $H_2{}^{17}O$ in frog striated muscle. *Biochim. Biophys. Acta* **343:** 399–408; 1974.

Clark, M.E., Burnell, E.E., Chapman, N.R., and Hinke, J.A.M. Water in barnacle muscle IV. Factors contributing to reduced self-diffusion. *Biophys. J.* **39:** 289–299; 1982.

Cleveland, G.G., Chang, D.C., Hazlewood, C.F., and Rorschach, H.D. Nuclear magnetic resonance measurement of skeletal muscle. Anisotropy of the diffusion coefficient of the intracellular water. *Biophys. J.* **16:** 1043–1053; 1976.

Cohen, S.M., Ogawa, S., Rottenberg, H., Glynn, P., Yamane, T., Brown, T.R., Shulman, R.G., and Williamson, J.R. $^{31}$P nuclear magnetic resonance studies of isolated rat liver cells. *Nature* **273:** 554–558; 1978.

Cohn, M., and Hughes, Jr., T.R. Nuclear magnetic resonance spectra of adenosine di- and triphosphate. II. Effects of complexing with divalent metal ions. *J. Biol. Chem.* **237:** 176–181; 1962.

Cohn, M., and Rao, B.D. $^{31}$P NMR of enzymatic reactions. *Bull. Magn. Resonance* **1**(No.1): 38–60; 1979.

Coles, B.A. Dual frequency proton spin relaxation measurements on tissues from normal and tumor bearing mice. *JNCI* **57:**(3): 389–393; 1976.

Colman, A., and Gadian, D.G. $^{31}$P nuclear magnetic resonance studies on the developing embryos of Xenopus laevis. *J. Biochem.* **61:** 387–396; 1976.

Conlon, T., and Outhred, R. Water diffusion permeability of erythrocytes using an NMR technique. *Biochim. Biophys. Acta* **288:** 354–361; 1978.

Cooke, R., and Wien, R. Nuclear magnetic resonance studies of intracellular water protons. *Annals N.Y.A.S.* **204:** 197–209; 1973.

Cooke, R., and Wien, R. The state of water in muscle tissue as determined by proton nuclear magnetic resonance. *Biophys. J.* **11:** 1002–1006; 1971.

Cooper, R.L., Chang, D.B., Young, A.C., Martin, C.J., Johnson, B.A. Restricted diffusion in biophysical systems: Experiment. *Biophys. J.* **14:** 161–177; 1974.

Cope, F.W. Complexing of sodium ions in myelinated nerve by nuclear magnetic resonance. *Physiol. Chem. Phys.* **2:** 545–550; 1970a.

Cope, F.W. Spin echo NMR evidence for complexing of $^{23}$Na in muscle, brain and kidney. *Biophys. J.* **10:** 843–858; 1970b.

Cope, F.W. Nuclear magnetic resonance evidence using $D_2O$ for structured water in muscle and brains. *Biophys. J.* **9:** 303–319; 1969.

Cope, F.W. NMR evidence for complexing of $^{23}$Na in muscle, kidney, brain, and by actomyosin. *J. Gen. Physiol.* **50:** 1353–1375; 1967.

Cope, F.W. Nuclear magnetic resonance evidence for the complexing of sodium ions in muscle. *PNAS USA* **54:** 225–227; 1965.

Cope, F.W., and Damadian, R. Pulsed nuclear magnetic resonance of potassium ($^{39}$K) of whole body live and dead newborn mice. Double oscillation frequencies in $T_1$ decay curves. *Physiol. Chem. Phys.* **11:** 143–149; 1979.

Cope, F.W., and Damadian, R. Nuclear magnetic resonance chemical shifts of potassium ions ($^{39}$K) on ion exchange resins and in muscle. *Physiol. Chem. Phys.* **9:** 461–471; 1977.

Cope, F.W., and Damadian, R. Biological ion exchanger resins. IV. Evidence for potassium association with fixed charges in muscle and brain by pulse NMR of $^{39}$K. *Physiol. Chem. Phys.* **6:** 17–25; 1974.

Cope, F.W., and Damadian, R. Cell potassium by $^{39}$K spin echo nuclear magnetic resonance. *Nature* **228:** 76–77; 1970.

Costello, A.J.R., Glonek, T., and Van-Wazer, J.R. Phosphorus-31 chemical shift variations with countereaction and ionic strength for the various ethyl phosphates. *Inorg. Chem.* **15:** 972–974; 1976.

Costello, A.J.R., Marshall, W.E., Omachi, A., and Henderson, T.O. Interactions between hemoglobin and organic phosphates investigated with phosphorus-31 nuclear magnetic resonance spectroscopy and ultrafiltration. *Biochim. Biophys. Acta* **427:** 481–491; 1976.

Cottam G., Vasek, A., and Lusted, D. Water proton relaxation rates in various tissues. *Res. Comm. Chem. Pathol. Pharmacol.* **4:** 495–502; 1972.

Cotton, G.L., Valentine, K.M., Yamaoka, K., and Waterman, M.R. The gelation of deoxyhemoglobin S in erythrocytes as detected by transverse water proton relaxation measurements. *Arch. Biochem. Biophys.* **162:** 487–492; 1974.

Cozzone, P.J., Nelson, D.J., and Jardetzky, O. Fourier transform phosphorous magnetic resonance study of ATP-calcium-G-actin complex. *Biochem. Biophys. Res. Comm.* **60:** 341; 1974.

Cresshull, I., Dawson, M.J., and Edwards, R.H.T. Human muscle analysed by $^{31}$P nuclear magnetic resonance in intact subjects. *J. Physiol.* **317:** 13P; 1981.

Crooks, L., Arakawa, M., Hoenninger, J., Watts, J., McRee, R., Kaufman, L., Davis, P.L., Margulis, A.R., and DeGroot, J. Nuclear magnetic resonance whole body images operating at 3.5 KGauss. *Radiology* **143:** 169–174; 1982.

Cyr, T.J., Derbyshire, W., Parsons, J.L., Blanshard, J.M.V., Lawrie, R.A. NMR investigations of water structure in frozen tropomyosin samples. *Trans. Faraday Soc.* **67:** 1887–1893; 1971.

Czeisler, J.L., Fritz, O.G., and Swift, T.J. Direct evidence from NMR studies for bound sodium in frog skeletal muscle. *Biophys. J.* **10:** 260–268; 1970.

Damadian, R., Goldsmith, M., and Minkoff, L. NMR in cancer. XVI. FONAR image of the live human body. *Physiol. Chem. Phys.* **9:** 97–100; 1977.

Damadian, R., Minkoff, L., Goldsmith, M., Stanford, M., and Koutcher, J. Field focusing nuclear magnetic resonance (FONAR): visualization of a tumor in a live animal. *Science, N.Y.* **194:** 1430–1432; 1976.

Damadian, R., and Cope, F.W. NMR in cancer V. Electronic diagnosis of cancer by potassium ($^{39}$K) nuclear magnetic resonance: Spin signatures and $T_1$ beat patterns. *Physiol. Chem. Phys.* **6:** 309–322; 1974.

Damadian, R., Zaner, K., and Hor, D. Human tumors detected by nuclear magnetic resonance. *PNAS USA* **71:** 1471–1473; 1974.

Damadian, R., and Cope, F. Potassium NMR relaxations in muscle, and brain and in normal E coli and a potassium transport mutant. *Physiol. Chem. Phys.* **5:** 511–514; 1973.

Damadian, R., Zaner, K., and Hor, D. Human tumors by NMR. *Physiol. Chem. Phys.* **5:** 381–402; 1973.

Damadian, R. Tumor detection by nuclear magnetic resonance. *Science* **171:** 1151–1153; 1971.

Davis, P.L., Kaufman, L., Crooks, L.E., and Miller, T.R. Detectability of hepatomas in rat livers by nuclear magnetic resonance imaging. *Invest. Radiol.* **16:** 354–359; 1981.

Dawson, M.J., Gadian, D.G., and Wilkie, D.R. Studies of the biochemistry of contracting and relaxing muscle by the use of $^{31}$P NMR in conjunction with other techniques. *Phil. Trans. R. Soc. Lond. B.* **289:** 445–455; 1980a.

Dawson, M.J., Gadian, D.G., and Wilkie, D.R. Mechanical relaxation rate and metabolism studied in fatiguing muscle by phosphorus nuclear magnetic resonance. *J. Physiol. Lond.* **299:** 465–484; 1980b.

Dawson, M.J., Gadian, D.G., Wilkie, D.R. Muscular fatigue investigated by phosphorus nuclear magnetic resonance. *Nature* **274:** 861–866; 1978.

Dawson, M.J., Gadian, D.G., and Wilkie, D.R. Contraction and recovery of living muscles studied by $^{31}$P nuclear magnetic resonance. *J. Physiol.* **267:** 703–735; 1977.

Dawson, M.J., Gadian, D.G., and Wilkie, D.R. Living muscle studied by $^{31}$P nuclear magnetic resonance. *J. Physiol.* **258:** 82–83P; 1976.

de Certaines, J., Herry, J.Y., Lancien, G., Benoist, L., Bernard, A.M., and Clech, G.L. Evaluation of human thyroid tumors by proton nuclear magnetic resonance. *J. Nucl. Med.* **23:** 48–51; 1982a.

de Certaines, J.D., Moulinoux, J.P., Benoist, L., Bernard, A.M., and Rivet, P. Proton nuclear magnetic resonance of regenerating rat liver after partial hepatectomy. *Life Sciences* **31:** 505–508; 1982b.

de Certaines, J., Bernard, A.M., Benoist, L., Rivet, P., Gallier, J., and Morin, P. Nuclear magnetic resonance study of cancer: Systemic effect on the proton relaxation times ($T_1$ and $T_2$) of human serum. *Can. Detect. Prevent.* **4:** 267–271; 1981.

de Certaines, J., Gallier, J., Lancien, G., and Bellossi, A. Nuclear magnetic resonance study of experimental Lewis lung carcinoma. *J. Clin. Hematol. Oncol.* **9**(3): 255–264; 1979.

Dehl, R.E. Collagen: Mobile water content of frozen fibers. *Science* **170:** 738; 1970.

De Kruijff, B. $^{13}C$ NMR studies on $^{13}C$ cholesterol incorporated in sonicated phosphatidylcholine vesicles. *Biochim. Biophys. Acta* **506:** 173–182; 1976.

De Layre, J.L., Ingwall, J.S., Malloy, C., and Fossel, E. Gated sodium 23 nuclear magnetic resonance images of an isolated perfused working rat heart. *Science* **212:** 935–936; 1981.

Derbyshire, W., and Parsons, J.L. NMR investigations of frozen muscle system. *J. Mag. Res.* **6:** 344–351; 1972.

Deslauriers, R., Ekiel, I., Kroft, T., Leveille, L., and Smith, I. NMR studies of malaria. *Tetrahedron* **38:** (in press); 1983.

Deslauriers, R., Ekiel, I., Byrd, R.A., Jarrell, H.C., and Smith, I.C.P. A $^{31}P$-NMR study of structural and functional aspects of phosphate and phosphonate distribution in Tetrahymena. *Biochimica et Biophysica Acta.* **720:** 329–337; 1982a.

Deslauriers, R., Ekiel, I., Kroft, T., and Smith, I.C.P. $^{31}P$ nuclear magnetic resonance of blood from mice infected with Plasmodium Berghei. *Biochimica et Biophysica Acta.* **721:** 449–457; 1982b.

Diegel, J.G., and Pintar, M.M. A possible improvement in the resolution of proton spin relaxation for the study of cancer at low frequency. *J. Natl. Cancer Inst.* **55:** 725–726; 1975a.

Diegel, J.G., and Pintar, M.M. Origin of the non exponentiality of the water proton spin relaxation in tissues. *Biophys. J.* **15:** 855–860; 1975b.

Doyle, D., and Barany, M. Quantitation of lactic acid in caffeine-contracted and resting frog muscle by high resolution natural abundance $^{31}C$ NMR. *FEBS Letts.* **140**(2): 237–240; 1982.

Doyle, D.D., Chalovich, J.M., and Barany, M. Natural abundance $^{13}C$ spectra of intact muscle. *FEBS Letts.* **131:**(1): 147–150; 1981.

Doyle, F.H., Pennock, J.M., Banks, L.M., McDonnell, M.J., Bydder, G.M., Steiner, R.E., Young, I.R., Clarke, G.J., Pasmore, T., and Gilderdale, D.J. Nuclear magnetic resonance imaging of the liver: Initial experience. *Amer. J. Radiol.* **138:** 193–196; 1982.

Drost-Hansen, W. Phase transition in biological systems: Manifestation of cooperative processes in vicinal water. *Ann. N.Y. Acad. Sci.* **204:** 100; 1973.

Duff, I.D., and Derbyshire, W. NMR investigations of frozen porcine muscle. *J. Magn. Reson.* **15:** 310–316; 1974.

Economou, J.S., Parks, L.C., and Saryan, L.A. Detection of malignancy by nuclear magnetic resonance. *Surg. Forum* **24:** 127–129; 1973.

Edwards, R.H.T. Physiological analysis of skeletal muscle weakness and fatigue. *Clin. Sci. Mol. Med.* **54:** 463–470; 1978.

Edwards, R.H.T., Dawson, M.J., Wilkie, D.R., Gordon, R.E., and Shaw, D. Clinical use of nuclear magnetic resonance in the investigation of myopathy. *The Lancet,* March 27: 725–731; 1982.

Edwards, R.H.T., and Wiles, C.M. Energy exchange in human skeletal muscle during isometric contraction. *Circulation Res.* **48**(suppl): 11–17; 1981.

Edzes, H.T., Ginzburg, M., Ginsburg, B.Z., and Berendsen, H.J.C. The physical state of alkali ions in a Halobacterium: Some NMR results. *Experentia* **33**(6): 732–734; 1977.

Eggleston, J.C., Saryan, L.A., and Hollis, D.P. Nuclear magnetic resonance investigations of human neoplastic and abnormal non neoplastic tissues. *Cancer Res.* **35:** 1326–1332; 1975.

Eisenstadt, M. NMR relaxation of protein and water protons in methemoglobin solutions. *Biophys. J.* **33:** 469–474; 1981.

Eisenstadt, M., and Fabry, M.E. NMR relaxation of the hemoglobin water proton spin system in red blood cells. *J. Magn. Reson.* **29:** 591–597; 1978.

Ekstrand, K.E., Dixon, R.L., Raben, M., and Ferree, C.F. Proton NMR relaxation in the peripheral blood of cancer patients. *Phys. Med. Biol.* **22:** 925–931; 1977.

Ernst, E., and Hazlewood, C.F. Inorganic constituents acting in bioprocessor. Part I: Water, Inorganic Perspectives in Biology and Medicine. **2:** 27–49; 1978.

Escanye, J.M., Canet, D., and Robert, J. Frequency dependence of water proton longitudinal NMR relaxation times in mouse tissues around the freezing transition. *Biochimica et Biophysica Acta* **762:** 445–451; 1983.

Evans, F.E., and Kaplan, N.O. Phosphorus-31 nuclear magnetic resonance studies of HeLa cells. *Proc. Natl. Acad. Sci. USA* **74:** 4909–4913; 1977.

Fabry, M.E., and Eisenstadt, M. Water exchange between red cells and plasma. Measurements by nuclear magnetic relaxation. *Biophys. J.* **15:** 1101–1110; 1975.

Farrar, T.C., and Becker, E.D. *Pulse and Fourier Transform NMR.* New York: Academic Press; 1971.

Fedotov, V.D., Miftakhutdinova, F.G., and Murtazin, S.F. Investigation of proton relaxation in live plant tissues by the spin echo method. *Biofizika* **14:** 873–882; 1969.

Finch, E., and Homer, L. Proton nuclear magnetic resonance relaxation measurements in frog muscle. *Biophys. J.* **14:** 907–913; 1974.

Finch, E.D., Harmon, J.F., and Muller, B.H. Pulsed NMR measurements of the diffusion constant of water in muscle. *Arch. Biochem. Biophys.* **147:** 229–310; 1971.

Floyd, R.A., Yoshida, T., and Leigh, J.S. Changes in tissue water proton relaxation rates during early phases of chemical carcinogenesis. *Proc. Natl. Acad. Sci. USA* **72:** 56–68; 1975.

Floyd, R.A., Leigh, J.S., Chance, B., and Miko, M. Time course of tissue water proton spinlattice relaxation in mice developing ascites tumor. *Cancer Res.* **34:** 89–91; 1974.

Forsslund, G., Odeblad, E., and Bergstrand, A. Proton magnetic resonance of human gingival tissue. *Acta Odont. Scand.* **20:** 121–124; 1962.

Fossel, E.T., Morgan, H.E., and Ingwall, J.S. Measurement

of changes in high-energy phosphates in the cardiac cycle using gated $^{31}P$ nuclear magnetic resonance. *Proc. Natl. Acad. Sci. USA* **77:** 3654–3658; 1980.

Fossel, E.T., and Ingwall, I.S. $^{31}P$ NMR on gated heart preparatoins. *Biophys. J.* **21:** 34a; 1978.

Fossel, E.T., and Solomon, A.K. Modulation of 2,3 diphosphoglycerate phosphorus-31-NMR resonance positions by red cell membrane shape. *Biochim. Biophys. Acta* **436:** 505; 1976.

Foster, K.F., Resing, H.A., and Garroway, A.N. Bounds on "bound water": transverse nuclear magnetic resonance relaxation in barnacle muscle. *Science, N.Y.* **194:** 324–326; 1976.

Frey, H.W., Knispel, R.R., Kruuv, J., Sharp, A.R., Thompson, R.T., and Pintar, M.M. Proton spin-lattice relaxation studies of non-malignant tissues of tumorous mice. *J. Natl. Cancer Inst.* **49:** 903–906; 1972.

Fritz, O.G., Jr., and Swift, T.J. The state of water in polarized and depolarized frog nerves. A proton magnetic resonance study. *Biophys. J.* **7:** 675–686; 1967.

Fruchter, R.G., Goldsmith, M., Boyce, I.G., Nicastri, A.D., Koutcher, J., and Damadian, R. Nuclear magnetic resonance properties of gynecological tissues. *Gynecol. Oncol.* **6:** 243–255; 1978.

Fullerton, G.D., Potter, J.L., and Dornbluth, N.C. NMR relaxation of protons in tissues and other macromolecular water solutions. *Mag. Res. Imaging* **1**(4): 209–226; 1982.

Fung, B.M., and McGaughy, T.W. Study of spin-lattice and spin-spin relaxation times of $^{1}H$, $^{2}H$ and $^{17}O$ in muscle water. *Biophys. J.* **28:** 293–304; 1979.

Fung, B.M. Proton and deuterium relaxation of muscle water over wide ranges of resonance frequencies. *Biophys. J.* **18:** 235–239; 1977a.

Fung, B.M. Carbon-13 and proton magnetic resonance of mouse muscle. *Biophys. J.* **19:** 315–319, 1977b.

Fung, B.M. Correlation of relaxation time with water content in muscle and brain tissues. *Biochim. Biophys. Acta* **497:** 317–322; 1977c.

Fung, B.M., Durham, D.L., and Wassil, D.A. The state of water in biological systems as studied by proton and deuterium relaxation. *Biochim. Biophys. Acta* **399:** 191–202; 1975a.

Fung, B.M., Wassil, D.A., Durham, D.L., Chesnut, R.W., Durham, N.N., and Berlin, K.D. Water in normal muscle and muscle with tumor. *Biochim. Biophys. Acta* **385:** 180–187; 1975b.

Fung, B.M. Non-freezable water and spin-lattice relaxation time in muscle containing a growing tumor. *Biochim. Biophys. Acta* **261:** 209–214; 1974.

Fung, B.M., and McGaughy, T.W. The state of water in muscle as studied by pulsed NMR. *Biochim. Biophys. Acta* **343:** 663–673; 1974.

Furuse, M., Saso, K., Motegi, Y., Inao, S., and Izawa, A. Studies on cerebrovascular diseases by means of nuclear magnetic resonance. *Proc. SMRM,* San Francisco; p. 132; 1983.

Furuse, M., Saso, K., Moteki, Y., Inao, S., Asai, H., Izawa, S., and Kasai, A. Proton density images and in vivo measurements of relaxation time in the human brains. Comparison of $T_1$ values in healthy volunteer and cerebrovascular disease. *Mag. Res. Imaging* **1**(4): 230–231; 1982.

Gadian, D.G. *Nuclear Magnetic Resonance and Its Application To Living Systems.* Oxford: Clarendon Press, 1982.

Gadian, D.G., and Radda, G.K. NMR studies of tissue metabolism. *Annu. Rev. Biochem.* **50:** 69–83; 1981.

Gadian, D.G., Radda, G.R., Brown, T.R., Chance, E.M., Dawson, M.J., Wilkie, D.R. The activity of creatine kinase in frog skeletal muscle studied by saturation transfer nuclear magnetic resonance. *Biochem. J.* **194:** 215–228; 1981a.

Gadian, D.G., Radda, G.K., Ross, B.D., Hockaday, J., Bore, P.J., Taylor, D., and Styles, P. Examination of a myopathy by phosphorus nuclear magnetic resonance. *Lancet* **II:** 774–775: 1981b.

Gadian, D.G., Radda, G.K., Richards, R.E., and Seeley, P.J. $^{31}P$ NMR in living tissue: The road from a promising to an important tool in biology. In: *Biological Application of Magnetic Resonance,* R.G. Shulman, ed. New York: Academic Press; pp. 463–470; 1979.

Gaggelli, E., Niccolai, N., and Valensin, G. $^{1}H$-NMR relaxation investigation of water bound to bovine rod outer segment disk membranes. *Biophys. J.* **37:** 559–561; 1982.

Gamsu, G., Webb, W.R., Sheldon, P., Kaufman, L., Crooks, L.E., Birnberg, F.A., Goodman, P., Hinchcliffe, W.A., and Hedgecock, M. Nuclear magnetic resonance imaging of the thorax. *Radiology* **147:** 473–480; 1983.

Gillies, R.J., and Benoit, A.G. NMR analysis of a cell division cycle mutant of *Saccharomyces cerevisiae. Biochimica et Biophysica Acta* **762:** 466–470; 1983.

Glasel, J.A. A study of water in biological systems by $^{17}O$ magnetic resonance spectroscopy. II. Relaxation phenomena in pure water. *PNAS, USA* **58:** 27–39; 1967.

Glonek, T., Burt, C.T., and Barany, M. NMR analysis of intact-tissue including several examples of normal and diseased human muscle determinations. In: *NMR in Medicine,* R. Damadian, ed. Berlin: Springer Verlag; pp. 121–134; 1981.

Glonek, T., Henderson, T.O., Hildebrand, R.L., and Myers, T.C. Biological phosphonates: Determination by phosphorus 31 nuclear magnetic resonance. *Science* **169:** 192–194; 1970.

Go, K.G., Kamman, R.L., Muskiet, F.A.J., Berendsen, H.J.C., and van Dijk, P. Studies on relaxation times of fatty tissue and cerebral white matter. *Proc. SMRM,* San Francisco, p. 140; 1983.

Goldsmith, M., Koutcher, J., and Damadian, R. NMR in cancer XI. Application of the NMR malignancy index to human gastro-intestinal tumors. *Cancer* **41:** 183–191; 1978.

Goldsmith, M., and Damadian, R. NMR in cancer VII. Sodium-23 magnetic resonance of normal and cancerous tissues. *Physiol. Chem. Phys.* **7:** 263; 1975.

Gonzalez-Mendez, R., Wemmer, D., Hahn, G., Wade-Jardetzky, N., and Jardetzky, O. Continuous-flow NMR culture system for mammalian cells. *Biochimica et Biophysica Acta* **720:** 274–280; 1982.

Gordon, R.E., Hancley, P.E., Shaw, D., Gadian, D.G., Radda, G.K., Styles, P., Bore, P.J., and Chan, L. Localization of metabolites in animals using $^{31}P$ topical magnetic resonance. *Nature* **287:** 736–738; 1980.

Gordon, R.E., Mallard, J.R., and Philip, J.F. *The Nuclear Resonance Effect in Cancer:* Chapter 16. R. Damadian, ed. Portland, Oregon: Pacific Publishing Co.; 1978.

Gore, J.C., Emery, E.W., Orr, J.S., and Doyle, F.H. Medical nuclear magnetic resonance imaging. I. Physical principles. *Investigative Radiology* **16:** 269–274; 1981.

Grove, T.H., Ackerman, J.J., Radda, G.K., and Bore, P.J. Analysis of rat heart *in vivo* by phosphorus nuclear magnetic resonance. *Proc. Natl. Acad. Sci. USA* **77**(1): 299–302; 1980.

Gunther, H. *NMR Spectroscopy: An Introduction.* New York: John Wiley & Sons, Ltd.; 1980.

Gupta, R.K., and Moore, R.D. $^{31}$P NMR studies of intracellular free $Mg^{+2}$ in intact frog skeletal muscle. *J. Biol. Chem.* **255:** 3987–3993; 1978.

Gwan Go, K., and Edzes, H.T. Water in brain edema. Observations by the pulsed nuclear magnetic resonance technique. *Arch. Neurol.* **32**(7): 462–465; 1975.

Hansen, J.R. Pulsed NMR study of water motility in muscle and brain tissue. *Biochim. Biophys. Acta* **230:** 482–486; 1971.

Haran, N., Malik, Z., and Lapidot, A. Water permeability changes studied by $^{17}$O nuclear magnetic resonance during differentiation of Friend leukemia cells. *Proc. Natl. Acad. Sci. USA* **76:** 3363; 1979.

Hardter, U. *Spin-Gitter Relaxationszeit: Messungen an Zellen und Geweben.* Dissertation, Eberhard-Karls-Universität, Tübingen, 1976.

Hazlewood, C.F., Kasturi, S.R., Dennis, L.W., and Yamanashi, W.S. Observations of chemical shift for intracellular water nuclei in cysts of the brine shrimp (Artemia). *Proc. SMRM,* San Francisco; p. 150; 1983.

Hazlewood, C.F., Yamanashi, W.S., Rangel, R.A.; and Todd, L.E. *In vivo* NMR imaging and $T_1$ measurements of water protons in the human brain. *Mag. Res. Imaging* **1:** 3–10; 1982.

Hazlewood, C.F., Beall, P.T., and Chang, D.C. (unpublished results); 1976.

Hazlewood, C.F., Cleveland, G., and Medina, D. Relationship between hydration and proton nuclear magnetic resonance relaxation times in tissues of tumor bearing and nontumor bearing mice: Implications for cancer detection. *J. Natl. Cancer Inst.* **52:** 1849–1853; 1974a.

Hazlewood, C.F., Chang, D.C., Nichols, B.L., and Woessner, D.E. Nuclear magnetic resonance transverse relaxation times of water protons in skeletal muscle. *Biophys. J.* **14:** 583–606; 1974b.

Hazlewood, C.F., Chang, D.C., Medina, D., Cleveland, G., and Nichols, B.L. Distinction between the preneoplastic and neoplastic state of murine mammary glands. *Proc. Natl. Acad. Sci. USA* **69:** 1478–1480; 1972.

Hazlewood, C.F., Nichols, B.L., Chang, D.C., and Brown, B. On the state of water in developing muscle: A study of the major phase of ordered water in skeletal muscle and its relationship to sodium concentration. *Johns Hopkins Med. J.* **128**(3): 117–131; 1971.

Hazlewood, C.F., Nichols, B.L., and Chamberlain, N.F. Evidence for the existence of a minimum of two phases of ordered water in skeletal muscle. *Nature* (London) **222:** 747–750; 1969.

Held, C., Noack, F., Pollak, V., and Melton, B. Protonenspinrelaxation und Wasserbeweglichkeit in Muskelgewebe. *Zeitschrifte für Naturforschung* **28c:** 59–63; 1973.

Heller, M., Moon, K.L., Helms, C.A., Schild, H., Chafetz, N.I., Rodrigo, J., Jergesen, H.E., and Genant, H.K. NMR imaging of femoral head necrosis. *Proc. SMRM,* San Francisco; p. 152; 1983.

Henderson, T.O., Costello, A.J.R., and Omachi, A. Phosphate metabolism in intact human erythrocytes. *Proc. Natl. Acad. Sci. USA* **71:** 2487–2490; 1974.

Herfkens, R.J., Higgins, C.B., Hricak, H., Lipton, M.J., Crooks, L.E., Sheldon, P.E., and Kaufman, L. Nuclear magnetic resonance imaging of atherosclerotic disease. *Radiology* **148:** 161–166; 1983a.

Herfkens, R.J., Sievers, R., Kaufman, L., Sheldon, P.E., Ortendahl, D.A., Lipton, M.J., Crooks, L.E., and Higgins, C.B. Nuclear magnetic resonance imaging of the infarcted muscle: A rat model. *Radiology* **147:** 761–764; 1983b.

Higgins, C.B., Herfkens, R., Lipton, M.J., Sievers, R., et al. NMR imaging of acute myocardial infarction. *Am. J. Cardiol.* **52:** 184–188; 1983.

Hindman, J.C., Svirmickas, A., and Wood, M. Relaxation processes in water. The spin–lattice relaxation of deuteron in $D_2O$ and oxygen $^1$17 in $H_2$ $^{17}$O. *J. Chem. Phys.* **54:** 621–634, 1971.

Hinshaw, W.S., Bottomley, P.A., and Holland, G.N. Radiographic thin-section image of the human wrist by nuclear magnetic resonance. *Nature* (London) **270:** 722–723; 1977.

Holland, G.N., Bottomley, P.A., and Hinshaw, W.S. $^{19}$F magnetic resonance imaging. *J. Mag. Res.* **28:** 133–137; 1978.

Hollis, D.P. Phosphorus NMR of cells, tissues and organelles. *Biol. Magn. Reson.* **2:** 1–44; 1980.

Hollis, D.P., Saryan, L.A., Eggleston, D., and Morris, H.P. Nuclear magnetic resonance studies of cancer. VI. Relationship among spin–lattice relaxation times, growth rate and water content of Morris hepatomas. *J. Natl. Cancer Inst.* **54:** 1469–1472; 1975.

Hollis, D.P., Saryan, L.A., and Economou, J.S. Nuclear magnetic resonance studies of cancer. V. Appearance and development of a tumor systemic effect in serum and tissues. *J. Natl. Cancer Inst.* **53:** 807–815; 1974.

Hollis, D.P., Economou, J.S., Parks, L.C., Eggleston, J.C., Saryan, L.A., and Czeisler, J.L. Nuclear magnetic resonance studies of several experimental and human malignant tumors. *Cancer Res.* **33:** 2156–2160; 1973.

Hopkins, A.L., Kirschner, D.A., and Bratton, C.B. Proton relaxation time measurements for bone tissue. *Proc. SMRM,* San Francisco; p. 161; 1983.

Hoult, D.I., Busby, S.J.W., Gadian, D.G., Radda, G.K., Richards, R.E., and Seeley, P.J. Observation of tissue metabolites using $^{31}$P nuclear magnetic resonance. *Nature* (London) **252:** 285–287; 1974.

House, C.R. *Water Transport in Cells and Tissues.* London: Edward Arnold Publishers; 1974.

Hricak, H., Filly, R.A., Margulis, A.R., Moon, K.L., Crooks, L.E., and Kaufman, L. Work in progress: Nuclear magnetic resonance imaging of the gallbladder. *Radiology* **147:** 481–484; 1983a.

Hricak, H., Williams, R.D., Moon, K.L., Jr., Moss, A.A., Alpers, C., Crooks, L.E., and Kaufman, L. Nuclear magnetic resonance imaging of the kidney: Renal masses. *Radiology* **147:** 765–772; 1983b.

Hricak, H., Crooks, L., Sheldon, P., and Kaufman, L. Nuclear magnetic resonance imaging of the kidney. *Radiology* **146:** 425–432; 1983c.

Huestis, W.H., and Raftery, M.A. Conformation and cooperativity in hemoglobin. *Biochemistry* **14:** 1886–92; 1975.

Huggert, A., and Odeblad, E. Proton magnetic resonance studies of some tissues and fluids of the eye. *Acta Radiologica* **51:** 385–392; 1959.

Iijima, N., Saitoo, S., Yoshida, Y., Fujji, N., Koike, T., Osanzi, K., and Hirose, K. Spin-echo nuclear magnetic resonance in cancerous tissue. *Physiol. Chem. Phys.* **5:** 431–435; 1973.

Inch, W.R., McCredie, J.A., Geiger, C., and Boctor, Y. Spin–lattice relaxation times for mixtures of water and gelatin or cotton compared with normal and malignant tissues. *J. Natl. Cancer Inst.* **53:** 689; 1974a.

Inch, W.R., McCredie, J.A., Knispel, R.R., Thompson, R.T., and Pintar, M.M. Water content and proton spin relaxation times for malignant and non-malignant tissues from mice and humans. *J. Natl. Cancer Inst.* **52:** 353–356; 1974b.

Ingwall, J.S. Phosphorus nuclear magnetic resonance spectroscopy of cardiac and skeletal muscles. *Am. J. Physiol.* **11:** H729–H744; 1982.

Jacobs, D.O., Albina, J.E., Settle, R.G., Wolf, G., and Rombeau, J.L. Nuclear magnetic resonance (NMR) and ethionine induced fatty liver infiltration. *Proc. SMRM,* San Francisco; p. 185; 1983.

Jacobus, W.E., Taylor, G.J.; Hollis, D.P., and Nunnally, R.L. Phosphorus nuclear magnetic resonance of perfused working rat hearts. *Nature* **265:** 756–768; 1977.

James, T., and Gillen, K.T. NMR relaxation time and self-diffusion constant of water in hen egg white and yolk. *Biochim. Biophys. Acta* **286:** 10–15; 1972.

Jardetzky, O., and Wertz, J.E. Detection of sodium complexes by nuclear spin resonance. *Am. J. Physiol.* **187:** 608–619; 1956.

Kainosho, M., Ajisaka, K., and Nakazawa, H., In situ analysis of the microbial fermentation process by natural abundance carbon 13 and phosphorus-31 NMR spectroscopy. *FEBS Lett.* **80:** 375–378; 1977.

Kamath, V., Rajagopalan, B., and Tio, F. Proton magnetic relaxation study of pleural fluids in vitro. *Proc. SMRM,* San Francisco; p. 189; 1983.

Kasturi, S.R., Seitz, P. Private communication (1982).

Kasturi, S.R., Chang, D.C., and Hazlewood, C.F. Study of anisotropy in nuclear magnetic resonance relaxation times of water protons in skeletal muscle. *Biophys. J.* **30:** 369–382; 1980.

Kasturi, S.R., Ranade, S.S., and Shah, S. Tissue hydration of malignant and uninvolved human tissues and its relevance to proton spin–lattice relaxation mechanisms. *Proc. Indian Acad. Sci.* Sect. B, **84b:** 60–74; 1976.

Kiricuta, I.C., and Simplacenu, V. Tissue water content and nuclear magnetic resonance in normal and tumor tissues. *Cancer Res.* **35:** 1164–1167; 1975.

Knispel, R.R., Thompson, R.T., and Pintar, M.M. Dispersion of proton spin–lattice relaxation in tissues. *J. Magn. Reson.* **14:** 44–51; 1974.

Koenig, S.H., Hallenga, K., and Shporor, M. Protein–water interaction studied by solvent $^{1}H$, $^{2}H$, and $^{17}O$ magnetic resonance, *PNAS USA* **72:** 2667–2671; 1975.

Koga, S., Echigo, A., and Oki, T. Nuclear magnetic resonance spectra of water in partially dried yeast cells. *Appl. Microbiol.* **14**(3): 466–467; 1966.

Koivula, A., Suominen, K., Timonen, T., and Kivinitty, K. The spin–lattice relaxation time in the blood of healthy subjects and patients with malignant disease. *Phys. Med. Biol.* **27**(7): 937–947; 1982.

Kopp, S.J., Glonek, T., and Greiner, J.V. Interspecies variations in mammalian lens metabolites as detected by phosphorus-31 nuclear magnetic resonance. *Science.* **215:** 1622–1625; 1982.

Koutcher, J.A., Zaner, K.S., and Damadian, R. $^{31}P$ as a nuclear probe for the diagnosis and treatment of malignant tissue. In: *NMR in Medicine,* R. Damadian, ed. Berlin: Springer-Verlag; pp. 101–119; 1981.

Koutcher, J.A., Goldsmith, M., and Damadian, R. NMR in cancer X. A malignancy index to discriminate normal and cancerous tissue. *Cancer* **41:** 174–182; 1978.

Koutcher, J.A., and Damadian, R. Spectral differences in the $^{31}P$ NMR of normal and malignant tissue. *Physiol. Chem. Phys.* **9:** 181–187; 1977.

Kushmerick, M.J., Meyer, R.A., and Brown, T.R. Phosphorus NMR of cat biceps and soleus muscles. In: *Proceedings of Fifth International Symposium of International Society for Oxygen Transport to Tissues,* H. I. Bicher, ed., 1983.

Labotka, R.J., Glonek, T., Hruby, M.A., and Honig, G.R. Phosphorus-31 spectroscopic determinations of the phosphorus metabolite profiles of blood components: Erythrocytes, reticulocytes, and platelets. *Biochem. Med.* **15:** 311–329; 1976a.

Labotka, R.J., Glonek, T., and Myers, T.C. Phosphorus-31 nuclear magnetic resonance studies on nucleoside phosphates in non-aqueous media. *J. Am. Chem. Soc.* **98:** 3699–3704; 1976b.

Lauterbur, P.C. Spatially-resolved studies of whole tissue, organs and organisms by NMR zeugmatography. In: *NMR in Biology,* R.A. Dwek, I.D. Campbell, R.E. Richards, R.J.P. Williams, eds. London: Academic Press; pp. 323–335; 1977.

Lauterbur, P.C. Image formation by induced local interactions: Examples employing NMR. *Nature* (London) **242:** 190–191, 1973.

Lee, S., Veis, A., and Glonek, T. Dentin phosphoprotein: an extracellular calcium-binding protein. *Biochemistry* **16:** 2971–2979; 1977.

Lewa, C.J., and Baczkowski, A. NMR relaxation times $T_1$ and $T_2$ versus the water content of normal and tumor animal tissues. *Acta Phys. Pol.* **AS0:** 865–870; 1976.

Lewa, C.J., and Zbytnlewski, Z. Magnetic transverse relaxation time of the protons in transplantable melanotic and amelanotic melanoma and in some inner organs of golden hamster (*Mesocricetus auratus*). *Waterhouse Bulletin du Cancer (Paris)* **63**(1): 69–72; 1976.

Lin-Er Lin, Shporer, M., and Civan, MM. *Amer. J. Physiol.* **243:** C76–C80; 1982.

Lindstrom, T.R., Koenig, S.H., Boussios, T., and Bertles, J.F. Intermolecular interactions of oxygenated sickle hemoglobin molecules in cells and cell-free solutions. *Biophys. J.* **16:** 679–689; 1976.

Lindstrom, T.R., and Koenig, S.H. Magnetic-field dependent water proton spin–lattice relaxation rates of hemoglobin solutions and whole blood. *J. Magn. Reson.* **15:** 344–353; 1974.

Ling, C.R., and Foster, M.A. Changes in NMR relaxation time associated with local inflammatory response. *Phys. Med. Biol.* **27:** 853–860; 1982.

Ling, C.R., and Foster, M.A. Comparison of NMR water proton $T_1$ relaxation times of rabbit tissues at 24 MHz and 2.5 MHz. *Phys. Med. Biol.* **25:** 748–751; 1981.

Ling, G.N., and Tucker, M. Nuclear magnetic resonance and water contents in normal mouse and rat tissues and in cancer cells. *J. Natl. Can. Inst.* **64:** 1199–1207; 1980.

Ling, G.N., Miller, C., and Ochsenfeld, M.M. The physical state of solutes and water in living cells according to the association-induction hypothesis. *Ann. N.Y. Acad. Sci.* **204:** 6–24; 1973.

Ling, G.N. A new model for the living cell: a summary of the theory and recent experimental evidence in its support. *Int. Rev. Cytol.* **26:** 1–61; 1969.

Ling, G.N., and Cope, F.W. Potassium ion: Is the bulk of intracellular K absorbed? *Science* **163:** 1335–1336; 1969.

Ling, G.N., Ochsenfeld, M.M., and Karreman, G. Is the cell membrane a universal rate limiting barrier to the movements of water between the living cell and its surrounding medium. *J. Gen. Physiol.* **50:** 1807–1820; 1967.

London, D.A., David, P.L., Williams, R.D., Crooks, L.E., Sheldon, P.E., and Gooding, C.A. Nuclear magnetic resonance imaging of induced renal lesions. *Radiology* **148:** 167–172; 1983.

Magnuson, J.A., and Magnuson, N.S. NMR studies of sodium and potassium in various biological tissues. In: *Physico Chemical State of Ions and Water in Living Tissues and Model Systems.* C.F. Hazlewood, ed. *Annals of the New York Academy of Sciences.* Vol. 204; p. 297; 1973.

Magnuson, J.A., Shelton, D.S., and Magnuson, N.S. A nuclear magnetic resonance study of sodium ion interaction with erythrocyte membranes. *BBRC* **39:** 279–281; 1970.

Mak, A., Smillie, L.B., and Barany, M. Specific phosphorylation at serine 283 of $\alpha$ tropomyosin from frog skeletal and rabbit skeletal and cardiac muscle. *PNAS USA* **75:** 3588–3594; 1978.

Mallard, J.R. Nuclear imaging. In: *Electronic Imaging,* T.P. McLean and P. Schagen, eds. London: Academic Press; pp. 415–459; 1979.

Mansfield, P., and Morris, P.G. *NMR Imaging in Biomedicine.* New York: Academic Press; 1982.

Mansfield, P., Morris, P.G., Ordidge, R.J., Pykett, I.L., Bangert, V., and Coupland, R.E. Human whole body imaging and detection of breast tumors by NMR. *Phil. Trans. R. Soc. Lond. B.* **289:** 503–510; 1980.

Mansour, T.E., Morris, P.G., Feeney, J., and Roberts, G.C.K. A $^{31}$P-NMR study of the intact liver fluke Fasciola Hepatica. *Biochimica et Biophysica Acta* **721:** 336–340; 1982.

Martinez, D., Silvidi, A.A., and Stokes, R.M. Nuclear magnetic resonance studies of sodium ions in isolated frog muscle and liver. *Biophys. J.* **9**(10): 1256–1260; 1969.

Mathur-DeVre, R. The NMR studies of water in biological systems. *Prog. Biophys. Molec. Biol.* **35:** 103–134; 1979.

Matthews, P.M., Williams, S.R., Seymour, A.M., Schwartz, A., Dube, G., Gadian, D.G., and Radda, G.K. A $^{31}$P-NMR study of some metabolic and functional effects of the inotropic agents epinephrine and ouabain, and the ionophore RO2-2985 (X537A) in the isolated, perfused rat heart. *Biochimica et Biophysica Acta* **720:** 163–171; 1982.

Matthews, R.M., and James, T.L. Effect of dichloromethane on sickle cells. An in vitro water proton magnetic resonance relaxation study. *Biochim. Biophys. Res. Comm.* **89:** 879; 1978.

McGilvery, R.W., and Murray, T.W. Calculated equilibria of phosphocreatine and adenosine phosphates during utilization of high energy phosphate by muscle. *J. Biol. Chem.* **249:** 5845–5850; 1974.

McLachlan, L.A. Cancer induced decrease in human plasma proton NMR relaxation rate. *Phys. Med. Biol.* **25:** 309–315; 1980.

McLachlan, L.A., and Hamilton, W. The effect of cancer on the spin–lattice relaxation time of mouse blood and tissue. *Biochim. Biophys. Acta* **583:** 119–128; 1979.

McLaughlin, A.C., Takeda, H., and Chance, B. Rapid ATP assays in perfused mouse liver by 31P NMR. *Proc. Natl. Acad. Sci. USA* **76**(11): 5445–5449; 1979.

Medina, D., Hazlewood, C.F., Cleveland, G., Chang, D.C., Spjut, H.J., and Moyers, R. Nuclear magnetic resonance studies on human breast dysplasias and neoplasms. *J. Natl. Cancer Inst.* **54:** 813–818; 1975.

Meyer, R.A., Kushmerick, M.J., and Brown, T.R. Application of $^{31}$P-NMR spectroscopy to the study of striated muscle metabolism. *Am. J. Physiol.* **242:** C1–C11; 1982.

Meyer, R.A., and Kushmerick, M.J. Fast-twitch and slow-twitch mammalian muscles perfused *in vitro* (Abstract). *Federation Proc.* **40:** 615a; 1981.

Miceli, M.V., Myers, T.C., Burt, C.T., and Henderson, T.O. The occurrence of aminoethylphosphonic acid in the planorbid snail helisoma SP. *Fed. Proc.* **37:** 1718a; 1978.

Michael, L.A., Seitz, P., McMillin-Wood, J., Chang, D.C., Hazlewood, C.F., and Entman, M.L. Mitochondrial water in myocardial ischemia: Investigation with nuclear magnetic resonance. *Science* **208:** 1267–1269; 1980.

Mild, K.H., James, T.L., and Gillen, K.T. Nuclear magnetic resonance relaxation time and self diffusion coefficient measurements of water in frog ovarian eggs. *J. Cell Physiol.* **80:** 155–158; 1972.

Minkoff, L., and Goldsmith, M. Nuclear magnetic resonance as a new tool in cancer research: human tumors by NMR. *Ann. N.Y. Acad. Sci.* **222:** 1048–1062; 1973.

Misra, L.K. Enhanced $T_1$ differentiation between normal and dystrophic muscles. *Magnetic Resonance Imaging* **2:** 33–36; 1984.

Misra, L.K., Narayana, P.A., Beall, P.T., Amtey, S.R., and Hazlewood, C.F. Developmental changes in $T_1$ relaxation times of muscle water protons in muscular dystrophy. *Proc. SMRM,* San Francisco; p. 236; 1983a.

Misra, L.K., Kasturi, S.R., Kundu, S.K., Harati, Y., Hazlewood, C.F., Luthra, M.G., Yamanashi, W.S., Munjaal, R.P., and Amtey, S.R. Evaluation of muscle degeneration in inherited muscular dystrophy by nuclear magnetic resonance techniques. *Mag. Res. Imaging* **1:** 75–80; 1983b.

Modak, S.G., Chaudhary, C.A., Kasturi, S.R., Phadke, R.S., Shah, S., and Ranade, S.S. Factors influencing the water proton relaxation in nuclear fractions from tissues of normal and tumor bearing animals. *Physiol. Chem. Phys.* **14:** 41–45, 1982.

Monoi, H. Nuclear magnetic resonance of tissue $^{23}$Na. *Biophys. J.* **14:** 645–651; 1974.

Monoi, H., and Katsukura, Y. NMR of $^{23}$Na in suspension of pig erythrocyte ghosts. *Biophys. J.* **16:** 979–981; 1976.

Moon, K.L., Jr., Hricak, H., Crooks, L.E., Gooding, C.A., Moss, A.A., Engelstad, B.L., and Kaufman, L. Nuclear magnetic resonance imaging of the adrenal gland: A preliminary report. *Radiology* **147:** 155–160; 1983a.

Moon, K.L., Jr., Genant, H.K., Hlems, C.A., Chafetz, N.I., Crooks, L.E., and Kaufman, L. Musculoskeletal applications of nuclear magnetic resonance. *Radiology* **147:** 161–171; 1983b.

Moon, K.L., Jr., David, P.L., Kaufman, L., Crooks, L.E., Sheldon, P.E., Miller, T., Brito, A.C., and Watts, J.C. Nuclear magnetic resonance imaging of a fibrosarcoma tumor implanted in the rat. *Radiology* **148:** 177–181; 1983c.

Moon, K.L., Jr., Moseley, M., Young, G., and Hricak, H. NMR relaxation characteristics of fasting and non-fasting bile in canines. *Proc. SMRM,* San Francisco; p. 245; 1983d.

Moon, R.B., and Richards, J.H. Determination of intracellular pH of $^{31}$P nuclear magnetic resonance. *J. Biol. Chem.* **248:** 7276–7278; 1974.

Moore, W.S., and Holland, G.N. Nuclear magnetic resonance imaging. *Br. Med. Bull.* **36:** 297–299; 1980.

Morariu, V.V., Kiricuta, I.C., and Hazlewood, C.F. Nuclear magnetic resonance investigation of the freezing of water in rat kidney tissues. *Physiol. Chem. Phys.* **10:** 517–524; 1978.

Morariu, V.V., and Benga, G. Evaluation of a nuclear magnetic resonance technique for the study of water exchange through erythrocytes membranes in normal and pathological subjects. *Biochim. Biophys. Acta* **469:** 301; 1977.

Naruse, S., Horikawa, Y., Tanaka, C., Hirakawa, K., Nishikawa, H., Shimizu, K., and Kiri, M. Fundamental interpretation of NMR-CT image: Comparison of relaxation time *in vitro* and NMR-CT image. *Proc. SMRM,* San Francisco; p. 252; 1983.

Navon, G., Ogawa, S., Shulman, R.G., and Yamane, T. $^{31}P$ nuclear magnetic resonance studies of Ehrlich ascites tumor cells. *PNAS USA* **74**(1): 87–91; 1977.

Navon, G., Navon, R., Shulman, R.G., and Yamane, T. Phosphate metabolites in lymphoid, Friend erythroleukemia, and Hela cells observed by high-resolution $^{31}P$ nuclear magnetic resonance. *Proc. Natl. Acad. Sci. USA* **75**(3): 891–895; 1978.

Needham, D.M. *The Biochemistry of Muscular Contraction in its Historical Development.* London: Cambridge University Press; 1971.

Neville, M.C., Patterson, C.A., Rae, J.L., and Woessner, D.E. Nuclear magnetic resonance studies and water ordering in the crystalline lens. *Science* **184:** 1072–1075; 1974.

Ngo, F.Q.H., Glassner, B.J., Bay, J.W., Dudley, A.W., and Neaney, T.F. $T_1$ and $T_2$ relaxation measurements of human brain tissues and NMR imaging optimizations. *Proc. SMRM,* San Francisco; p. 264; 1983.

Nicolay, K., Hellingwerf, K.J., Kaptein, R., and Konings, W.N. Carbon-13 nuclear magnetic resonance studies of acetate metabolism in intact cells of Rhodopseudomonas Sphaeroides. *Biochimica et Biophysica Acta* **720:** 250–258; 1982.

Nishikawa, H., Yamada, S., Yoshizaki, K., and Watari, H. Phosphorus-31 NMR study on perfused ventral nerve cords of crayfish. *Proc. Japan Acad.(B)* **54:** 297–303; 1978.

Nunnally, R.L., and Bottomley, P.A. Assessment of pharmacological treatment of myocardial infarction by phosphorus-31 NMR with surface coils. *Science* **211:** 177–180; 1981.

Odeblad, E., Ingelman-Sundberg, A., Åsberg, K., et al. Proton magnetic relaxation times $T_1$ and $T_2$ for normal types of cervical secretions. *Acta Obstet. and Gynecol.* in press.

Odeblad, E., and Ingelman-Sundberg, A. Proton magnetic resonance studies on the structure of water in the myometrium. *Acta Obstet. Gynecol. Scand.* **44:** 117–125; 1966.

Odeblad, E. Proton magnetic resonance studies on the physical state of proteins in the vaginal epithelium. *Acta Obst. Gynec. Scand.* **44:** 595–608; 1965.

Odeblad, E. and Westin, B. Proton magnetic resonance of human milk. *Acta Radiologica,* **49,** 389–392; 1958.

Odeblad, E., and Bryhn, U. Proton magnetic resonance of human cervical mucus during the menstrual cycle. *Acta Radiol.* **47:** 314–320; 1957.

Odeblad, E., Bahr, B.N., and Lindstrom, G. Proton magnetic resonance of human red blood cells in heavy water exchange experiments. *Arch. Biochem. Biophys.* **63:** 221–225; 1956.

Ogawa, S., Boens, C.C., and Lee, T. A $^{31}P$ nuclear magnetic resonance study of the pH gradient and the inorganic phosphate distribution across the membrane in intact rat liver mitochondria. *Arch. Biochem. Biophys.* **210**(2): 740–747; 1981.

Ogawa, S., Rottenberg, H., Brown, T.R., Shulman, R.G., Castillo, C.L., and Glynn, P. High resolution $^{31}P$ nuclear magnetic resonance study of rat liver mitochondria. *Proc. Natl. Acad. Sci. USA* **75:** 1796–1800; 1978.

Oldfield, E., and Allerhand, A. Studies of individual carbon sites of hemoglobins in solution by natural abundance carbon 13 nuclear magnetic resonance spectroscopy. *J. Biol. Chem.* **250**(16): 6403–6407; 1975.

O'Neill, I.K., and Richards, C.P. Biological $^{31}P$-NMR spectroscopy. In: *Annual Reports on NMR Spectroscopy.* **10**(A): 133–236; 1980.

Outhred, R.K., and George, E.P. A nuclear magnetic resonance study of hydrated systems using the frequency dependence of the relaxation processes. *Biophys. J.* **13:** 83–96; 1973a.

Outhred, R.K., and George, E.P. Water and ions in muscles and model systems. *Biophys. J.* **13:** 97–103; 1973b.

Paalme, T., Olivson, A., and Vilu, R. $^{13}C$ NMR study of $CO_2$-fixation during the heterotrophic growth in Chlorobium thiosulfatophilum. *Biochimica et Biophysica Acta* **720:** 311–319; 1982a.

Paalme, T., Olivson, A., and Vilu, R. $^{13}C$-NMR study of the glucose synthesis pathways in the bacterium chlorobium thiosulfatophilum. *Biochimica et Biophysica Acta* **720:** 303–310; 1982b.

Packer, K.J. The effects of diffusion through locally inhomogeneous magnetic fields on transverse nuclear spin relaxation in heterogeneous systems. Proton transverse relaxation in striated muscle tissue. *J. Magn. Reson.* **9:** 438–443; 1973.

Packer, K.J., and Sellwood, T.C. Proton magnetic resonance studies of hydrated *Stratum corneum.* Part 1. Spin–lattice and transverse relaxation. *J. Chem. Soc. Faraday Trans.* **74:** 1579; 1978a.

Packer, K.J., and Sellwood, T.C. Proton magnetic resonance studies of hydrated *Stratum corneum.* Part 2. Self-diffusion. *J. Chem. Soc. Faraday Trans.* **74:** 1592; 1978b.

Parrish, R.G., Kurland, R.J., Janese, W.W., and Bakay, L. Proton relaxation rates of water in brain and brain tumors. *Science* **183:** 433–439; 1974.

Pearson, R.T., Duff, I.D., Derbyshire, W., and Blanshard, J.M.V. An NMR investigation of rigor in porcine muscle. *Biochim. Biophys. Acta* **362:** 188–200; 1974.

Peemoeller, H., Shenoy, R.K., Pintar, M.M., Kydon, D.W., and Inch, W.R. Improved characterization of healthy and malignant tissue by NMR line-shape relaxation correlations. *Biophys. J.* **38:** 271–276; 1982.

Peemoeller, H., Pintar, M.M., and Kydon, D.W. Nuclear magnetic resonance analysis of water in natural and deuterated mouse muscle above and below freezing. *Biophys. J.* **29:** 427–435, 1980.

Peemoeller, H., and Pintar, M.M. Nuclear magnetic resonance multiwindow analysis of proton local fields and magnetization distribution in natural and deuterated mouse muscle. *Biophys. J.* **28:** 339–345; 1979.

Peemoeller, H., Schreiner, L.J., Pintar, M.M., Inch, W.R., and McCredie, J.A. Proton $T_1$ study of coverage parameter changes in tissues from tumor-bearing mice. *Biophys. J.* **25:** 203–212; 1979.

Pieper, G.M., Tood, G.L., Wu, S.T., Salhany, J.M., Clayton, F.C., and Eliot, R.S. Attenuation of myocardial acidosis by propranolol during ischemic arrest and reperfusion: Evidence with $^{31}P$ nuclear magnetic resonance. *Cardiovasc. Res.* **14:** 646–653; 1980.

Polak, J.F., Neirinckx, R.D., Garnic, J.D., and Adams, D.F. Dependence of bulk susceptibility and $T_1$ of rat myocardium on the tissue water content. *Proc. SMRM,* San Francisco; p. 282; 1983.

Pollard, H.B., Shindo, H., Creutz, C.E., Pazoles, C.J., and Cohen, J.S. Internal pH and state of ATP in adrenergic chromaffin granules determined by phosphorus-31 NMR spectroscopy. *J. Biol. Chem.* **254:** 1170–72; 1979.

Raaphorst, G.P., Law, P., and Kruuv, J. Water content and spin–lattice relaxation times of cultured mammalian cells subjected to various salt, sucrose, or DMSO solutions. *Physiol. Chem. Physics* **10:** 177–182; 1978.

Raaphorst, G.P., Kruuv, J., and Pintar, M.M. Nuclear magnetic resonance study of mammalian cell water. *Biophys. J.* **15:** 391–402; 1975.

Radda, G.K. Dynamic properties of biological membranes. *Trans. R. Soc. London, Sec. B* **272:** 159–171; 1975.

Radda, G.K., and Seeley, P.J. Recent studies on cellular metabolism by nuclear magnetic resonance. *Annu. Rev. Physiol.* **41:** 749–769; 1979.

Ranade, S.S., Smita, S., Advani, S.H., and Kasturi, S.R. Pulsed nuclear magnetic resonance studies of human bone marrow. *Physiol. Chem. Phys.* **9:** 297–299; 1977.

Ranade, S.S., Shah, S., Korgaonkar, K.S., Kasturi, S.R., Chaughule, R.S., and Vijayaraghavan, R. Absence of correlation between spin-lattice relaxation times and water content in human tumor tissues. *Physiol. Chem. Phy.* **8:** 131–135; 1976.

Ranade, S.S., Chaughule, S., Kasturi, S.R., Nad Karni, J.S., Talwalkar, G.V., Wash, U.V., Korgaonkar, K.S., and Vijayaraghavan, R. Pulsed nuclear magnetic resonance studies on human malignant tissues and cells *in vitro. Indian J. Biochem. Biophys.* **12:** 229–238, 1975.

Rangel-Guerra, R.A., Perez-Payan, H., and Todd, L.E. Nuclear magnetic resonance in the bipolar affective disorders. *Mag. Res. Imaging* **1**(4): 229–230; 1982.

Ratkovic, S., Bacic, G., Radenovic, C., and Vucinic, Z. Water in plants: A review of some recent NMR studies concerning the state and transport of water in leaf, root, and seed. *Studia Biophysica* **91:** 9–18; 1982.

Ratkovic, S., NMR studies of water in biological systems in different levels of their organization. *Scientia Yugoslavica* **7**(1–2): 19–54; 1981.

Ratkovic, S., and Bacic, G. Water exchange in *Nitella* cells: A PMR study in the presence of paramagnetic $Mn^{+2}$ ions. *Bioelectrochemistry and Bioenergetics* **7:** 405; 1980.

Ratkovic, C., Rusov, C., and Stojanovic, D. Rous sarcoma studied by $^{1}H$ nuclear magnetic resonance. *Acta Biol. Med. Exp.* **3:** 31; 1978.

Ratkovic, S., and Sinadinovic, J. Nuclear magnetic resonance studies on the thyroid gland. II. On the state of tissue water in normal thyroid gland. *Studia Biophys.* **63:** 25–39; 1977.

Ratkovic, S., and Rusov, C. Magnetic relaxation of water protons in the detection of tissue changes induced by erythroleukosis. *Period. Biol.* **76:** 19–23; 1974.

Ratner, A.V., Carter, E. A., Wands, J. R., and Pohost, G.M. Characterization of changes in proton NMR relaxation parameters in models of hepatic steatosis with and without necrosis. *Proc. SMRM,* San Francisco; p. 290; 1983.

Reisin, I.L., Rotunno, C.A., Corchs, L., Kowalewski, V., and Cereijido, M. The state of sodium in epithelial tissues as studied by nuclear magnetic resonance. *Physiol. Chem. Phys.* **2:** 171–179; 1970.

*Report of the Task Group on Reference Man,* International Commission on Radiological Protection, No. 23. New York: Pergamon Press; 1981.

Resing, H.A., Foster, K.R., and Garroway, A.N. "Bound water" in barnacle muscle as indicated in nuclear magnetic resonance studies. *Science, N.Y.* **198:** 1180–1182; 1977.

Richards, T., Budinger, T.F., Wesbey, G., and Engelstad, B. Evaluation of heavy ion radiation damage in the rat brain using proton NMR imaging. *Proc. SMRM,* San Francisco; p. 298; 1983.

Roos, A., and Boron, W.F. Intercellular pH. *Physiol. Rev.* **61:** 296–434; 1981.

Rorschach, H.E., Chang, D.C., Hazlewood, C.F., and Nichols, B.L. The diffusion of water in striated muscle. *Ann. N.Y. Acad. Sci.* **204:** 444–452; 1973.

Ross, B.D., Radda, G.K., Gadian, D.G., Rocker, G., Esiri, M., and Falconer-Smith, J. Examination of a case of suspected McArdle's syndrome by $^{31}P$ nuclear magnetic resonance. *N. Engl. J. Med.* **304:** 1338–1343; 1981.

Ross, R.J., Thompson, J.S., Kim, K., and Bailey, R.A. Nuclear magnetic resonance imaging and evaluation of human breast tissue: Preliminary clinical trials. *Radiology* **143**(1): 195–206; 1982a.

Ross, R.J., Thompson, J.S., Kim, K., and Bailey, R. Nuclear magnetic resonance evaluation of a leiomyosarcoma. *Mag. Res. Imaging* **1:** 87–90; 1982b.

Rotunno, C.A., Kowalewski, V., and Cereijido, A.M. Nuclear spin resonance evidence for complexing of sodium in frog skin. *Biochim. Biophys. Acta* **135:** 170–173; 1967.

Rustgi, S.N., Peemoeller, H., Thompson, R.T., Kydon, D.W., and Pintar, M.M. A study of molecular dynamics and freezing phase transition in tissues by proton spin relaxation. *Biophys. J.* **22:** 439–452; 1978.

Saks, V.A., Rosentraukh, V., Smirnov, V.N., and Chazov, E.I. Role of creatine phosphokinase in cellular function and metabolism. *Can. J. Physiol. Pharmacol.* **56:** 691–706; 1978.

Salhany, J.M., Pieper, G.M., Wu, S., Todd, A.L., Clayton, F.C., and Eliot, R.S. $^{31}P$-nuclear magnetic resonance measurements of cardiac pH in perfused guinea pig hearts. *J. Mol. Cell. Cardiol.* **11:** 601–610; 1979.

Salhany, J.M., Yamane, T., Shulman, R.G., and Ogawa, S. High resolution $^{31}P$ NMR studies of intact yeast cells. *Proc. Natl. Acad. Sci. USA* **72:** 4966–4970; 1975.

Sandhu, H.S., and Friedman, G.B. Proton spin–lattice relaxation time study in tissues of the adult newt Taricha granulosa (Amphibia: Urodele). *Med. Phys.* **5**(6): 513–517; 1978.

Saryan, L.A., Hollis, D.P., Economou, J.S., and Eggleston, J.C. Nuclear magnetic resonance studies of cancer. IV. Correlation of water content with tissue relaxation times. *J. Natl. Cancer Inst.* **52:** 599–602; 1974.

Schara, M., Sentjurc, M., Auersperg, M., and Golouh, R. Characterization of malignant thyroid gland tissue by magnetic resonance methods. *Br. J. Cancer.* **29:** 483–486; 1974.

Schmidt, K., Breitmaier, E., Aeikens, B., Zieger, K.H., and Knutel, B. Spin–Gitter Relaxationszeit der Protonen des Zellwassers in normalen und tumorosen Geweben des Menschen. *Z. Krebsforsch* **80:** 209–222; 1973.

Seeley, P.J., Sehr, P.A., Gadian, D.G., Garlick, P.B., and Radda, G.K. In: *NMR in Biology,* R.A. Duck, I.D. Campbell, R.E. Richards, R.J.P. Williams, eds. New York: Academic Press; pp. 247–275; 1977.

Seeley, P.J., Busby, S.J.W., Gadian, D.G., Radda, G.K., and Richards, R.E. *Biochem. Soc. Trans.* **4:** 62–64; 1976.

Sehr, P.A., Radda, G.K., Bore, P.J., and Sells, R.A. A model kidney transplant studied by phosphorus NMR. *Biochem. Biophys. Res. Commun.* **77:** 195–202; 1977.

Seitz, P.K., Chang, D.C., Hazlewood, C.F., Rorschach, H.E., and Clegg, J.S. The self-diffusion of water in artemia cysts. *Arch. Biochem. Biophys.* **210**(2): 517–524; 1981.

Sentjurc, M., Schara, M., Auersperg, M., Golouh, R., and Lamovec, J. Magnetic resonance in the diagnosis of cancer. *Radiol. Ingosl.* **4:** 319–327; 1974.

Shah, S.S., Ranade, S.S., Phadke, R.S., and Kasturi, S.R. Significance of water proton spin–lattice relaxation times in normal and malignant tissues and their subcellular fractions. *Mag. Res. Imaging* **2:** 91–104; 1982.

Shporer, M., and Civan, M.M. Pulsed nuclear magnetic resonance study of $^{39}K$ within Halobacteria. *J. Membr. Biol.* **33:** 385–400; 1977.

Shporer, M., Haas, M., and Civan, M. Pulsed nuclear magnetic resonance study of $^{17}O$ from $H_2{}^{17}O$ in rat lymphocytes. *Biophys. J.* **16:** 601–611; 1976.

Shporer, M., and Civan, M.M. NMR study of $^{17}O$ from $H_2{}^{17}O$ in human erythrocytes. *Biochim. Biophys. Acta* **385:** 81–87; 1975.

Shporer, M., and Civan, and M.M. Effects of temperature and field strength on the NMR relaxation times of $^{23}Na$ in frog striated muscle. *Biochim. Biophys. Acta.* **354:** 291–304; 1974.

Shulman, R.G., Brown, T.R., Ugurbil, K., Ogawa, S., Cohen, S.M., and Den Holladner, J.A. Cellular applications of $^{31}P$ and $^{13}C$ nuclear magnetic resonance. *Science* **205:** 160–166; 1979.

Sinadinovic, J., Ratkovic, S., Kraincanic, M., and Jovanic, M. Relationship of biochemical and morphological changes in rat-thyroid and proton spin-relaxation of the tissue water. *Endokrinologie* **69:** 55–66; 1977.

Small, W.C., McSweeney, M.B., Bernardino, M.E., and Goldstein, J.H. Multiple component in vitro relaxation studies in human breast and liver tissue. *Proc. SMRM,* San Francisco; p. 342; 1983.

Smith, F.W. Two years clinical experience with NMR imaging. *Applied Radiology,* Vol. 12, 3: 29–42; 1983.

Smith, F.W., Reid, A., Hutchinson, J.S., and Mallard, J.R. Nuclear magnetic resonance, imaging of the pancreas. *Radiology* **142:** 677–680; 1982.

Stadelmann, E. Permeability of the plant cell. *Ann. Rev. Plant Physiol.* **20:** 585–597; 1969.

Stout, D.G., Steponkus, P.L., Bustard, L.D., and Cotts, R.M. Water permeability of Chlorella cell membranes by nuclear magnetic resonance. Measured diffusion coefficients and relaxation times. *Plant Physiol.* **62:** 146–149; 1978.

Stout, D.G., Cotts, R.M., and Steponkus, P.L. The diffusional water permeability of Elodea leaf cells as measured by nuclear magnetic resonance. *Can. J. Botany* **55:** 1623–1627; 1977.

Sussman, M.V., and Chin, L. Nuclear magnetic resonance spectrum changes accompanying rigor mortis. *Nature* **211:** 414–416; 1966.

Swift, T.I., and Barr, E.M. An oxygen magnetic resonance study in frog skeletal muscle. *NYAS* **204:** 191–196; 1973.

Swift, T.I., and Fritz, O.G. A proton spin-echo study of the state of water in frog nerves. *Biophys. J.* **9:** 54–59; 1969.

Takhavieva, D.G., Rokityanskii, V.I., and Yakimov, Y.V. Changes in the proton relaxation time and water content of bone tissue during conservation by cold. *Byulleten Ekoperimental nor Biologii i Meditainy* **77**(2): 92–94; 1974.

Tanaka, K., Yamada, Y., Shimizer, T., Sano, F., and Abe, Z. Fundamental investigation (in vitro) for a non invasive method of tumor detection by nuclear magnetic resonance. *Biotelemetry* **1:** 337–350; 1974.

Tanner, J.E. Intracellular diffusion of water. *Arch. Biochem. Biophys.* **224:** 416–428; 1983.

Tanner, J.E. Self diffusion of water in frog muscle. *Biophys. J.* **28:** 107–116; 1979.

Tanner, J.E. Magnetic resonance in colloid and interface science. In: ACS Symposium No. 34, H.A. Reising, C.G. Wade, eds. *Am. Chem. Soc.* **16:** 1976.

Tennvall, J., Biorklund, A., Moller, T., Olsson, M., Persson, B., and Akerman, M. NMR-relaxation-times studies in malignant tumours and normal tissues of the human thyroid gland. *Proc. SMRM,* San Francisco; p. 356; 1983.

Thickman, D.I., Kundel, H.L., and Wolf, G. Nuclear magnetic resonance characteristics of fresh and fixed tissue: The effect of elapsed time. *Radiology* **148:** 183–185; 1983.

Thompson, B.C., Waterman, M.R., and Cottam, G.L. Evaluation of the water environment in deoxygenated sickle cells by longitudinal and transverse water proton relaxation rates. *Arch. Biochem. Biophys.* **166:** 193–200; 1975.

Thompson, R.T., Knispel, R.R., and Pintar, M.M. A study of the proton exchange in tissue water by spin relaxation in the rotating frame. *Chem. Phys. Lett.* **22:** 335–337; 1973.

Torchia, D.A., Lyerla, J.R., and Quattrone, A.J. A $^{13}C$ magnetic resonance study of the helix and coil stages of the collagen peptide $\alpha$1-CB1. In: *Peptides, Polypeptides and Proteins,* E.R. Blount, ed. New York: Wiley and Sons; pp. 436–448; 1974.

Trantham, E.C. *Quasi-elastic neutron scattering study of water in agarose gels.* Ph.D. Dissertation, Rice University, Houston, Texas; 1980.

Udall, J.N., Alvarez, L.A., Chang, D.C., Soriano, H., Nichols, B.L., and Hazlewood, C.F. Effects of cholera enterotoxin on intestinal tissue water as measured by nuclear magnetic resonance (NMR) spectroscopy. II. *Physiol. Chem. Phys.* **9:**(1): 13–20; 1977.

Udall, J.N., Alvarez, L.A., Nichols, B.L., and Hazlewood, C.F. The effects of cholera enterotoxin on intestinal tissue water as measured by nuclear magnetic resonance (NMR) spectroscopy. *Physiol. Chem. Phys.* **7:** 533–539; 1975.

Ugurbil, K., Guernsey, D.L., Brown, T.R., Glynn, P., Tobkes, N., and Edelman, I.S. $^{31}P$ NMR studies of intact anchorage-dependent mouse embryo fibroblasts. *PNAS USA* **78:** 4843–4847; 1981.

Ugurbil, K., Rottenberg, H., Glynn, P., and Shulman, R.G. $^{31}P$ nuclear magnetic resonance studies of bioenergetics and glycolysis in anaerobic Escherichia coli cells. *Proc. Natl. Acad. Sci. USA* **75:** 2244–2249; 1978.

Ujeno, Y. Nuclear magnetic resonance relaxation times for tissues of mice given radioprotective compounds. *Physiol. Chem. Phys.* **12:** 271–275; 1980.

Valensin, G., Gaggelli, E., Tiezzi, E., Valensin, P.E., Bianchi Bandelli, M.L., and di Cairano, M.L. Virus-cell interactions: Nuclear magnetic resonance behavior of intracellular water. *Microbiologica* **5:** 195–205; 1982.

Valensin, P.E., Bianchi Bandinelli, M.L., and Di Cairano, M.L. Proton spin–lattice relaxation rates in erythrocytes adsorbed with hemagglutinating viruses. *Biophysical Chemistry* **14:** 357–362; 1981.

Valikanov, G.A., and Volkov, V.Y. *Biophysics* **23:** 820–823; from Biofizika **23:** 813–816; 1979.

Venable, W.C., Miller, D.M., Yopp, J.H., and Zimmerman,

J.R. Pulsed NMR study of water in an obligately halophilic blue-green algo. *Physiol. Chem. Phys.* **10:** 405–414; 1978.

Villey, D., and Martin, G. $^{1}$H-NMR spectra of muscle, red cells, brain, liver, and kidney in acutely hyponatremic rats. *Physiol. Chem. Physics* **6:** 339; 1974.

Vucelic, D., Macura, S., Neskovic, B., Djuric, L.J., and Ajdaric, Z. Proton relaxation of malignant and embryonic mouse cells. *Studia Biophysica,* **68:** 179–185; 1978.

Wagh, U.V., Kasturi, S.R., Chaughule, R.S., Smita, S.S., and Ranade, S.S. Studies on proton spin–lattice relaxation time ($T_1$) in experimental cell cultures. *Physiol. Chem. Phys.* **9:** 167–174; 1977.

Walter, J.A. Nuclear magnetic resonance and the state of water in cells. *Prog. Biophys. Mol. Biol.* **23:** 1–20; 1971.

Walter, J.A., and Hope, A.B. Proton magnetic resonance studies of water in slime mold Plasmodia. *Austr. J. Biol. Sci.* **24:** 497–507; 1971.

Weisman, I.D., Bennett, L.H., Maxwell, L.R., Woods, D.B., and Burk, D. Recognition of cancer *in vivo* by NMR. *Science* **178:** 1288–1290; 1972.

Wesbey, G., Goldberg, H.I., Mayo, J., Moon, K., Stark, D., and Moss. A. In-vivo proton NMR imaging characterization of the normal human liver and spleen. *Proc. SMRM,* San Francisco; p. 373; 1983.

Westover, C.J., and Dresden, M.H. Collagen hydration. *Biochim. Biophys. Acta* **365:** 389–399; 1974.

White, J.P., Kuntz, I.D., and Cantor, C.R. Studies on the hydration of *Escherichia coli* ribosomes by nuclear magnetic resonance. *J. Mol. Biol.* **64:** 511–514; 1971.

Wiles, C.M., Young, A., Jones, D.A., and Edwards, R.H.T. Muscle relaxation rate and energy turnover in hyper and hypo-thyroid patients. *Clin. Sci.* **57:** 375–384; 1979.

Wilkie, D.R. Shortage of chemical fuel as a cause of fatigue: Studies by nuclear magnetic resonance and bicycle ergometry. In: *Human Muscle Fatigue: Physiological Mechanisms,* R. Porter, J. Whelan, eds. London: Pitman Medical; 1981.

Williams, E.S., Kaplan, J.I., Thatcher, F., Zimmermann, G., and Knoibel, S.B. Prolongation of proton spin relaxation times in regionally ischemic tissue from dog hearts. *J. Nucl. Med.* **51**(5): 449–453; 1980.

Winkler, M.W., Matson, G.B., Hershey, J.W.B., and Bradbury, E.M. $^{31}$P-NMR study of the activation of the sea urchin egg. *Exptl. Cell Res.* **139:** 217–222; 1982.

Wu, S.T., Pieper, G.M., Salhany, J.M., and Eliot, R.S. $^{31}$P-NMR and multiequilibria analysis of intracellular free $Mg^{+2}$ in perfused and ischemic arrested guinea pig hearts (Abstract). *Biophys. J.* **33:** 26a; 1981.

Yeh, H.J.C., Brinkley, F.J., Jr., and Becker, E.D. Nuclear magnetic resonance studies in intracellular sodium in human erythrocytes and frog muscle. *Biophys. J.* **13:** 56–71; 1973.

Yoshikawa, K., Matsushita, K., and Ohsaka, A. $^{1}$H-NMR spectroscopy in aqueous mediums. Examination of experimental conditions with human urine as a model sample. *Physiol. Chem. Phys.* **14:** 385–389; 1982.

Yoshikawa, K., and Ohsaka, A. $^{1}$H and $^{13}$C NMR spectroscopic study of rat organs. *Physiol. Chem. Phys.* **12:** 515–518; 1980.

Yoshikawa, K., and Ohsaka, A. Nuclear magnetic resonance of rat skin. *Physiol. Chem. Phys.* **11:** 185–188; 1979.

Yoshizaki, K., Seo, Y., Nishikawa, H., and Morimoto, K. Application of pulsed-gradient $^{31}$P NMR on frog muscle to measure the diffusion rates of phosphorus compounds in cells. *Biophys. J.* **38:** 209–211; 1982.

Yoshizaki, K., Nishikawa, H., Yamade, S., Morimoto, T., and Watari, H. Intracellular pH measurement in frog muscle by means of $^{31}$P nuclear magnetic resonance. *Japan J. Physiol.* **29:** 211–225; 1979.

Yoshizaki, K. Phosphorus nuclear magnetic resonance studies of phosphorus metabolites in frog muscle. *J. Biochem. (Tokyo)* **84:** 11–18; 1968.

Zaner, K.S., and Damadian, R. NMR in cancer. VIII. Phosphorus-31 as a nuclear probe for malignant tumors. *Physiol. Chem. Phys.* **9**(4–5): 473–483; 1977.

Zaner, K.S., and Damadian, R. NMR in cancer IX. The concept of cancer treatment by NMR: A preliminary report of high resolution NMR or phosphorous in normal and malignant tissues. *Physiol. Chem. Phys.* **7:** 437–451; 1975a.

Zaner, K.S., and Damadian, R. Phosphorus-31 as a nuclear probe for malignant tumors. *Science* **189:** 729–731; 1975b.

Zipp, A., James, T.L., Kuntz, I.D., and Shohet, S.B. Water proton magnetic resonance studies of normal and sickle erythrocytes. Temperature and volume dependence. *Biochim. Biophys. Acta* **428:** 291–303; 1976a.

Zipp, A., Kuntz, I.D., and James, T.L. An investigation of "bound" water in frozen erythrocytes by proton magnetic resonance spin-lattice, spin-spin, and rotating frame spin-lattice relaxation time measurements. *J. Magn. Reson.* **24:** 411; 1976b.

Zipp, A., Kuntz, I.D., Rehfeld, S.J., and Shohet, S.B. Proton magnetic resonance studies of intracellular water in sickle cells. *FEBS Lett.* **34**(1): 9–12; 1974.

# CHAPTER 10

# *SUPPLEMENTAL INFORMATION*



## 10.1. A SHORT BIBLIOGRAPHY ON NUCLEAR MAGNETIC RESONANCE OF WATER


1. Abragam, A. *The Principles of Nuclear Magnetism.* London: Oxford University Press, 1961.
2. Andrasko, J., and Forsen, S. NMR study of rapid water diffusion across lipid bilayers in dipalmitoil lecithin vesicles. *Biochem. Biophys. Res. Comm.* **60:** 813; 1974.
3. Andrew, E.R. *Nuclear Magnetic Resonance.* London: Cambridge University Press, 1958.
4. Berendsen, H.J.C. Nuclear magnetic resonance study of collagen hydration. *J. Chem. Phys.* **36:** 3297–3305; 1962.
5. Berendsen, H.J.C. Specific interactions of water with biopolymers. In: *Water: A Comprehensive Treatise,* Vol. 5, F. Franks, ed. New York: Plenum Press. pp. 293–349; 1975.
6. Bienkiewicz, K.J., Berendsen, H.J.C., and Andree, P.J. Properties of water in native and modified collagen. *Ann. Soc. Chim. Polonorum* **51:** 149–158; 1977.
7. Blears, D.J., and Danyluk, S.S. Proton wide-line nuclear magnetic resonance spectra of hydrated proteins. *Biochim. Biophys. Acta* **154:** 17–27; 1968.
8. Blicharska, O.B., Florkowski, Z., Hennel, J.W., Held, G., and Noack, F. Investigation of protein hydration by proton spin relaxation time measurements. *Biochim. Biophy. Acta* **207:** 381; 1970.
9. Bloch, F. Nuclear induction. *Phys. Rev.* **70:** 460; 1946.
10. Bloembergen, N., Purcell, E.M., and Pound, R.V. Relaxation effects in nuclear magnetic resonance absorption. *Phys. Rev.* **73:** 679; 1948.
11. Brnjas-Kraljevic, J., Pifat, G., and Maricic, S. Quaternary structure of hemoglobin. Its hydration and selfassociation. A proton magnetic relaxation study. *Physiol. Chem. Physics* **11:** 371; 1979.
12. Bystrov, G.S., Nikolaev, N.I., and Romanenko, G.I. Temperaturnaja zavisimost jadernoj relaksacii protonov vody v nekotoryh tkanjah. *Biofizika* **18:** 484; 1973.
13. Canet, D., Goulon-Ginet, C., and Marchal, J.P. Accurate determination of parameters for $^{17}O$ in natural abundance by Fourier transform NMR. *J. Magn. Reson.* **22:** 537; 1976.
14. Carr, H.Y., and Purcell, E.M. Effect of diffusion on free precession in nuclear magnetic resonance experiments. *Phys. Rev.* **94:** 630; 1954.
15. Child, T.F. Pulsed NMR study of molecular motion and environment of sorbed water on cellulose. *Polymer* **13:** 259; 1972.
16. Civan, M.M., Achlama, A.M., and Shporer, M. The relationship between the transverse and longitudinal nuclear magnetic resonance relaxation rates of muscle water. *Biophys. J.* **21:** 127–136; 1978.
17. Cooper, R.L., Chang, D.B., Young, A.C., Martin, C.J., and Ancker-Johnson, B. Restricted diffusion in biophysical systems: Experiment. *Biophys. J.* **14:** 161; 1974.
18. Davis, C.M. Jr., and Litovitz, T.A. Two-state theory of the structure of water. *J. Chem. Phys.* **42:** 2563; 1965.
19. Dehl, R.E. The effect of salts on the NMR spectra of $D_2O$ in collagen fibers. *Biopolymers* **12:** 2329; 1973.
20. Dehl, R.E., and Hoeve, C.A.J. Broad-line NMR study of $H_2O$ and $D_2O$ in collagen fibers. *J. Chem. Phys.* **50:** 3245–3251; 1969.
21. Dwek, R.A. *Nuclear Magnetic Resonance (NMR) in Biochemistry: Application to Enzyme Systems.* Oxford: Clarendon Press; 1973.
22. Edzes, H.T., and Berendsen, H.J.C. The physical state of diffusible ions in cells. *Ann. Rev. Biophys. Bioeng.* **4:** 265; 1975.
23. Edzes, H.T., and Samulski, E.T. The measurement of cross-relaxation effects in the proton NMR spin–lattice relaxation of water in biological systems: hydrated collagen and muscle. *J. Magn. Reson.* **31:** 207–229; 1978.
24. Edzes, H.T., and Sanulski, E.T. Cross relaxation and spin diffusion in the proton NMR of hydrated collagen. *Nature* (London) **265:** 521–523; 1977.
25. Eisenberg, D., and Kauzmann, W. *The Structure and Properties of Water.* Oxford: Clarendon Press; 1969.
26. Ellis, G.E., and Packer, K.J. Nuclear spin-relaxation studies of hydrated elastin. *Biopolymers* **15:** 813–832; 1976.

27. Emsley, J.W., Feeney, J., and Sutcliffe, L. *High Resolution Nuclear Magnetic Resonance Spectroscopy,* Vols. I and II. London: Pergamon Press; 1965.
28. Fabry, M.E., Koenig, S.H., and Schillinger, W.E. Nuclear magnetic relaxation dispersion in protein solutions. IV. Proton relaxation at the active site of carbonic anhydrase. *J. Biol. Chem.* **245:** 4256; 1970.
29. Farrar, T.C., and Becker, E.D. *Pulse and Fourier Transform NMR. Introduction to Theory and Methods.* New York: Academic Press; 1971.
30. Finch, E.D., and Schneider, A.S. Mobility of water bound to biological membranes. A proton NMR relaxation study. *Biochim. Biophys. Acta* **406:** 146; 1975.
31. Franks, F. *Water: A Comprehensive Treatise,* Vols. 1–5. New York: Plenum Press; 1972–1975.
32. Franks, F. Solvation interactions of proteins in solution. *Phil. Trans. R. Soc. B* **278:** 89–96; 1977.
33. Fuller, M.E., and Brey, W.S. Jr. Nuclear magnetic resonance study of water sorbed on serum albumin. *J. Biol. Chem.* **243:** 243–280; 1968.
34. Fung, B.M., and Trautmann, P. Deuterium NMR and EPR of hydrated collagen fibers in the presence of salts. *Biopolymers* **10:** 391–397; 1971.
35. Fung, B.M., Witschel, J. Jr., and McAmis, L.L. The state of water on hydrated collagen as studied by pulsed NMR. *Biopolymers* **13:** 1767–1776; 1974.
36. Gillen, K.T., Douglass, D.C., and Hoch, M.J.R. Self-diffusion in liquid water to $-31°C$. *J. Chem. Phys.* **57:** 5117; 1972.
37. Glasel, J.A. A study of water in biological systems by $^{17}O$ magnetic resonance spectroscopy. II. Relaxation phenomena in pure water. *Proc. Natl. Acad. Sci. USA* **58:** 27–33; 1967.
38. Grigera, J.R., and Mascarenhas, S. A model for NMR, dielectric relaxation and electret behaviour of bound water in proteins. *Studia Biophysica* **73:** 19; 1978.
39. Grosch, L., and Noack, F. NMR relaxation investigation of water mobility in aqueous bovine serum albumin solutions. *Biochim. Biophys. Acta* **453:** 218–232; 1976.
40. Hahn, E.L. Spin echoes. *Phys. Rev.* **80:** 580; 1950.
41. Hallenga, K., and Koenig, S.H. Protein rotational relaxation as studied by solvent $^1H$ and $^2H$ magnetic relaxation. *Biochemistry* **15:** 4255–4264; 1976.
42. Hansen, J.R., and Lawson, K.D. Magnetic relaxation in ordered systems. *Nature* (London) **225:** 542–543; 1970.
43. Haran, N., and Shporer, M. Study of water permeability through phospholipid vesicle membranes by $^{17}O$ NMR. *Biochim. Biophys. Acta* **426:** 638; 1976.
44. Hilton, B.D., Hsi, E., and Bryant, R.G. $^1H$ nuclear magnetic resonance relaxation of water on lysozyme powders. *J. Am. Chem. Soc.* **99:** 8483–8490; 1977.
45. Hindman, J.C., Zielen, A.J., Svirmickas, A., and Wood, M. Relaxation processes in water. The $^1H$ and oxygen-17 in $H_2{}^{17}O$. *J. Chem. Phys.* **54:** 621; 1971.
46. Hindman, J.C., Svirmickas, A., and Wood, M. Relaxation processes in water. A study of the proton spin–lattice relaxation time. *J. Chem. Phys.* **59:** 1517; 1973.
47. Hindman, J.C. Proton resonance shift of water in the gas and liquid states. *J. Chem. Phys.* **44:** 4582; 1966.
48. Hinshaw, W.S., Bottowley, P.A., and Holland, G.N. Radiographic thin section image of the human wrist by nuclear magnetic resonance. *Nature* **270:** 722; 1977.
49. Horowitz, S.B., and Paine, P.L. Reference phase analysis of free and bound intracellular solutes. II. Isothermal and isotopic studies of cytoplasmic sodium, potassium and water. *Biophys. J.* **25:** 45; 1979.
50. Hoult, D.I. Zeugmatography: A criticism of the concepts of a selective pulse in the presence of a field gradient. *J. Magn. Reson.* **26:** 165–167; 1977.
51. Hsi, E., and Bryant, R.G. Nuclear magnetic resonance relaxation in frozen lysozyme solutions. *J. Am. Chem. Soc.* **97:** 3220; 1975.
52. Hutchinson, J.M.S., Sutherland, R., Mallard, J.R., and Foster, M.A. In: *NMR in Biology,* R.A. Dwek, I.D. Campbell, R.E. Richards, R.J.P. Williams, eds. London: Academic Press; pp. 368–369; 1977.
53. Inglefield, P.T., Lindblom, K.A., and Gottlieb, A.M. Water binding and mobility in the phosphatidylcholine (cholesterol) water lammellar phase. *Biochim. Biophys. Acta* **419:** 196; 1976.
54. Jacobsen, B., Anderson, W.A., and Arnold, T. A proton magnetic resonance study of the hydration of deoxyribonucleic acid. *Nature* (London) **173:** 772–773; 1954.
55. Kalk, A., and Berendsen, H.J.C. Proton magnetic relaxation and spin diffusion in proteins. *J. Magn. Reson.* **24:** 343–366; 1976.
56. Kimmich, R., and Noack, F. Kernmagnetische Relaxation in Proteinlosungen. *Z. Naturforsch.* **25a:** 299; 1970.
57. Klose, G., and Stelzner, F. NMR investigations of the interaction of water with lecithin in benzene solutions. *Biochim. Biophys. Acta* **363:** 1–8; 1974.
58. Knispel, R.R., and Pintar, M.M. Temperature dependence of the proton exchange time in pure water by NMR. *Chem. Phys. Lett.* **32:** 238; 1975.
59. Koenig, S.H., Bryant, R.G., Hallenga, K., and Jacob, G.S. Magnetic cross-relaxation among protons in protein solutions. *Biochemistry* **17:** 4348–4358; 1978.
60. Koenig, S.H., Hallenga, K., and Shporer, M. Protein-water interaction studied by solvent $^1H$, $^2H$ and $^{17}O$ magnetic relaxation. *Proc. Natl. Acad. Sci. USA* **72:** 2667–2671; 1975.
61. Kruger, G.J., and Helcke, G.A. Proton relaxation of water absorbed on protein. In: *Magnetic Resonance and Relaxation,* R. Blinc, ed. Amsterdam: North Holland; 1136; 1968.
62. Krynicki, K. Proton spin–lattice relaxation in pure water between 0°C and 100°C. *Physica* **32:** 167; 1966.
63. Kumar, A., Welti, D., and Ernst, R.R. NMR Fourier zeugmatography. *J. Magn. Reson.* **18:** 69–83; 1975.
64. Kuntz, I.D. Hydration of macromolecules. III. Hydration of polypeptides. *J. Am. Chem. Soc.* **93:** 514–516; 1971a.
65. Kuntz, I.D., and Kauzmann, W. Hydration of proteins and polypeptides. *Adv. Protein Chem.* **28:** 239–345; 1974.
66. Kuo, A.L., and Wade, C.G. Lipid lateral diffusion by pulsed nuclear magnetic resonance. *Biochemistry* **18:** 2300; 1979.
67. Lahajnar, G., Zupancic, I., Miljkovic, L.J., and Rupprecht, A. The structure of water sorbed in oriented DNA samples. *Period. Biol.* **80:** 135; 1978.
68. Lauterbur, P.C. Image formation by induced local interactions: Examples employing NMR. *Nature* (London) **242:** 190–191; 1973.
69. Leung, H.K., Steinberg, M.P., Wei, L.S., and Nelson, A.I. Water binding of macromolecules determined by pulsed NMR. *J. Food Sci.* **41:** 297; 1976.

70. Lindstrom, T.R., and Koenig, S.H. Magnetic-field-dependent water proton spin–lattice relaxation rates of hemoglobin solutions and whole blood. *J. Magn. Reson.* **15:** 344; 1974.
71. Ling, G.N. In: *A Physical Theory of the Living State: The Association-Induction Hypothesis.* New York: Blaisdell; 1962.
72. Look, D.C., and Lowe, I.J. Nuclear magnetic dipole–dipole relaxation along the static and rotating magnetic fields. *J. Chem. Phys.* **44:** 2995; 1966.
73. Lubas, B., and Wilczok, T. NMR study on molecular mobility of DNA molecules in solution. *Biopolymers* **10:** 1267–1276; 1971.
74. Lucken, E.A.C. *Nuclear Quadrupole Coupling Constants.* London: Academic Press; p. 52; 1969.
75. Lynch, L.J., and Webster, D.S. The use of NMR techniques in the study of protein hydration. *J. Polymer Sci.* **49:** 43; 1975.
76. Lynch, L.J., Marsden, K.H., and George, E.P. NMR of absorbed systems. I. A systematic method of analyzing NMR relaxation time data for a continuous distribution of nuclear correlation times. *J. Chem. Phys.* **51:** 5673; 1969.
77. Mansfield, P., and Pyckett, I.L. Biological and medical imaging by NMR *J. Magn. Reson.* **29:** 355; 1978.
78. Mansfield, P., and Maudsley, A.A. Line scan proton spin imaging in biological structures by NMR. *Phys. Med. Biol.* **21:** 847; 1976.
79. Marchi, R.P., and Eyring, H. Application of significant structure theory to water. *J. Chem. Phys.* **68:** 221; 1964.
80. Mathur-DeVré, R., and Bertinchamps, A.J. The effects of $\gamma$-irradiation on the hydration characteristics of DNA and polynucleotides. II. An NMR study of mixed $H_2O/D_2O$ frozen solutions. *Radiat. Res.* **72:** 181–189.
81. Mears, P. The mechanism of water transport in membranes. *Phil. Trans. R. Soc. London B* **278:** 113; 1977.
82. Meiboom, S. Nuclear magnetic resonance study of the proton transfer in water. *J. Chem. Phys.* **34:** 375; 1961.
83. Meiboom, S., and Gill, D. Modified spin–echo method for measuring nuclear relaxation times. *Rev. Sci. Instrum.* **29:** 688; 1958.
84. Migchelsen, C., and Berendsen, H.J.C. Proton exchange and molecular orientation of water in hydrated collagen fibers. An NMR study of $H_2O$ and $D_2O$. *J. Chem. Phys.* **59:** 296–305; 1973.
85. Migchelsen, C., and Berendsen, H.J.C., Rupprecht, A. Hydration of DNA. Comparison of nuclear magnetic resonance results for oriented DNA in the A, B and C form. *J. Molec. Biol.* **37:** 235–237; 1968.
86. Migchelsen, C., and Berendsen, H.J.C. Deuteron magnetic resonance on hydrated collagen. In: *Magnetic Resonance and Relaxation,* R. Blinc, ed. Amsterdam: North-Holland; pp. 761–766; 1967.
87. Mullen, K., and Pregosin, P.S. *Fourier Transform NMR Techniques: A Practical Approach.* London: Academic Press; 1976.
88. Nemethy, G., and Scheraga, H.A. Structure of water and hydrophobic bonding in proteins. I. A model for the thermodynamic properties of liquid water. *J. Chem. Phys.* **36:** 3382; 1962.
89. Oakes, J. Nuclear magnetic resonance relaxation studies of the state of water in native bovine serum albumin solutions. *J. Chem. Soc. Faraday Trans. I* **72:** 216; 1976.
90. Ohgushi, M., Nagayama, K., and Wada, A. Dextrane-magnetite: A new relaxation reagent and its application to $T_1$ measurements in gel systems. *J. Magn. Reson.* **29:** 599; 1978.
91. Outhred, R.K., and George, E.P. A nuclear magnetic resonance study of hydrated systems using the frequency dependence of the relaxation processes. *Biophys. J.* **13:** 82–96; 1973a.
92. Packer, K.J. The dynamics of water in heterogeneous systems. *Phil. Trans. R. Soc. B* **278:** 59–87; 1977.
93. Pope, J.M., and Cornell, B.A. A pulsed NMR study of lipids, bound water and sodium ions in macroscopically oriented lecithin/water lyotropic liquid crystal model membrane systems. *Chem. Phys. Lipids* **24:** 27; 1979.
94. Ramirez, J.E., Cavanaugh, J.R., and Purcell, J.M. Nuclear magnetic resonance studies of frozen aqueous solutions. *J. Phys. Chem.* **78:** 807; 1974.
95. Reeves, L.W. The study of water in hydrate crystals by nuclear magnetic resonance. In: *Progress in Nuclear Magnetic Resonance Spectroscopy,* J.W. Emsley, J. Feeney, L.H. Sutcliffee, eds. Oxford: Pergamon Press. **4:** 193; 1969.
96. Rupprecht, A. Hydration of DNA, a wide line NMR study of oriented DNA. *Acta Chem. Scand.* **20:** 582–585; 1966.
97. Schneider, A.S., Middaugh, C.R., and Oldewurtel, M.D. Role of bound water in biological membrane structure: Fluorescence and infrared studies. *J. Supramol. Structure* **10:** 265; 1979.
98. Shaw, D. *Fourier Transform NMR Spectroscopy.* Amsterdam: Elsevier; 1976.
99. Slichter, C.P. *The Principles of Magnetic Resonance with Examples from Solid State Physics.* New York: Harper; 1963.
100. Smith, D.W.G., and Powles, J.G. Proton spin–lattice relaxation in liquid water and liquid ammonia. *Mol. Phys.* **10:** 452; 1966.
101. Stejskal, E.O., and Tanner, J.S. Spin diffusion measurements: Spin–echoes in the presence of a time-dependent field gradient. *J. Chem. Phys.* **42:** 288; 1965.
102. Stillinger, F.H. Theoretical approaches to the intermolecular nature of water. *Phil. Trans. R. Soc. Lond. B* **278:** 97; 1977.
103. Sykes, B.D., Hull, W.E., and Snyder, G.H. Experimental evidence for the role of cross-relaxation in proton nuclear magnetic resonance spin–lattice relaxation time measurements in proteins. *Biophys. J.* **21:** 137–146; 1978.
104. Szent-Gyorgyi, A. *Bioenergetics.* New York: Academic Press; 1957.
105. Tanner, J.E., and Stejskal, E.O. Restricted self-diffusion of protons in colloidal systems by pulsed-gradient, spin–echo method. *J. Chem. Phys.* **49:** 1768; 1968.
106. Texter, J. Nucleic acid-water interactions. *Prog. Biophys. Molec. Biol.* **33:** 83–97; 1978.
107. Tricot, Y., and Niederberger, W. Water orientation and motion in phospholipid bilayers: A comparison between $^{17}O$ and $^{2}H$-NMR. *Biophys. Chem.* **9:** 195; 1979.
108. Volkov, V.Y., and Velikanov, G.A. Izuchenie transporta vody v membranah kletok drozhzhej impulsnim

metodom yadernogo magnitnogo rezonanca. *Biofizika* **24:** 77; 1979.

109. Walmsley, R.H., and Shporer, M. Surface induced NMR line splittings and augmented relaxation rates in water. *J. Chem. Phys.* **68:** 2584–2590; 1978.
110. Wieslander, A., Ulmius, J., Lindblom, G., and Fontell, K. Water binding and phase structures for different *Acholeplasma laidlawii* membrane lipids studied by deuteron nuclear magnetic resonance and X-ray diffraction. *Biochim. Biophys. Acta* **512:** 241; 1978.
111. Woessner, D.E. Temperature dependences of nuclear-transfer and spin relaxation phenomena of water absorbed on silica gel. *J. Chem. Phys.* **39:** 2783; 1963.
112. Zimmerman, J.R., and Brittin, W.E. Nuclear magnetic resonance studies in multiple phase systems: Lifetime of a water molecule in an adsorbing phase on silica gel. *J. Phys. Chem.* **61:** 1328; 1957.
113. Zupancic, I., Lahajnar, G., Rutar, V., Miljkovic, L., and Rupprecht, A. Proton NMR and self-diffusion of absorbed water in oriented DNA. In: *Magnetic Resonance and Related Phenomena,* 1979.

## 10.2. GLOSSARY OF PHYSICAL TERMS FOR NUCLEAR MAGNETIC RESONANCE

**absorption line (in NMR):** A line observed in a nuclear magnetic resonance (NMR) spectrum corresponding to a single transition or a set of transitions arising from absorption of the incident radio-frequency radiation by the nuclear spins.
See also *FWHM, integral, line amplitude, line shape.*

**acquisition delay time:** The time between the end of the radio-frequency pulse and the beginning of data acquisition in an NMR experiment.

**acquisition rate:** The number of data points acquired per second. Also known as sampling rate or digitizing rate.

**acquisition time:** The time during an NMR experiment in which data are acquired and digitized. Acquisition time = $Nt_d$, where $N$ is the number of data points sampled and $t_d$ is the dwell time.

**adiabatic fast passage:** A method of conducting an NMR experiment in which either the static field or the radio-frequency field is swept while the other is held constant. This sweeping can be done either slowly or quickly, compared with the spin–lattice proton relaxation time ($T_1$). If the sweep is done quickly, it is *adiabatic.* (An adiabatic process is an irreversible thermodynamical process carried out with no change of entropy, i.e., without any change in the magnetization in this case of fast passage.) This method can be used to invert the spins in a continuous-wave NMR experiment. (It is similar to the 180° pulse in pulsed NMR.)

**algorithm:** A set of well-defined rules for the solution of a problem in a finite number of steps. For example, a complete statement of an arithmetical procedure for evaluating $e^x$.

**alternating current (AC):** An electrical current in a circuit that reverses direction alternately, at regular intervals of time. Compare *direct current.*

**angular frequency** (angular velocity): Rate of change of angular motion over time about an axis. In the centimeter-gram-second (cgs) system, the unit is radians/s. If the angle described in time $t$ s is $\Theta$ radians, the angular frequency, $\omega = \Theta t$ radians/s. $\omega = 2\pi\nu$, where $\nu$ is frequency in Hz or cycles/s.

**angular momentum** (quantity of motion): For a particle of mass $m$, moving with a velocity $v$, the angular momentum relative to a point $O$, is the vector product of the position vector **r** of the particle relative to $O$ and linear momentum $m\mathbf{v}$. That is, $\mathbf{L} = \mathbf{r} \times m\mathbf{v}$, where **L** is the angular momentum vector. For atomic and nuclear particles, this momentum is quantized.

**anisotropic motion:** Motion of an atomic or molecular species about one preferred axis, e.g., rotation about one axis in a molecule.

**anisotropy:** Change of a physical parameter with directions in a substance, e.g., variation of the self-diffusion coefficient of water and spin–spin relaxation time ($T_2$) with the static magnetic field direction being along or perpendicular to muscle fiber orientation.

**aperture time:** The time interval during which the signal is received by the sampling device in signal processing, e.g., the aperture time of the boxcar integrator is the duration of the gate pulse during which the signal is allowed into the integrator. The aperture time is generally a small fraction of the dwell time one uses in typical FT NMR experiments.

**automatic frequency control:** A technique by which the microwave frequency is kept in tune with the resonance frequency of the ESR cavity. This is necessarily used in ESR spectrometers.

**axial symmetry:** In some systems, there may be one unique axis about which there is symmetry and the other two perpendicular axes are equivalent, e.g., the hyperfine tensor A can be defined in terms of $A_\parallel$ and $A_/$ when there is axial symmetry.

**bandwidth:** The frequency range in hertz, e.g., filter bandwidth implies the range of frequencies that the filter will allow to pass through with less than 3db (50%) attenuation in power by the band pass filter.

**bimodular cavity:** A special type of microwave cavity used in electron spin resonance experiments. This type of cavity can be simultaneously subjected to two microwave frequencies.

**Bloch equations:** Phenomenological equations of motion developed by Bloch for the macroscopic magnetization used in nuclear magnetic resonance techniques.

**bohr magneton:** The unit of atomic magnetic moment denoted by $\beta$ or $\mu_\beta$. It is equal to $eh/4\pi m_e c$ where $e$ is the electron charge, $h$ is Planck's constant, $m_e$ is the electron rest mass and $c$ is the speed of light. It is the magnetic moment of a single electron spin and its value is $0.927 \times 10^{-20}$ erg/gauss (cgs units) or $9.27 \times 10^{-24}$ (joules/tesla, SI units).

**Boltzmann distribution:** Population distribution of a system of particles in thermal equilibrium. The number of particles in energy level $E$ at absolute temperature $T$ (°K) is proportional to $e^{-E/kT}$ where $e$ is the base of the Napierian logarithms and $\boldsymbol{k}$ is Boltzmann's constant. More specifically in the case of nuclear spins, the number of nuclear spins in parallel ($N_+$) and anti-parallel ($N_-$) orientations with respect to the magnetic field is given by $N = Ne^{-\mu H/RT}$ where $\mu$ is the magnetic moment, $H$ is the magnetic field in gauss and $\boldsymbol{k}$ is Boltzmann's constant. This small excess of nuclear spins in the lower energy state (orientation parallel to the static field) gives rise to the resonance phenomenon.

**bolus:** A quantity of opaque medium or radiopharmaceutical introduced into an artery at one time. In NMR method of blood flow measurement, the bolus consists of a certain volume of blood whose protons have been tipped by the rf pulse.

**boxcar integrator:** A type of signal processing equipment used to improve the signal-to-noise ratio in pulsed NMR experiments. It is usually a single channel or dual channel averaging device.

**broad band decoupling:** A technique used in high resolution NMR spectroscopy in which nuclei of the same isotope but possibly different chemical shifts are decoupled simultaneously from a heteronucleus, e.g., in $^{13}C$ NMR spectroscopy, the coupling of all protons to the $^{13}C$ nuclei is removed by broad band decoupling.

**capacitor:** Capacitor is an electrical device for storing electrical charge. Capacitance of the capacitor indicates the amount of charge stored in it when its potential is raised by one volt.

$$C = Q/V$$

$C$ is capacitance (in farads)
$Q$ is charge in coulombs
$V$ is the potential in volts

**Carr-Purcell sequence:** A sequence of rf pulses consisting of 90°-$\tau$-180°-$\tau$-180°-$\tau$-180° used to measure the spin–spin relaxation time, $T_2$. This method reduces the contributions due to molecular diffusion unlike the case of the 90°-$\tau$-180° spin–echo method.

**Carr-Purcell-Meiboom-Gill sequence:** A modification of the Carr-Purcell sequence, designed to correct any inaccuracies in the setting of the 180° pulses. This method gives the most accurate $T_2$ value. This method is the same as the Carr-Purcell sequence except that a phase shift of 90° is introduced into the 180° pulses.

**carrier:** It is a wave having at least one characteristic (amplitude, phase or frequency) that may be varied from a known reference value by another physical function, e.g., in AM radio broadcasts, it is the amplitude of the rf carrier that is modulated by the audio signal whereas in FM radio broadcasts it is the frequency of the rf carrier that is modulated by the audio signal.

**CAT:** (a) Computer of average transients used for signal averaging in CW NMR experiments. (b) In radiological context, Computed Axial Tomography: A diagnostic imaging modality using x-rays.

**cavity wavemeter:** Device for measuring microwave frequency used in ESR spectroscopy.

**chemical shift:** This specifies the position of a line in a high resolution NMR spectrum with reference to the position of a standard reference line. This shift arises due to the shielding of the applied magnetic field by the electrons surrounding the particular nucleus. Hence provides information about chemical structure. It is defined as

$$\delta = \frac{\Delta\nu}{\nu_R} \times 10^6 \text{ (ppm)}$$

where $\nu_R$ is the frequency with which the reference substance is in resonance at the applied magnetic field and $\Delta\nu$ is the frequency difference between the reference substance and that particular nucleus in the substance whose chemical shift is being determined at the same field. The sign of $\Delta\nu$ is to be chosen such that shifts to the high frequency side of the reference are positive.

**coherence (phase coherence):** If two signals of the same frequency or nearly the same frequency have the same phase, they are said to be phase coherent.

**computer limited spectral resolution:** It is the spectral width divided by the number of data points.

**conformation:** It is the three-dimensional structure of a molecule. The angles of rotation about the different chemical bonds with molecules make up the conformation of the molecule. More than one conformation of the molecule can be allowed.

**conformational change:** This is the change of the conformation of the molecule defined above. This conformational change can be either gross or small. An example of a gross change is the unfolding of a protein or uncoiling of a DNA molecule, etc. Small changes in the conformation of big molecules can occur in the binding of a ligand, such as the binding of a substrate to an enzyme.

**continuous wave NMR:** The type of NMR technique in which the spin system is continuously excited by a radiofrequency radiation of small strength. The rf field strength is such that the populations of the nuclear Zeeman energy levels are never equal. Absorption of the rf energy by the spin system is detectable by sweeping either the field or the frequency of the rf through the resonance.

**correlation time ($\tau_c$):** It is the characteristic "average" time between two molecular orientations for a molecule undergoing some kind of motion through interaction of a particular type, e.g., dipolar interaction correlation time, such as diffusional correlational time, rotational correlational time, etc.

**covalent bond:** A type of chemical bond which involves transfer of electrons between the atoms forming the molecule.

**cryomagnet:** Superconducting magnet which can operate without external electrical power once the field is established. At very low temperatures (less than 23 °K) some substances become superconducting, i.e., zero resistance. A cryomagnet needs to be kept at these very low temperatures by using liquid helium.

**cryoshims:** The superconducting correction coils (shim coils) used in the superconducting magnet for adjustment of the homogeneity of the field. See also *shim coils.*

**cryostat:** A multiwalled vacuum insulated vessel used for storing liquid helium and liquid nitrogen to maintain low temperatures, e.g., 4.2 °K. The superconducting magnet is operated at constant temperature of 4.2 °K and is located inside the liquid helium bath.

**C W NMR:** See *continuous wave NMR.*

**dead time:** Time during which receiver is unable to receive the signal in a pulsed NMR spectrometer.

**decibel (db):** One-tenth of a bel: the number of decibels denoting the nature of the two amounts of power being ten times the logarithm to the base 10 of this ratio; db is the commonly used abbreviation for decibel. With $P_1$ and $P_2$ designating two amounts of power and $n$, the number of decibels denoting their ratio,

$$n = 10 \log_{10} (P_1/P_2) \text{ decibels}$$

For current and voltage ratio the corresponding expressions are,

$$n = 20 \log_{10} (I_1/I_2) \text{ decibels}$$

$$n = 20 \log_{10} (V_1/V_2) \text{ decibels}$$

where $I_1/I_2$ and $V_1/V_2$ are the current and voltage ratios respectively.

**dewar:** Double-walled, vacuum insulated vessel used for storing cold or hot liquids.

**diamagnetism:** Many substances when they are subjected to a magnetization force, become magnetized in such a direction as to oppose that force, i.e., exhibiting negative susceptibility. This property is referred to as diamagnetism, e.g., aluminum is diamagnetic. See also *ferromagnetism, paramagnetism.*

**dielectric absorption:** Strong microwave electrified absorption by polar liquids, e.g., water.

**digital computer:** Operates on digital data; performs arithmetic and logical operations.

**diode:** An electronic device which allows current to flow in only one direction in a circuit. A diode can convert AC current to DC current.

**diode detector:** A type of detection used in radios and NMR spectrometers using simple diodes. They are non-linear under some conditions. They are amplitude detectors and carry no phase information. See also *phase sensitive detector.*

**dipolar correlation time:** It is the characteristic "average" time taken by a spin system to change between two molecular orientations by "through space" interactions with neighboring spins.

**dipolar interactions:** Magnetic interaction between two spins possessing magnetic moments. This interaction is due to the magnetic field produced by one spin acting on the other.

**direct current (DC):** An electric current in a circuit which does not change its direction or polarity with time. See *alternating current.*

**d-orbital:** Electron orbital or wave function which has an orbital angular momentum of ½. Electrons in the d-orbital give rise to paramagnetism of transitional metal ions, e.g., 3d electrons in iron group transformation metals.

**D S S:** Sodium salt of 2,2 dimethylsilapentane 5-sulphonic acid. NMR reference substance used for aqueous samples. $^1$H and $^{13}$C NMR reference.

**dwell time:** The time between the beginning of sampling one data point and the beginning of sampling of the next successive point in the FID.

**echo:** See *spin–echo.*

**electromagnet:** A type of magnet which is energized by electrical power.

**electron spin–echo spectroscopy:** Technique exactly similar to nuclear spin–echo spectroscopy except that one observes electron spin–echos. In this technique, pairs of microwave pulses of high strength are applied to the electron spins in a paramagnetic sample in the same way as in a nuclear spin–echo experiment and an electron spin–echo is observed.

**electron spin–lattice relaxation time ($T_1$):** The characteristic time constant taken by the electron spins to reach thermal equilibrium through interaction with the fluctuating local electrical fields surrounding these spins (lattice).

**electron spin–spin relaxation time ($T_2$):** The characteristic time constant taken by the electron spins to lose phase coherence, i.e., return to the equilibrium value through interaction with neighboring spins in an ESR experiment. It is inversely related to the line widths of the resonance and its origin is similar to the "uncertainty broadening" observed in other forms of spectroscopy.

**energy level:** The stable energy states that an atom or molecule can take up or the nuclear or electron spins take up in the presence of the applied magnetic field. The lowest energy states are occupied while the higher energy states are partially occupied. Only discrete sets of energy levels exist according to quantum mechanical principles.

**E S R:** Electron spin resonance, i.e., a magnetic resonance experiment carried out on unpaired electron spins.

**exponential function:** An exponential function is a function of the type $e^{Ax}$ where $e$ is the base of the Naperian logarithm ($e$ has the value of 2.718) and $A$ is a constant, which has the inverse dimensions of $x$. For example, (i), the rate of decay of radioactivity is exponential with respect to time. If the activity at time $t = 0$ is $A_0$, then the activity at any later time will be given by $A = A_0 e^{-\lambda t}$, where $\lambda$ is a characteristic decay constant of that specific radioactive nuclei. (ii) The charge $Q$ on a capacitor of capacitance $C$, which is being charged through a resistor $R$ will build up to its full value $Q_0$ exponentially with respect to time as follows $Q = Q_0 e^{t/RC}$. (iii) The FID signal in a time domain NMR experiment may decay exponentially with a time constant $T_2$. $M(t) = M_0 e^{-t/T_2}$ where $M_0$ is the size of the signal at $t = 0$. Note the following characteristics of the exponential functions: (a) the exponent ($Ax$) in the exponential function must be dimensionless e.g., $RC$ must have the same units (dimension) as time.
(b) The increase (decrease) of the exponential function is relatively fast for values of $x < A$, where $x = 1/A$, the function has grown to approximately 3 times the value (decreased to approximately ⅓ of the value) at the beginning. For $x \gg A$ (i.e., 5 or 6 times) the function changes very little.
(c) If the growth or decay is due to two or more independent processes each being exponential, then the total growth (or decay) can still be represented as an exponential function.

**external lock (NMR):** In a high resolution NMR technique, an NMR signal from another substance other than the actual substance under investigation is used to control the field-frequency ratio of the spectrometer. If the material giving rise to the locking signal is located outside the sample tube, it is referred to as an external lock.

**external reference (NMR):** In high resolution NMR spectroscopy, the chemical shifts are measured with reference to the position of a selected standard material in the NMR spectrum. If this reference compound is not dissolved in the same phase as the sample under investigation, it is referred to as an external reference.

**fast Fourier transform (FFT):** FFT is a computational algorithm first developed by Cooley and Tukey which reduces the computational time required for Fourier analysis by digital computer. For a function which does not have a closed form Fourier transform and which has $N$ separate digital data points, the computational time required to determine the $N$ sinusoidal amplitudes, the digital computer requires $N^2$. However, in the Cooley and Tukey FFT algorithm this time becomes proportional to $N$ $\mathrm{Log}_2 N^2$ computations, thus reducing the total time.

**ferromagnetism:** The phenomenon whereby certain materials exhibit a high degree of magnetism in weak fields and possess a very high permeability. Ferromagnetic materials exhibit spontaneous magnetization, residual magnetism, and are subject to hysteresis. This behavior is markedly dependent on whether the temperature is above or below that of the Curie point. See also *diamagnetism* and *paramagnetism*.

**field gradient:** See *gradient*.

**field sweep:** One mode of observing the NMR spectrum by systematically varying (sweeping) the applied static field and keeping the frequency of the rf radiation constant. By varying the magnetic field strength, NMR transitions of different energies can be brought into resonance successively. In this kind of experiment, the NMR spectrum consists of signal intensity versus the magnetic field strength.

**filling factor:** That fractional volume of the NMR coil which is filled by the sample to be investigated.

**filter (electronic):** Electronic device which lets through only certain frequencies and is characterized by a bandwidth ($\Delta\nu$) or window.

**filtered back projection:** An algorithm used in NMR and CT scanning techniques to reconstruct an image from a set of projection data.

**first derivative:** The instantaneous rate of change of a function with respect to the variable. Two special interpretations:
(1) As the slope of a curve.
(2) As the speed of a moving particle.

**flat cell:** A special type of cell which is made of quartz and has flat rectangular faces (~0.3 mm apart) used in ESR spectroscopy for the observation of ESR signals in aqueous samples.

**foldover:** A term used in pulsed Fourier NMR spectroscopy when single phase detection is used. Because of this type of detection, a single spectral line can be detected at two different frequencies symmetrically located at about the carrier frequency ($\nu_0$). The process of reflecting the line at for example $\nu_0 + \Delta\nu$ to $\nu_0 - \Delta\nu$ is called foldover.

**FONAR:** One type of imaging technique called Field Focusing NMR, developed by Damadian in which the field is focused at a point on the object called a saddle point. In this technique, the volume of interest can be localized with the aid of an inhomogeneous magnetic field which produces a saddle-shaped profile.

**Fourier transform (FT):** A mathematical technique by which it is possible to transform a signal that is observed as a function of intensity versus time (time domain) to a signal that is observable as intensity versus frequency consisting of sine and cosine waves. This technique is used in conventional high resolution NMR as well as in imaging techniques. Mathematically speaking, a function $f(t)$ can usually be expressed as a Fourier series, i.e., an infinite series of cosines and sines;

$$f(t) = \sum_{n=0}^{\infty} A_n \cos\left(\frac{n\pi}{T}\right)t + \sum_{n=0}^{\infty} B_n \sin\left(\frac{n\pi}{T}\right)t$$

This expansion is valid only over the region $-T < t < T$. When we carry out the mathematical operations of Fourier analysis, we customarily deal not with Fourier series, but with related integrals that can cope with functions in which the variable $t$ is not restricted to the range $-T < t < T$, but can go to infinity. Then, the Fourier transform of $f(t)$ is defined as follows:

$$F(\omega) = \int_{-\infty}^{\infty} f(t)\, e^{-i\omega t} dt$$

When $\omega$ is an angular frequency ($\omega = 2\pi f$) and $i = \sqrt{-1}$. Also $f(t) = 1/2\pi \int_{-\infty}^{\infty} F(\omega) e^{i\omega t} d\omega$.

**free induction decay (FID):** Transient nuclear signal induced in the NMR coil after an rf pulse has excited the nuclear spin system in pulsed NMR techniques. This is referred to as free induction decay signal because the signal is induced by the free precession of the nuclear spins around the static field after the rf pulse has been turned off.

**free radical:** A molecule containing an unpaired electron and it is generally chemically reactive. Some free radicals may not be stable at room temperature.

**frequency (NMR):** Characteristic quantity of a magnetic or electrical field which is changing its direction alternatively at a regular interval. It is equal to the reciprocal of the period of this periodic motion. It is measured in hertz (Hz) which is one cycle per second.

**frequency sweep (NMR):** One mode of observing the NMR spectrum by systematically varying the frequency of the radio-frequency (rf) radiation and keeping the static field strength constant. This can also be done by varying the frequency of the modulation side band of the applied radio-frequency field. By varying the frequency of the applied rf radiation, NMR transitions of different energies can be brought into resonance successively. In this kind of experiment, the NMR spectrum consists of signal intensity versus frequency of the applied rf.
See also *field sweep*.

**FWHM:** Full width at half-maximum ($\Delta\nu_{1/2}$). The line width of a spectral line is usually referred to as FWHM.

**gauss:** Unit of magnetic flux density (frequently referred to as unit of magnetic field strength). The two are numerically equivalent in air.

1kG = 1000 gauss and 1 tesla = 10,000 gauss.

**gaussian line shape:** Shape of a spectral line whose height changes as a function of frequency in the frequency domain as

$$I(\nu) = I_{\max} \exp\left[\frac{-(\nu_0 - \nu)^2}{2\sigma^2}\right]$$

where $\nu_0$ is the frequency at the center of the line and the full line width at half-maximum is given by $\Delta\nu = 2.36\sigma$.

$$I(t) = I(0) \exp\left(\frac{-\sigma^2 t^2}{2}\right)$$

The characteristic decay time for $I(t)$ can be defined as the time $T_2$ required for $I(t)$ to decay to $1/e$ of $I(0)$. Then $T_2 = \sqrt{2}/\sigma$ so that the full width at half-height in the frequency domain is $3.34/T_2$.

**gradient:** The change in the value of a quantity with a change in a given variable. For example, if the magnetic field is not uniform but is changing continuously in the $x$ direction, we would say that there exists a magnetic field gradient in the $x$-direction.

**g-value:** A characteristic number which denotes the size of the magnetic moment of a paramagnetic species. The ESR spectrum of a paramagnetic species occurs at a position in the frequency domain determined by the g-value of the species. The g-value of the free electron is 2.0023.

**gyromagnetic ratio ($\gamma$):** See *magnetogyric ratio*.

**Hall probe:** A device used for measuring magnetic fields. A Hall probe is used with wide line NMR and ESR instruments for stabilizing the magnetic field produced by the electromagnets.

**hamiltonian:** The hamiltonian ($\mathcal{H}$) is a mathematical operator used in quantum mechanical treatment of some phenomena (e.g., magnetic resonance). It represents the sum of the kinetic and potential energies of a particle or a system. It can be represented as a function of momentum and position coordinates of the particle. It can be expressed as $\mathcal{H} = p^2/2m + V(r)$, where $p$ is the momentum operator, $V(r)$ is the potential energy as a function of position operator $r$, and $m$ is the mass of the particle.

**Helmholtz coils:** Two circular parallel coils of the same radius separated by a distance equal to their radius. $\Delta\nu_{1/2} = 1/\pi T_2^*$, where $T_2^*$ is the characteristic time constant for the decay of the FID signal. They produce a region of constant magnetic field on either side of the mid-point between the two coils.

**hertz:** Frequency unit, 1 Hz is 1 cycle per second. 1 KHz = $10^3$ Hz = 1000 Hz; 1 MHz = $10^6$Hz = 1000 KHz. 1 GHz = $10^9$ Hz = $10^3$ MHz.

**heteronuclear decoupling:** A technique used in high resolution NMR spectroscopy for elimination of spin–spin interactions between nuclei of different types, irradiating with high rf power at the resonance frequency of the heteronucleus, e.g., coupling between $^{31}P$ and $^{1}H$ spins is eliminated by irradiating at the $^{1}H$ NMR frequency of the protons couples to $^{31}P$ nucleus under observation.

**heteronuclear lock:** A lock signal obtained from a different nucleus other than the one that is being observed in high resolution NMR spectrometer, e.g., in $^{1}H$, $^{13}C$, and $^{31}P$ high resolution FT NMR spectrometers, a deuterium lock is commonly used.

**high resolution NMR spectroscopy:** An NMR spectrometer that is capable of producing very narrow NMR lines for the nucleus of a given isotope, i.e., lines with widths that are less than the majority of chemical shifts and coupling constants for that isotope.

**homogeneity spoil pulse:** One of the techniques used in pulsed Fourier transform NMR spectroscopy. In this technique, a temporary deterioration of the homogeneity of the static magnetic field ($B$) is deliberately introduced in a pulsed mode.

**homonuclear decoupling:** A technique used in high resolution NMR spectroscopy for the elimination of spin–spin interactions between the nuclei of the same type by irradiation with high rf power at the resonance frequency of one of them. Very useful in the assignment of lines in NMR spectra, e.g., by the decoupling of one proton from another adjacent to it in a molecule.

**homonuclear lock:** A lock signal in high resolution NMR spectrometer which is obtained from the same nuclide that is being observed.

**hyperfine splitting:** Splittings in the lines of an ESR spectrum arising due to the interaction of the nuclei in the vicinity of an unpaired electron with the unpaired electron under observation. This parameter can be used for determining the structure of a free radical or in identifying the ligands of a paramagnetic ion.

**inductance:** Inductance of a coil is equal to the magnetic flux through the coil when unit current flows through it. It is represented by $L$. When current, $I$, flows through the coil, a magnetic field is produced around the coil. The energy stored in the coil is given by $\frac{1}{2} LI^2$, where $L$ is the inductance of the coil.

**integral (NMR):** A quantitative measure of the relative intensities of the NMR signals. The integral is given by the area of the spectral line and is usually presented as a step function in which the heights of the steps are proportional to the intensities of the resonance lines, if integration is done in the analogue mode. If the integration is done digitally, it is obtained by summing the amplitudes of the digital data points that define the envelope of each NMR line. The results of those summations are usually displayed either as a normalized total number of digital counts for each line or as a step function superimposed on the spectrum.

**interaction broadening:** ESR lines in an ESR spectrum can be broadened by magnetic interactions between paramagnetic molecules. This interaction is dependent on the concentration of the paramagnetic species.

**interface:** That part which is intended to provide the connection between one item of equipment to another. The specification of an interface should include the characteristics of signals (amplitude, duration, timing) and of the signal path (e.g., impedance, timing) and the nature of any control sequences.

**internal reference (NMR):** A reference substance that is dissolved in the same phase as the sample. The primary internal reference for proton spectra and $^{13}C$ spectra of non-aqueous solutions is tetramethylsilane (TMS). A concentration of 1% is recommended. For proton spectra of aqueous solutions, the recommended internal reference is sodium salt of 2,2,3,3,-tetradeutro-4,4-dimethyl-4-silapentanoic acid (TSP-d4). For $^{13}C$ spectra of aqueous solutions, secondary standards such as dioxane have been found satisfactory.

**inversion recovery sequence:** A type of pulse sequence in which the magnetization is inverted by means of a 180° rf pulse and recovery from this is monitored by means of a 90° rf pulse applied after a time $\tau$. This sequence is commonly used for measurement of $T_1$ (*in vivo* and *in vitro*) in biological systems.

**isosbestic points:** The points of intersection in a spectrum observed when one species is converted to a spectrally distinct species.

**isotopes:** Atoms which have the same atomic numbers (i.e., the same number of electrons) but different mass numbers (atomic weights). Therefore, isotopes are the same element but have different atomic weights, and, hence, different nuclear spins.

**isotropic:** If a property is the same in every direction, for example while the proton NMR spectrum of liver tissue is isotropic, that of muscle may depend on the direction of the fibers with respect to the applied magnetic field. Contrast with *anisotropy* where the property is not the same in all directions, e.g., anisotropy of $T_2$ in muscle tissue.

**isotropic motion:** Motion which is equally probable in all directions, e.g., random rotational motion of a molecule about the three axes of a molecule.

**isotropic spectrum (NMR or ESR):** ESR or NMR spectrum in which the molecular motion is so fast that any anisotropy in the spectrum is averaged out, e.g., ESR spectrum of a spin label in a liquid or NMR spectrum of bulk water.

**J-spectra:** In high resolution Fourier transform NMR spectroscopy, Fourier transformation of the Carr-Purcell spin–echo sequence gives spectra called J-spectra consisting of lines which have their natural line width and are separated only by coupling constants. Chemical shifts do not appear in the spectra. J-spectra are useful for measurement of very small coupling constants.

**Jeener-Broekaert phase shifted-pulse pair sequence:** A type of pulse sequence used in pulsed NMR experiments to measure $T_{1D}$, the spin–lattice relaxation time in the dipolar field. In this method, pulse sequence used is $90°–\tau_1–(45°)_0–\tau_2–(45°)_{90}$ –echo. The subscripts 0 and 90 refer to 0° and 90° phase shifts introduced into the rf pulse, $\tau_1$ and $\tau_2$ are time delays between the pulses.

**kilo (k):** Kilo implies one thousand, e.g., kilohertz implies one thousand hertz. In computers K is used to denote 1024 bytes ($2^{10}$), e.g., 32K memory.

**klystron:** Electronic device used to generate microwave radiation.

**Larmor equation:** The equation defining the resonance condition in magnetic resonance phenomena. The Larmor equation is $\omega_0 = \gamma B_0$ where $\omega_0$ is the Larmor frequency in radiations per second, $\gamma$ is the gyromagnetic ratio and $B_0$ is the magnetic field (induction) strength.

**Larmor frequency:** The frequency ($\omega_0$) of Larmor precession. $\omega_0$ is the angular frequency of Larmor precession expressed in radians per second and $\omega_0 = 2\pi \nu_0$, where $\nu_0$ is the frequency expressed in hertz. $\pi$ is 3.1416.

**Larmor precession:** The precessional motion of the orbit of a charged particle which is subjected to a magnetic field. The precession occurs about the direction of the field. For an electron revolving about a nucleus, the angular velocity of the Larmor precession is given by $eH/2mc$ where $e$ is the charge, $m$ is the mass, $c$ is the speed of light, and $H$ is the magnetic field strength.

**ligand:** An atom or small molecule binding to an atom or molecule of interest. The binding can be either covalent binding (e.g., nitrogens or oxygen to a metal ion) or non-covalent binding of a small molecule to a big molecule (e.g., binding of a substrate to an enzyme).

**line amplitude:** The height of the spectral line in an NMR spectrum as measured from the baseline.

**line intensity:** Area under the spectral line in an NMR spectrum. Under certain conditions, the intensity is an index of the number of nuclei giving rise to this resonance line.

**line scanning:** One type of NMR imaging technique in which the spin density distribution is sampled one line at a time. A line is scanned sequentially through the sample to obtain the complete image. This method is also referred to as the "sensitive line" or "multiple sensitive point" imaging.

**line shape:** Shape of a spectral line, i.e., the variation of the height of the line as a function of frequency. See also *gaussian line shape, lorentzian line shape.*

**line width (NMR):** The width of a line in a spectrum (e.g., NMR spectrum). This width is defined as the separation between the points of half-maximum height. The line width of an NMR spectral line may be expressed in hertz or gauss. Typical proton NMR line widths are of the order of a few Hz and in solids are of the order of a few kilohertz. In liquids line width ($\Delta\nu_{1/2}$) is related to $T_2^*$, $\Delta\nu_{1/2} = 1/\pi T_2^*$.

**lock signal:** The NMR signal used to control the field frequency ratio of the spectrometer generally in high resolution NMR spectroscopy. See also *external lock.* In most of the modern FT NMR high resolution instruments, a deuterium signal from the solvent is used for lock signal.

**longitudinal relaxation time ($T_1$):** Another name for spin–lattice relaxation time ($T_1$). See *spin–lattice relaxation time* for definition.

**lorentzian line shape:** Spectral line with narrow peak and long tails. It is the shape of a spectral line whose height varies as a function of frequency domain as $I(\nu) = I_{max}/[1 + a^2(\nu_0 - \nu)]^{-1}\sqrt{2}$, where $\nu_0$ is the frequency of line center and the line width is $\Delta\nu = 2/a$. In time domain it can be expressed as $I(t) = I(0)\exp(-t/T_2)$. The full width at half-maximum is $2/T_2$. Note that the Fourier transform of the lorentzian line is an exponential function.

**macroscopic magnetization vector:** In a sample which has individual microscopic nuclear magnetic moments, the net magnetic moment is the sum of all these individual magnetic moments. The net magnetic moment is treated as vector in the classical picture of magnetic resonance phenomenon. This magnetic moment vector is called the magnetization vector or magnetic moment vector, $M_0$.

**magic angle:** The angle 54.7° is called the magic angle because $3\cos^2\Theta - 1 = 0$ for this value of $\Theta$, e.g., the dipolar coupling contains the terms $3\cos^2\Theta - 1$, where $\Theta$ is the angle between the internuclear vector and the applied field. Hence for an angle $\Theta = 54.7°$, the dipolar interaction vanishes.

**magnet:** Any body which has the power of attracting iron.

**magnetic dipole:** North and south poles of a magnet separated by a finite distance constitutes a magnetic dipole moment. An electrical current in a circular loop can create a magnetic dipole moment.

**magnetic field:** The region in the neighborhood of a magnetized body in which magnetic forces can be detected. Also used interchangeably to refer to magnetic induction or magnetic intensity produced by an electromagnet, permanent magnet, or a superconducting magnet.

**magnetic flux ($\Phi_B$):** Through a closed figure (e.g., a circular or rectangular loop): The product of the area of the figure and the average component of magnetic induction normal to the area, i.e., the surface integral of the magnetic induction normal to the surface. The SI unit is the Weber (Wb) and the CGS unit, the Maxwell, is equal to $10^{-8}$ Wb: $\Phi = \overline{S}\overline{B}ds$, where $\overline{B}$ is the magnetic induction and $\overline{S}$ is the surface area.

**magnetic induction: (magnetic flux density) (*B*):** May be defined as the magnetic flux per unit area at right angles to the flux or as a product of the magnetic intensity and permeability. The magnetic induction is related to the magnetic field strength $H$ and the magnetization $M$ through the relationship $B = \mu_0(H + M)$, where $\mu_0$ is the permeability of free space. The unit of $B$ is in terms of magnetic flux density; the SI unit is the tesla ($1\,Wb/m^2$) and the CGS unit is the gauss (1 Maxwell/cm$^2$ = $10^{-4}$ tesla.)

**magnetic moment:** For a flat current loop of $N$ turns, each carrying a current $I$ and situated in a magnetic induction field $\overline{B}$, the magnetic moment may be defined through the torque relationship: $\overline{\tau} = \overline{\mu} \times \overline{B}$. Here $\overline{\mu} = Ni\,\overline{A}$, where $\overline{A}$ represents the area vector whose direction is perpendicular to the coil plane in the sense given by the right hand rule.

**magnetic moment (atomic):** The magnetic moment arising from the magnetic moment of the nucleus and the (relatively large) magnetic moment of the electrons.

**magnetic moment (electron):** The magnetic moment associated with the electron spin on the one hand and the orbital motion on the other. The moment of a single electron spin is one Bohr magneton. Electrons possess about 1000 times larger magnetic moment than the largest of the nuclear moments.

**magnetic moment (nuclear):** The magnetic moment associated with the nucleus as a consequence of their inherent charge and spin. It is related to the nuclear angular momentum $L$ through the gyromagnetic ratio constant ($\bar{\mu} = \gamma L = \gamma I\hbar$, where $I$ is the spin angular momentum vector). The nucleus may be viewed as a bar magnet spinning about its north–south axis.

**magnetic pole:** A convenient fiction for describing certain magnetic phenomena. It denotes the points of a magnet from which the magnetic field appears to diverge or towards which it appears to converge. The pole strength of a magnet is the magnetic moment divided by the distance between the poles.

**magnetic resonance:** That form of absorption spectroscopy in which the transitions between the energy levels in which different orientations of an electron or nuclear magnetic moment are induced when the frequency of the radiation is equal to the separation between the energy levels, i.e., in resonance. For nuclear magnetic resonance, the rf radiation is used while for electron resonance microwave radiation is used.

**magnetic shielding:** The reduction of the magnetic field "seen" by a nucleus in an atom or molecule due primarily to screening by the electron cloud of the molecule.

**magnetic susceptibility:** A characteristic quantity of a substance denoting the intensity of the magnetization produced in it by an applied magnetic field. Paramagnetic substances are characterized by positive susceptibility while diamagnetic substances have negative susceptibilities.

**magneticogyric ratio ($\gamma$):** Also called *gyromagnetic ratio.* A constant for each nucleus given by the ratio of the nuclear magnetic moment to the angular momentum. It is also equal to $\mu/I\hbar$, where $\mu$ is the magnetic moment, $I$ is the spin quantum number and $\hbar$ is Planck's constant divided by $2\pi$. It also appears in the Larmor equation relating the resonant frequency (Larmor frequency) to the magnetic induction field for protons $\gamma/2\pi = 4.26$ MHz/kg (or 42.6 MHz/tesla).

**magnetometer:** Device used for measurement of magnetic field strength, e.g., proton magnetometer which uses proton NMR frequency for measuring the magnetic field strength.

**megahertz (MHz):** $10^6$ Hz or 1000 KHz.

**Meiboom-Gill sequence:** See *Carr-Purcell-Meiboom-Gill sequence.*

**microsecond:** One millionth part of a second, i.e., $10^{-6}$ s.

**microwaves:** Electromagnetic radiation with frequencies in the range of 1000 MHz. For example, ESR is observed at 10 GHz, i.e., $10^{10}$ KHz (~3 cm wavelength).

**millisecond:** One thousandth part of a second, i.e., $10^{-3}$ s.

**modulation:** Super position of a time varying quantity on to a steady quantity, e.g., frequency modulation, magnetic field modulation, amplitude modulation, etc. In magnetic field modulation, a time varying magnetic field, varying at a particular frequency, is superimposed on the static magnetic field.

**modulation side bands:** Bands introduced into NMR absorption spectrum by modulation of the resonance signal. This can be done either by the modulation of the static magnetic field or by the amplitude or frequency modulation of the applied rf radiation.

**molecular orbital:** Wave function of an electron in a molecule. A molecular orbital is usually a linear combination of the electron wave functions of the constituent atoms of the molecule.

**nanosecond:** One nanosecond is equal to $10^{-9}$ s.

**NMR (nuclear magnetic resonance):** The resonant absorption of electromagnetic energy by a system of atomic nuclei situated in a magnetic field ($H_0$). The frequency, $\omega_0$, of the magnetic resonance is the same as the frequency of the Larmor precession of the nuclei in the magnetic field and is proportional to the strength of the field. Thus,

$$\omega_0 = \gamma H_0$$

where $\gamma$ is a characteristic constant, called the magnetogyric ratio, for a given nucleus.

**NMR absorption band:** A region of the spectrum in which a detectable signal exists and passes through one or more maxima.

**NMR absorption line:** A single transition or a set of degenerate transitions is referred to as a line.

**NMR integral:** A quantitative measure of the relative intensities of NMR signals, defined by the areas of the spectral lines.

**NMR line width:** The full width, expressed in hertz (Hz) of an observed NMR line at one-half maximum height (FWHM).

**NMR spectral resolution:** The width of a single line in the spectrum which is known to be sharp, such as, TMS or benzene. This definition includes sample factors as well as instrumental factors.

**nuclear angular momentum:** The concept of angular momentum is applied to a particle (nucleus) or a system of particles (atoms, molecules) that spins about an axis (or behaves as though it does) as well as a particle or system of particles that revolves in an orbit.

**nuclear magneton:** The unit of nuclear magnetic moment, given by $eh/4\pi mc$, where $e$ is the electronic charge, $h$ is Planck's constant, $m$ is the rest mass of a proton, and $c$ is the speed of light. Its value is $5.05 \times 10^{-24}$ erg/gauss, i.e., about 1/1840 that of a Bohr magneton.

**Nyquist frequency:** The sampling theorem states that to record a spectrum unambiguously with frequencies as high as $\nu$, one must sample the FID at a rate of at least $2\nu$ (Nyquist frequency).

**orbital magnetic moment:** Magnetic moment of a particle which is associated with its orbiting motion. This magnetic moment is in addition to the spin magnetic moment.

**paramagnetism:** The property shown by many substances of becoming magnetized in the direction of the applied field, but not retaining this directional magnetization when the field is removed.

**partially relaxed Fourier transform (PRFT) NMR:** A set of multiline FT spectra obtained from an inversion-recovery sequence and designed to provide information on spin–lattice relaxation times.

**phase (wave):** If the peaks of the signal waves coincide (in position and in time), then the two are exactly in phase. If the peak of one coincides with the trough of the other, then the signals are exactly out of phase.

Phase is the fractional part of a period through which the quantity has advanced from an arbitrary origin.

**phase sensitive detector:** A filter which allows only those signals modulated at a certain frequency to pass by comparing the phase of the signal with that of the modulated source.

**planar imaging:** An NMR imaging technique where data is obtained from an entire plane rather than from a line or a point.

**point scanning:** An NMR imaging technique where data is gathered point by point throughout the region of interest.

**potential energy:** Energy of a particle by virtue of its position, e.g., orientation within a magnetic field.

**precession:** A rotation of the spin axis produced by a torque applied about an axis mutually perpendicular to the spin axis and the axis of the resulting motion.

**probe:** The part of an NMR spectrometer which contains the sample and rf coils.

**projection profile:** One-dimensional projection of a given quantity such as spin-density.

**pulse:** To apply for a specified period of time a perturbation (e.g., a radio-frequency field) whose amplitude envelope is normally rectangular.

**pulse amplitude:** The radio-frequency field, $H_1$, in gauss.

**pulse flip angle:** The angle (in degrees or in radians) through which the magnetization is rotated by a specific pulse.

**pulse interval:** The time between two pulses of a sequence.

**pulse phase:** The phase of the radio-frequency field as measured relative to chosen axis in the rotating coordinate frame.

**pulse sequence:** A set of defined pulses and time spacings between these pulses.

**pulse width:** The duration of a pulse.

**Q-factor:** A measure of energy loss for a coil. The value of the reactance of either the inductor or capacitor at the resonant frequency of a series-resonant circuit, divided by the series resistance in the circuit, is called the $Q$ (quality factor) of the circuit.

**quantum:** A discrete energy packet. Electromagnetic radiation of frequency $\nu$ does not have continuously variable energy, but is made of packets of energy $h\nu$, where $h$ is Planck's constant.

**quantum mechanics:** The set of physical laws which universally apply to certain micro systems and predict that their properties (such as magnetic moment, energy) cannot have a continuous set of values but must have a set of discrete, discontinuous values.

**quenching:** Loss of superconductivity in a superconducting magnet.

**radio-frequency (rf):** Electromagnetic frequencies that are used in radio transmission and reception, 100 KHz to 500 MHz.

**radio-frequency bandwidth:** The range of radio-frequency within which the performance with respect to a given characteristic falls between specified limits.

**radio-frequency pulse:** A short burst of radio-frequency electromagnetic radiation. Rotation of the magnetization vector can be caused by controlling the duration and amplitude of the rf pulse.

**relaxation:** A process by which atoms or molecules in an excited state return to their ground state (or lower energy state).

**relaxation rate:** Spin–lattice relaxation rate is defined as $1/T_1$ or $R_1$, and spin–spin relaxation rate is defined at $1/T_2$ or $R_2$, where $T_1$ and $T_2$ are spin–lattice and spin–spin relaxation times, respectively.

**relaxation time:** Spin–lattice: A measure of the time taken for the spin population to return to its equilibrium value through interactions with the fluctuating internal electric fields which surround it (the lattice). It is characterized by the exponential time constant $T_1$ given by $M_z = M_0(1 - e^{-t/T_1})$, where $M_z$ is the $Z$ component of magnetization and the field is in the $Z$ direction, $M_0$ is the equilibrium value of the magnetization. Note the following:

| $t$ | $M_0$ |
|---|---|
| 0 | 0 |
| $T_1$ | $0.632\ M_0$ |
| $2T_1$ | $0.865\ M_0$ |
| $3T_1$ | $0.950\ M_0$ |
| $4T_1$ | $0.982\ M_0$ |
| $5T_1$ | $0.993\ M_0$ |

Thus, in about 5 times the relaxation time, the magnetization is recovered back to its equilibrium value, $M_0$.

Spin–spin: A measure of the time to lose phase coherence, i.e., return to equilibrium through interactions with neighboring spins. It is inversely related to the linewidths of the resonance and its origin is similar to the uncertainty broadening observed in other forms of spectroscopy. It is also characterized by an exponential time constant ($T_2$) as follows:

$$M_{xy} = M_0 e^{-t/T_2}$$

where $M_{xy}$ is the transverse component of the magnetization.

**resistive magnet:** An electromagnet that uses wires with sufficient electrical resistance so that there is considerable loss of power in terms of generation of heat in the conducting wires rather than conversion into magnetic field. Present-day large size imaging resistive magnets do not produce fields greater than 2 kG.

**resonance:** The phenomenon of amplification of an oscillation of a system by an external oscillation of exactly equal period. In NMR experiments, the frequency of the rf waves have to match the Larmor frequency of the nuclei in the magnetic field to create resonance phenomenon. Energy is then transferred from the rf field to the nucleus.

**rotating frame:** In NMR experiments, the effects of external rf field on the precessing nuclear magnetic moments are studied in detail. As a reference, a coordinate system is chosen with its $Z$ axis in the same direction as the applied magnetic field. If the motion of the precessing moments is viewed from the coordinate system which is also rotating around the magnetic field at the same Larmor frequency and in the same direction as the precessing nuclear moments, the motion is much simpler.

**saturation:** The state of containing or exhibiting the maximum amount of substance, energy, field, etc., or the action of bringing about that state; e.g., if the rates of upward and downward energy-level transitions are equal, no net energy can be absorbed and the system is said to be saturated.

**selection rule:** The energy of atomic or molecular systems, or energy of systems of their constituent nuclei, is quantized. The transition between these energy levels is governed by certain rules called selection rules.

**selective irradiation:** Rf excitation so designed that only a limited spatial region of the sample is excited.

**sensitive line imaging:** An imaging method in which a given characteristic, such as spin density, is determined along a line and then the procedure is repeated sequentially throughout the sample.

**sensitive point imaging:** An imaging method in which a given characteristic, such as spin density, is measured at predetermined points through the volume of interest.

**sequence delay time:** It is the time between the last pulse of a pulse sequence and the beginning of the next identical pulse sequence. It is the time allowed for the nuclear spin system to recover its magnetization and is equal to the sum of the acquisition delay time, data acquisition time, and the waiting time.

**sequence repetition time:** This is the time between the beginning of a pulse sequence and the beginning of the succeeding identical pulse sequence.

**shim coils:** Coils used to produce small field gradients to correct field inhomogeneities.

**shimming:** The process of optimization of magnetic field homogeneity in an NMR instrument by adjusting currents through the shim coils.

**signal averaging:** Technique of improving signal-to-noise ratio by adding up repeated scans through the same region of interest. The noise tends to average because of its random nature whereas the signal reinforces itself.

**signal-to-noise ratio:** In general, the ratio of the value of the signal to that of the noise.
(a) This ratio is usually in terms of peak values in the case of impulse noise and in terms of the root-mean-square values in the case of random noise.
(b) Where there is a possibility of ambiguity, suitable definitions of the signal and noise should be associated with the term: e.g., peak-signal to peak-noise ratio; root-mean-square signal to root-mean-square noise ratio; peak-to-peak signal to peak-to-peak noise ratio, etc.
(c) This ratio is usually expressed in decibels.

**skin-effect:** The decrease in the depth of penetration of an electrical current in a conductor as the frequency of the current increases. The skin depth is the depth below the surface at which the current density has decreased to $1/e$ of its value at the surface.

**spectral width:** In NMR, the frequency range represented without foldover.

**spectrum:** Plot of absorption (or emission) of energy as a function of energy, wavelength, or frequency. The sharp dips (or peaks) are referred to as spectral lines.

**spin:** A property of an elementary particle, such as an electron or proton that determines its angular momentum and magnetic moment. The spin is quantized and may have the value zero (in which case no magnetic moment), or whole or half-integral numbers. The spin of an electron or of a proton is ½.

**spin–coupling:** The interaction between the magnetic moments of two neighboring nuclei.

**spin density:** The number of resonating nuclei in a unit volume.

**spin diffusion:** The diffusion of magnetic moments due to actual movement of the associated molecule and/or chemical exchange.

**spin–echo:** The reappearance of a somewhat weaker NMR signal arising from rephasing of the components of magnetization in the $x,y$ plane.

**spin–echo sequence:** The sequence 90°–$\tau$–180°. Images produced using such sequences have strong $T_2$ dependence.

**spin–lattice relaxation time ($T_1$):** (Also called the *longitudinal relaxation time.*) The exponential time constant that characterizes the growth or decay of the component of magnetization parallel to the external field. The physical mechanism involved in this process is the interaction of the nucleus with its entire surroundings (lattice). See also *relaxation rate, spin–lattice relaxation time.*

**spin–lattice relaxation time in the rotating frame ($T_{1\rho}$):** NMR experiments can be conducted in such a fashion that the magnetization relaxes not in the longitudinal or transverse direction relative to the direction of the large applied external field, but in the direction of a much smaller applied magnetic field. Such relaxation viewed in the rotating frame is characterized by an exponential time constant known as $T_{1\rho}$. This technique is sensitive to ultra-slow motions.

**spin–spin broadening:** Increased line width due to interactions between neighboring dipoles.

**spin–spin relaxation time ($T_2$):** (Also called transverse relaxation time.) The exponential time constraint that characterizes the decay of confinement of magnetization perpendicular to the external field. This decay results from interaction at the nuclei with its immediate neighboring nuclei.

**spin–spin splitting:** Splittings in the lines of an NMR spectrum arising from the interaction of the nuclear magnetic moment with those of neighboring nuclei.

**spin–warp imaging:** A planar imaging method that uses time-dependent magnetic field gradients to encode NMR signals with phase and intensity information. The method was primarily developed by the group at Aberdeen, Scotland.

**spinning sidebands:** Bands, paired symmetrically about a principal band, arising from spinning of the sample in a field that is inhomogeneous at the sample position.

**steady-state free precession (SSFP):** An NMR technique in which the interpulse interval is much shorter compared to $T_1$ or $T_2$.

**sweep rate:** The rate, in Hertz-per-second (Hz) at which the applied radio-frequency is varied to produce an NMR spectrum.

**$T_1$:** See *spin–lattice relaxation time.*

**$T_2$:** See *spin–spin relaxation time.*

**$T_{1\rho}$:** See *spin–lattice relaxation time in the rotating frame.*

**tailored excitation:** RF excitation designed to excite only a selected region of the sample.

**tesla:** Unit of magnetic induction or flux density ($B$) in the SI system and is equal to 1 Weber per $m^2$ ($Wbm^2$) or 10,000 gauss.

**thermal equilibrium:** When a system and its surroundings are at the same temperature, there is no tendency of any heat flow to or from its surroundings. Such a system is said to be in thermal equilibrium with its surroundings.

**time constant:** Characteristic response time of a system. The longer the time constant, the longer it takes for the system to respond. (The response is usually exponential in time.)

**TMS:** Tetramethylsilane—an NMR reference compound.

**torque:** The turning moment exerted by a tangential force acting at a distance from the axis of rotation. It is equal to the product of the force and the distance in question.

**tuning:** To make radiation frequency match the natural "resonance" frequency of a given component such as the rf coil.

**unpaired electron:** An atomic or molecular electron whose spin is not paired with the oppositely-directed spin of another electron in the atom or molecule.

**viscosity:** The resistance to fluid flow set up by sheer stresses within the flowing liquid.

**waiting time:** The time between the end of data acquisition after the last pulse of a sequence and the initiation of a new sequence. (Some NMR systems place the waiting time prior to the initiation of the first pulse of the sequence in order to ensure equilibrium at the beginning of the first sequence.)

**wave function:** A function giving the probability distribution of electric charge (e.g, electron) in space. For atoms, it is a wave-mechanical description of their stationary states.

**wave guide:** A system consisting of a metal tube, a dielectric tube or rod, or a single wire, for the transmission of electromagnetic energy. Only certain waves can be transmitted by a particular guide.

**Zeeman effect:** Quantization of orbital angular momentum of atoms can be observed experimentally by placing atoms in a uniform magnetic field. The normal energy levels of the atom, i.e., in the absence of magnetic field, are split into multiple levels in the presence of magnetic field since each slightly different quantized angular momentum state has a different energy associated with it in the presence of the magnetic field. One can "observe" such splitting by looking at transitions between the shift levels. Such quantization of orbital angular momentum in the presence of a uniform magnetic field and its observation by looking at transitions between different energy states is called Normal Zeeman Effect.

**Zero field splitting:** Refers to the removal of degeneracy, i.e., an energy level is further split, in the absence of an applied magnetic field. This is due to the presence of the internal crystalline electric field.

**Zeugmatography:** A type of NMR imaging making use of magnetic field gradients to induce local interactions. In this technique the gradient field restricts the interaction of the object with the radio-frequency field to a limited region thus correlating the position of the region with the strength of the gradient field in that region. Lauterbur proposed the word "Zeugmatography" for this technique coining the word from the Greek term for "that which is used for joining".

## 10.3. GLOSSARY OF BIOLOGICAL AND BIOMEDICAL TERMS FOR NMR USERS

**abdomen:** Portion of the body between the diaphragm and the pelvis, containing the soft organs—liver, intestines, stomach, spleen, kidneys,

**abiotic:** Characterized by an absence of life.

**abnormal:** Contrary to the usual structure, position, condition, or rule.

**abscess:** Localized pocket of pus caused by death of tissue.

**abscissa:** The horizontal coordinate of a graph.

**absorption:** The uptake of substances into or across tissues.

**acellular:** Not made up of cells or containing no cells.

**acid:** Any compound of an electronegative element with dissociable hydrogen atoms.

**actomyosin:** The system of actin and myosin protein filaments responsible for muscle contraction.

**adenoma:** A benign epithelial tumor derived from glandular tissue. The numerous types are prefixed by cell of origin, e.g., chief-cell adenoma, papillary adenoma, fibroadenoma, tubular adenoma.

**adenosinetriphosphatase (ATPase):** An enzyme that degrades ATP to ADP (adenosinediphosphate) and Pi (inorganic phosphate).

**adenosinetriphosphate (ATP):** An energy storing compound found in all cells.

**adenosis:** Any disease of a gland.

**adipose:** Fat present in fat storage or adipose tissue.

**adolescence:** Period of life from appearance of secondary sex characteristics to full growth.

**adoral:** Toward or near the mouth.

**adrenal gland:** Gland situated on top of kidney, produces epinephrine.

**adrenalopathy:** Any disease of the adrenal glands.

**adsorption:** Attachment of one substance to the surface of another.

**adventitia:** Loose connective tissue covering an organ.

**aerobe:** A microorganism that can live in the presence of free oxygen; aerobic.

**agar:** Dried mucilaginous polymer extracted from algae, used as a gel for growth of microorganisms, extended polymer structure.

**albumin:** A common protein found in cells and fluids of plants and animals; used to balance serum osmotic pressure.

**aldosterone:** An electrolyte-water regulating steroid hormone secreted by the adrenal cortex.

**algae:** Unicellular plants accounting for 90% of the earth's photosynthetic production of oxygen.

**alimentary:** Pertaining to the organs of digestion.

**aliquot:** A small sample of the whole.

**alveolus:** Small sac-like volume of the lung, location of oxygen exchange.

**ambi-:** Prefix signifying on both sides; ambidextrous.

**amenorrhea:** Absence or stoppage of normal menstrual flow.

**amentia:** A congenital lack of mental ability.

**amino acid:** Acid containing -$NH_2$ group with COOH.

**amnion:** Thin tough membrane containing the fetus; amnionic fluid—fluid surrounding fetus.

**amorphous:** Having no definite form, shapeless.

**anabolism:** A process by which cells convert simple substances into complex compounds; opposite of catabolism.

**anaerobe:** Organism which grows in the absence of oxygen; anaerobic.

**anatomy:** Science of the structure of the animal body and its parts.

**androgen:** Substance that possesses masculinizing activity.

**anemia:** Lower than normal level of red blood cells or alteration in red cell oxygen carrying capacity.

**anesthesia:** Loss of feeling or sensation, chemically induced unconsciousness.

**aneurysm:** Dilatation of an arterial wall due to weakness.

**angina:** Sharp pain in the chest.

**angio-:** Referring to a blood vessel.

**angioma:** A tumor whose cells tend to form a blood vessel or related to blood vessels.

**angiomyosarcoma:** A tumor made up of elements of angioma and smooth muscle.

**anomalous:** Irregular; deviation from natural case; applied often to hereditary defects.

**anorexia:** Lack of appetite and metabolic decline.

**anoxia:** Absence or lack of oxygen.

**anterior:** Front or forward part of an organ.

**antibiotic:** A chemical produced by a microorganism which can inhibit the growth of other microorganisms.

**antibody:** An immunoglobulin protein that specifically combines with an antigen due to its structural recognition.

**anticoagulant:** Substance which prevents the coagulation or clotting of blood.

**aorta:** Major blood vessel leading from the heart to other organs.

**aplasia:** Lack of development of an organ or tissue.

**aplastic:** Pertaining to lack of an organ or tissue.

**aqueous:** Watery; pertaining to water.

**arteriolosclerosis:** Thickening of the walls of the smaller arterioles.

**artery:** Blood vessel carrying blood away from the heart to other organs.

**ascites:** Accumulation of serous fluid in the abdominal cavity.

**aseptic:** Free from infection or germs, antiseptic.

**atrium:** The upper chambers of the heart leading into the ventricles.

**atrophy:** Wasting away, loss of size in an organ or tissue.

**axon:** Extension of the nerve into the body; axoplasm—cytoplasm of the nerve axon.

**bacterium:** Microorganism of the class Schizomycetes, usually on the order of 1 $\mu$ in diameter, spherical or rod shaped.

**barbital:** Chemical 5,5 diethylbarbituric acid, used as a long acting hypnotic and sedative, derivative sodium pentabarbital or nembutal.

**basophil:** Structure, cell, or histological element staining easily with basic dyes; immature erythrocytes staining blue or gray.

**bigeminy:** Occurring in pairs; bigeminal pulse pair.

**bilateral:** Having two sides; symmetrical procedure such as bilateral adrenalectomy—removal of both adrenal glands.

**bile:** Fluid secreted by the liver, stored in the gall bladder, and secreted through the bile duct into the small intestine to aid in the digestion of fats.

**bilirubin:** Bile pigment; breakdown product of hemoglobin, causes yellow color in serum and tissues.

**bioassay:** Determination of the activity of a chemical in live animals or cultured cells.

**biogenesis:** The origin of life; the theory that organisms can originate only from organisms already living.

**biology:** The science that deals with the phenomena of life.

**biophysics:** The science dealing with the application of physical methods and theories to biology.

**biopsy:** Removal and examination of a portion of an organism or tissue, used to establish a diagnosis; usually a small piece of tissue.

**bladder:** Membranous sac or organ used to hold urine; sometimes other sacs such as gall bladder.

**blast:** Immature stage of cell development; precursor of differential cell type; i.e., lymphoblast, fibroblast, neuroblast.

**blood:** The fluid circulating through the heart, arteries, capillaries, and veins, carries nutrients and oxygen to body cells, blood groups defined by antigens on red cells surfaces.

**blood plasma:** All liquid portion of the blood without cells.

**blood serum:** Liquid portion of the blood plasma left after clotting of fibrinogen; has slightly lower protein concentration than plasma.

**bone:** Hard form of connective tissue that makes up the supporting skeleton of vertebrates; consists of an organic matrix of collagen and cells, and mineral portion made of calcium phosphate and calcium carbonate, matrix in dynamic turnover.

**brachium:** Referring to the arm; i.e., brachial artery.

**brain:** The portion of the central nervous system within the skull; also called encephalon, cerebrum; complex organ with many types of tissues.

**breast:** Anterior portion of the chest; in mammals, also refers to milk producing glands.

**bronchiole:** Fine subdivision of the airways of the lungs; bronchoscope—instrument for examining the interior of the lung.

**bubo:** An enlarged and inflamed lymph node due to infections such as plague and tuberculosis.

**bursa:** A sac or cavity filled with fluid to provide lubrication; often at a joint such as the knee; bursitis—inflammation of the bursa.

**bypass:** Auxiliary flow route; replacement of major blood vessels by synthetic tubes.

**cachexia:** A profound state of ill health and malnutrition; used with organ to denote origin—i.e., thyroid cachexia—ill health due to malfunction of the thyroid gland.

**cadaver:** A dead body; generally a dead human body.

**calcification:** Process by which an organic tissue becomes hardened by a deposit of calcium salts.

**calix:** A cup shaped organ or cavity, i.e., renal calices, calyx.

**callus:** Localized hyperplasia of the epidermis due to pressure or friction.

**cancer:** A cellular tumor whose natural course is fatal. Cancer cells unlike benign tumor cells exhibit the properties of invasiveness and spread to other organs. There are two broad categories termed:
*Carcinoma*—meaning a malignant new growth of epithelial cells that infiltrate surrounding tissues.
*Sarcoma*—referring to a tumor made up of a substance, e.g., embryonic connective tissue. For further definition these suffixes are preceded by the name of the tissue or organ they are associated with; i.e., adeno—carcinoma of a gland; melanotic—carcinoma (abbreviated melanoma) of the melanin pigment-producing cells; epithelial—carcinoma of the epithelium (abbreviated epithelioma); lympho—sarcoma of the lymphatic glands; asteo—sarcoma of the pigment-producing cells; fibro—sarcoma of the fibroblasts which are collagen-producing cells. Cancers described by these suffixes are considered to be malignant and life threatening.

**cannula:** A tube inserted into a vessel, duct, or cavity for the infusion or removal of fluid; cannulate—to introduce a tube.

**capillary:** Smallest diameter blood vessel; greatest exchange of oxygen and nutrients at this level.

**capsula:** Connective or fatty tissue surrounding an organ.

**carcinogen:** Any cancer-producing substance; often referring to chemicals that cause changes in cell properties.

**carcinoma:** A malignant new growth of cells from the epithelial layer (lining of organs, vessels, cavities) that tends to invade surrounding tissues and gives rise to metastases; organ name precedes to indicate type.

**cardiac:** Pertaining to the heart.

**caries:** Molecular decay or death of a bone; refers to decay of teeth.

**cartilage:** Fibrous connective tissue; infiltration with calcium minerals leads to bone development.

**casein:** The principal phosphoprotein of milk.

**castrate:** To deprive of gonads rendering the individual incapable of reproduction.

**catabolism:** The breakdown of complex compounds into simple chemicals during metabolism or after death.

**cataract:** Opacity of the crystalline lens of the eye.

**catheter:** A tube for introducing or withdrawing fluids from a cavity of the body; especially for withdrawing urine from the bladder through the natural pathway.

**cauterize:** To seal by burning.

**cecum:** The first portion of the large intestine.

**celiac:** Pertaining to the abdomen; celioma—a tumor of the abdomen.

**cell:** The unit of living tissues; in higher forms consists of a nucleus containing genetic material, and a cytoplasmic space.

**cellulose:** Carbohydrate polymer forming the skeleton of plants.

**cephalo:** Pertaining to the head.

**cerebellum:** Portion of the brain concerned with coordination of movements.

**cerebrum:** Main portion of the brain occupying upper cranial cavity; two hemispheres associated with conscious actions.

**cervix:** Term denoting connection of two parts of a functional organ; in female mammals the cervix connects the uterus and vagina.

**chancre:** A sore or lesion usually caused by an infection.

**chemoreceptor:** A protein structure capable of combination with specific chemicals and transfer of information of their presence to the organism; often on cell surfaces.

**chemotaxis:** Movement of an organism toward a higher chemical concentration or in response to a chemical presence.

**chimera:** Organism which develops from combined portions of different types of tissues.

**chondrocyte:** Cartilage cell; chondrocarcinoma—tumor type.

**chromatin:** Readily stainable nuclear material consisting of DNA, histones, and associated proteins.

**chromosomes:** Structure involved in the transmission of genetic information (DNA) during cell division.

**chyme:** The semifluid liquid material produced during the digestion of food.

**cilium:** A fine hairlike projection of the surface of a cell.

**cirrhosis:** Liver disease with loss of lobular architecture, fibrosis, and loss of liver function.

**cisterna:** A closed reservoir for lymph or other body fluid.

**clastic:** Causing or undergoing a division into parts.

**cloaca:** A common term of fecal, urinary, or reproductive discharge.

**clone:** The progeny of a single cell.

**clot:** A semi-solid mass of blood or lymph; coagulate of blood.

**coarctation:** Stricture or contraction, often refers to major arteries; coarctation of the aorta.

**coel-:** Denoting cavity or space; sometimes cel.

**colic:** Pertaining to the colon; intense abdominal pain.

**colitis:** Inflammation of the colon.

**collagen:** The main fibrous protein of skin, tendon, bone, cartilage, and connective tissue; collagenase—an enzyme which breaks up collagen.

**collateral:** Secondary or accessory; alternate passage.

**colloid:** A suspension of fine particles which does not settle out; glutinous or like glue.

**colon:** Part of the large intestine from the cecum to the rectum; used inaccurately to refer to all of the large intestine.

**colostrum:** Thin yellow milky fluid secreted by the mammary gland after birth, containing immunoglobulins to transfer immunity from mother to infants.

**coma:** State of unconsciousness from which a patient cannot be aroused.

**comedocarcinoma:** Intraductal cancer of the breast.

**complement:** A blood protein that is part of a complex immune response.

**concentric:** Having a common center.

**continence:** Ability to refrain; incontinence—inability to refrain from desire or urging—inability to control the release of urine.

**contuse:** To bruise.

**convex:** Having a rounded elevated surface.

**cor:** Muscular organ maintaining circulation; heart.

**cord:** Any long rounded flexible anatomical structure; umbilical cord of the fetus; spermatic cord.

**cornea:** Transparent anterior portion of the fibrous tunic of the eye.

**cornification:** Conversion to the hard protein keratin.

**corpus:** A discrete mass of material; human corpse.

**corpuscle:** Small mass or body; formed elements of the blood; after cell type.

**cortex:** The external anatomical layer; outer surface of the brain and other organs.

**coxa:** Used to denote hip joint.

**cranio-:** Denoting relationship to the head or skull; i.e., cranial, cranium.

**crenation:** Formation of notching on edge of a cell.

**crest:** A projection or ridge surmounting a bone or organ; also crista.

**crevice:** A longitudinal fissure.

**cryobiology:** Science dealing with the effect of low temperatures on biological systems.

**crypto-:** Denoting hidden or concealed; as in crypt.

**culture:** Propagation of microorganisms or cells in defined media.

**cyanosis:** Blue discoloration of tissues due to lack of oxygen caused by excess reduced hemoglobin in the blood.

**cyst:** Closed cavity or sac filled with liquid or semi-solid material; may be due to infection or presence of foreign material.

**cytogenetics:** Science of cellular constituents concerned with heredity and chromosomes.

**cytology:** Study of cells, their origin, structure and chemical properties.

**cytoplasm:** The protoplasm of the cell exclusive of the nucleus; everything outside the nucleus.

**cytoskeleton:** Filamentous network of proteins in the cytoplasm.

**cytosol:** Liquid portion of the cytoplasm.

**cytotoxic:** Deadly to cells.

**dacryo-:** Pertaining to tears or crying.

**death:** Cessation of life; legally—"the irreversible cessation of total cerebral function, spontaneous function of the respiratory system, and spontaneous function of the circulatory system."

**debride:** To remove foreign material or contaminated tissue from wound with a scalpel.

**deferent:** Conveying away from a center.

**dehydration:** Removal of water from a substance, condition when water content of the body falls below normal.

**demi-:** Denoting one half.

**denaturation:** Destruction of the usual nature of a substance; a change in the tertiary or secondary structure of protein which causes it to lose its unique properties.

**dens:** Referring to teeth.

**dentin:** Hard substance of the tooth; similar to bone.

**derma:** Denoting skin; dermatitis—inflammation of the skin.

**deuterium:** Isotope of hydrogen whose nucleus contains one proton and one neutron.

**development:** The process of growth and differentiation from the egg until adulthood.

**dexter:** Denoting the right hand or the right side.

**diabetes:** Disorders characterized by excessive urine excretion; familiar type due to faulty function of the pancreas and insulin imbalance.

**diagnosis:** The *art* of distinguishing one disease from another; the determination of the cause of a disease.

**dialysis:** Process of separating materials by their relative permeability through a membrane.

**diaphragm:** Muscular membrane separating the thorax and abdomen; involved in breathing.

**diaphysis:** Portion of long bone between the articulate ends; shaft.

**diarrhea:** Abnormal frequency and fluidity of fecal material.

**diathermy:** Heating of tissue by high-frequency electromagnetic radiation.

**differentiation:** Process of acquiring individual fixed characteristics for a cell or tissue; cancers are sometimes characterized by an apparent reversal of this process in specific cells.

**digit:** A finger or toe.

**diploid:** Having two sets of chromosomes.

**dissection:** To cut apart into sections.

**distal:** Remote; opposite of proximal.

**diuresis:** Increased secretion of urine.

**diverticulum:** A closed pouch or sac in an organ.

**dorsum:** Referring to the back; dorsal, dorsa.

**duct:** A passage with well-defined walls or a tube for excretions or secretions.

**duodenum:** First or proximal portion of the small intestine; up to jejunum.

**dura mater:** Outermost, toughest of three membranes covering the brain; "hard mother."

**dysplasia:** Abnormality of development; in pathology alteration in size, shape, and organization of adult cells.

**dystrophia:** Dystrophy, defective or faulty function; muscular dystrophy, degeneration of muscle structure, sometimes hereditary.

**ear:** The organ of hearing.

**ectasia:** Dilation, expansion, or distension.

**ectoderm:** The outermost of three primary germ layers of the embryo; origin of epidermis, nails, hair, glands, nervous system, ear, membranes of mouth and anus.

**edema:** Presence of abnormal amounts of fluid in intercellular spaces of the body.

**elastin:** Protein of elastic connective tissue; elasto-fibroma-tumor consisting of elastin and fibrous elements.

**electrocardiogram (EKG, ECG):** Graphic tracing of the electrical activity of the heart.

**electroencephalogram (EEG):** Graphic tracing of the electrical potentials of the brain.

**embolism:** Blocking of a blood vessel by a clot or air bubble.

**enamel:** White compact hard outer surface of the teeth.

**encephalitis:** Inflammation of the brain.

**encephalopathy:** Any degenerative disease of the brain.

**endocarditis:** Inflammation of the lining of the heart.

**endoceliac:** Inside one of the body cavities.

**endocrinology:** Study of the endocrine system.

**endometrium:** Mucous membrane lining the uterus.

**endothelium:** Layer of epithelial cells lining the heart, lymph vessels, and serous cavities of the body.

**enzyme:** Protein catalyst of biochemical reactions.

**epidermis:** Outermost nonvascularized layer of the skin.

**epiphysis:** Wide end of long bones just before the joint; usual region of growth.

**epithelium:** Covering internal and external surfaces of the body; single layer of cells.

**erythroblast:** Immature red blood cell.

**erythrocyte:** Mature red blood cell; in humans red cells are biconcave discs without nuclei.

**esophagus:** Muscular passage from the pharynx to the stomach.

**exo-:** Referring to outside.

**familial:** Affecting more members of a family than would be expected by chance.

**fascia:** A sheet of fibrous tissue lying under the skin or inside organs.

**fat:** Adipose tissue; reserve supply of fatty acids, yellow, white, and brown fat.

**fecal:** Referring to excrement.

**femur:** Large bone of the thigh.

**fetus:** Unborn offspring after major structures have been defined from the embryo.

**fever:** Elevation of body temperature.

**fibroblast:** Connective tissue cell; immature form of collagen-producing cells.

**fibroma:** A tumor consisting of connective tissue.

**fimbria:** Fringe, border, or edge.

**fissure:** Term for cleft or groove in organ.

**fistula:** Abnormal passage between two body cavities.

**flux:** Excessive flow or discharge.

**follicle:** Small sac or cavity.

**fossa:** Hollow or depressed area; especially in bone.

**fovea:** Term for small pit on the surface of an organ.

**fundus:** The bottom or base of an organ.

**gallbladder:** Pear-shaped reservoir for bile on the underside of the liver; gallstone—hard lump of cholesterol.

**gamete:** One of two cells, male and female, that must unite to assure reproduction.

**ganglion:** Term for mass of nerve cells located outside the central nervous system; a knotlike mass or tumor.

**ganglioside:** Class of galactose containing cerebrosides found in tissues of the central nervous system.

**gastric:** Pertaining to the stomach.

**gel:** Colloid which is firm although it contains large amounts of water; often a crosslinked polymer.

**generic:** Public or common name for a chemical.

**genetics:** Science of heredity; chemical and physical nature of genes.

**genitalia:** The reproductive organs.

**genu:** The knee; anatomical structure which bends like a knee.

**geny:** Related to the jaw.

**germ:** A pathogenic or harmful microorganism.

**gero-:** Referring to old age; geriatrica; gerontology.

**gestation:** Period of development before birth.

**gibbous:** Convex; humped, protuberant.

**gingiva:** Mucous membrane of the mouth.

**gland:** Group of cells specialized to secrete or excrete biochemical materials.

**glaucoma:** Opacity of the lens of the eye.

**glia:** The neuroglia; gluelike structure; glio—related to a gluey substance.

**glioma:** Tumor of the brain or spinal cord.

**globule:** A small spherical mass; small spherical bodies in cells or secretions.

**globulin:** A class of protein insoluble in water but soluble in saline; important in the immune response.

**glomerulus:** Cluster of blood vessels or nerve fibers, glomus—a ball of arterioles and veins.

**glossa:** The tongue.

**glottis:** Vocal apparatus of the larynx.

**gluteal:** Pertaining to the buttocks.

**gnathic:** Pertaining to the jaw or cheek.

**goiter:** Enlargement of the thyroid gland in the neck.

**gon:** Denoting relation to seed or semen; or relationship to the knee.

**gout:** Excess uric acid crystal accumulation in the joints.

**graft:** Tissue or organ for transplantation to another site or another individual.

**granule:** Small particle of grain; non-soluble particles in cells; beadlike masses of tissue.

**granuloma:** Tumor-like mass of granulated tissue.

**gravid:** Pregnant.

**growth:** A normal process of increase in mass; or an abnormal tissue mass, normal or malignant cells.

**gut:** Intestine or bowel.

**gyneco-:** Denoting relationship to women or the female sex.

**gyrus:** Infolding of the surface of the brain.

**halo:** A luminous or colored circle seen in the light around an object.

**hamarto-:** Denoting a defect.

**hamartoma:** Benign tumor composed of overgrowth of normal cells.

**haplo-:** Form meaning single or simple.

**haploid:** Having one set of chromosones.

**hapten:** Factor important to the immune response.

**head:** The upper extremity of the body; casing for the brain and sense organs.

**heart:** Organ which maintains the circulation of the blood.

**HeLa cells:** From the name of the patient from whom these human cervical carcinoma cells were derived; possibly Henrietta Lacks or Helen Lane.

**helio:** Denoting relation to the sun; heliopathia—pathology due to sunlight.

**hematocrit:** Volume percentage of red cells in whole blood; 46% in males, 43% in females.

**hematoma:** Localized collection of clotted blood in an organ usually due to a break in the wall of a blood vessel.

**hematuria:** Blood in the urine.

**heme:** Nonprotein iron protoporphyrin ring compound of hemoglobin and other respiratory pigments.

**hemi-:** Meaning half.

**hemilateral:** Affecting only one side.

**hemiplegia:** Paralysis of one side of the body.

**hemoglobin:** Oxygen-carrying pigment of red blood cells.

**hemostat:** A small surgical tool for clamping off blood vessels.

**heparin:** A mucopolysaccharide which renders blood incoagulable.

**hepatitis:** Inflammation of the liver; many causes—often viral.

**hepatoma:** Tumor of the liver; usually carcinoma.

**hepatoportal:** Pertaining to the portal or venous side of the liver circulation.

**hernia:** Protrusion of a loop of an organ through an abnormal opening; weakness of the intestinal wall allowing protrusion.

**herpes:** Any inflammatory skin disease characterized by sores; also the virus causing inflammation.

**hetero-:** Denoting a relationship to another; heterosexual.

**hiatus:** Term for gap, cleft, or opening; opening in diaphragm through which aorta and thoracic duct pass.

**hidradenoma:** General term for tumors of the skin.

**hilus:** Denoting depression or pit on that part of the organ where blood vessels and nerves enter.

**histo-:** Referring to tissue; histology—study of tissues, microscopic anatomy.

**holo-:** Relationship to the whole.

**homeo-:** Remaining the same; homeostasis—ability of the body to strive to remain the same in composition and structure.

**hormone:** Chemical regulatory substance made by the body.

**host:** An animal or plant that harbors or nourishes another organism or parasite.

**humerus:** Bone that extends from the shoulder to the elbow.

**humor:** Fluid materials from the body.

**hyaline:** Glassy and transparent; hyaloid membrane.

**hybrid:** Animal or plant produced from parents different in kind; cell made by fusion of two cell types.

**hydranencephaly:** Almost complete replacement of cerebral hemispheres by cerebrospinal fluid in developing brain.

**hydro-:** Referring to water.

**hydrocephalus:** Accumulation of fluid in the cranium, but brain may be completely formed.

**hydrops:** Accumulation of serous fluid in tissues or a body cavity; dropsy.

**hygroma:** A sac or cyst filled with fluid; a watery tumor.

**hyper-:** Above, beyond, excessive.

**hyperplasia:** Excessive multiplication of normal cells in a normal arrangement; hyperplastic.

**hypertrophy:** Enlargement of an organ due to increased size of individual cells; not due to hyperplasia.

**hypo-:** Beneath, below, under.

**hypoplasia:** Incomplete development of an organ because of too few cells.

**hypoxia:** Low oxygen content or tension; insufficient for normal metabolism.

**iatro-:** Denoting relationship to a physician or to medicine.

**idiotype:** Genotype; inherited characteristics.

**ileum:** Distal portion of the small intestine; parts—duodenum, jejunum, ileum.

**image:** A picture or conception with more or less likeness to objective reality.

**immunity:** Response to antigenic or foreign material in the body; protection against disease.

**impatency:** Being closed or obstructed.

**implant:** Material inserted or grafted to the body.

**impressio:** Indentation produced on the surface of one organ by the pressure of another.

**incision:** A cut produced by a sharp instrument.

**inclusion:** A particle in the cell cytoplasm.

**incontinent:** Unable to control excretory functions.

**infarct:** Area of dead tissue due to blood clot blocking the circulation.

**infection:** Invasion and multiplication of microorganisms in the body.

**inflammation:** Localized response to redness and swelling due to trauma or infection.

**infra-:** Meaning to be situated below or beneath.

**infracostal:** Below the ribs.

**infusion:** The steeping of a substance or drug in water to extract its medicinal benefit.

**inguen:** The junction between the abdomen and thigh.

**innate:** Inborn, hereditary, congenital; present at birth.

**innocuous:** Harmless.

**in ovo:** In the egg; experiments on the egg.

**in situ:** In the natural or normal place, in the body.

**insufflation:** The act of blowing powder, vapor, gas or air into a body cavity.

**in tela:** In tissue; especially to staining of tissues.

**inter-:** Prefix meaning between two separate elements.

**intercellular:** Between two separate cells.

**interstitial:** Between the parts of a tissue; connective tissue.

**intestine:** Organ extending from the stomach to the anus.

**intima:** Innermost part of a structure.

**intra-:** Occurring within an element; intracellular, inside one cell.

**intraabdominal:** Within the abdomen.

**intravenous:** Inside a vein.

**in vacuo:** In a vacuum.

**in vitro:** Observable in a test tube; removed from the natural environment; excised.

**in vivo:** Within the living body.

**iris:** Circular pigmented portion of the eye perforated by the central dark pupil.

**irrigation:** Washing with a stream of water or saline.

**iso-:** Prefix meaning equal, alike, or same.

**isotonic:** A solution in which body cells can be bathed without shrinking or swelling.

**jaundice:** Disease characterized by deposition of yellow hemoglobin breakdown products in skin and mucous membranes.

**jecur:** Liver.

**jejunum:** Middle portion of the small intestine.

**joint:** Articulation of two bones coming together.

**jugal:** Pertaining to the cheek.

**junction:** Where two organs came together; also junctura.

**juxtaposition:** Opposing one another.

**kalium:** Potassium; hypokalemia—low potassium.

**karyoplast:** The nucleus of a cell.

**karyosome:** Irregular lumps of chromatin in the nucleus.

**keloid:** Elevated, progressively enlarging scar due to excessive amounts of collagen synthesis during repair.

**keno-:** Meaning empty.

**keratin:** Scleroprotein; principal component of hair, nails, and tooth enamel.

**kidney:** Organ which filters the blood and produces urine.

**kino-:** Form pertaining to movement; kinology—kinesiology, study of movement.

**labio-:** Referring to the lips.

**labyrinth:** System of interconnecting cavities or canals.

**lac:** Referring to milk.

**lamella:** Thin leaf or plate; arrays of membranes.

**lamina:** Thin flat plate or layer; composite structure.

**lance:** To cut with a lancet or small sharp knife.

**laparo-:** Denoting the loin or flank.

**larynx:** Upper part of the windpipe between the mouth and the trachea.

**latero:** Meaning the side.

**lavage:** The wash from inside an organ; or to wash out toxic material.

**leio-:** Meaning smooth.

**leiomyoma:** Benign tumor from smooth muscle.

**lemma:** A covering layer designed to be shed.

**lens:** Clear biconvex part of eye through which light passes.

**lepto-:** Meaning slender, thin, or delicate.

**lesion:** Any pathological or traumatic disorder of tissue or loss of function: sore, cut, opening.

**leucocyte:** White blood cell or corpuscle; numerous functional types; leukocytes.

**leukemia:** Progressive malignant disease of the blood-forming organs; excessive and distorted proliferation of leucocytes and their precursors.

**leukoencephalitis:** Inflammation of the white substance of the brain.

**leukopenia:** Reduced white cells in the blood.

**leukosis:** Proliferation of leukocyte-producing tissue.

**levo-:** Meaning to the left.

**lien:** The spleen.

**ligament:** Band of fibrous tissue connecting bones.

**lingua:** The tongue.

**linitis:** Inflammation of the gastric cellular tissue.

**lipedema:** Accumulation of fat and fluid in subcutaneous tissue.

**lipo:** Related to fats or lipids.

**lipoma:** Benign tumor composed of mature fat cells.

**litho:** Referring to a stone or calculus.

**liver:** Large dark red glandular organ in the upper abdomen; hepar—suffix related to liver, multifunctional.

**livid:** Discolored, lead colored, black and blue.

**lobe:** Well-defined pendant portion of an organ.

**lobotomy:** Incision or removal of a portion of an organ.

**lobular:** Having small lobes.

**locus:** Place, site, specific location.

**lumen:** The cavity or channel within an organ or gland.

**lung:** Organ of respiration.

**lymph:** Transparent slightly yellow fluid found in the lymphatic system; filtrate of plasma through capillary walls.

**lymphangioma:** A tumor of newly formed lymph spaces.

**lymphedema:** Accumulation of interstitial lymph fluid due to obstruction of the lymphatic channels.

**lymphocyte:** A mononuclear white blood cell which participates in the immune response.

**lymphoma:** Neoplastic disorder of lymphoid tissue: e.g., Hodgkin's disease. Lymphomas are usually malignant; lymphosarcomas.

**lyse:** To cause disintegration of cells.

**lysosome:** Cell organelle containing lytic enzymes.

**macro-:** Meaning large.

**macrophage:** Large phagocytic cells occurring in the walls of blood vessels and connective tissue; eats foreign particles.

**macula:** Stain, spot; area distinguished by color or appearance.

**magnum:** Meaning great.

**mal:** Disease; bad condition; sickness.

**malignant:** Becoming progressively worse and leading to death; invasive and metastasizing.

**malum:** Evil or disease.

**mamma:** The breast; milk-secreting gland on the chest, mammary.

**mammography:** X-ray of the mammary gland.

**mandibula:** Mandible, lower jawbone.

**manus:** The hand.

**marrow:** Soft tissue filling the bone cavities.

**mass:** Lump or body made of cohering particles; sometimes refers to a tumor.

**mater:** Mother; dura mater—hard mother; pia mater—tender mother; membranes covering the brain.

**maxilla:** Bone of the upper jaw.

**mazo-:** Denoting the breast.

**meatus:** A general term for an opening or passageway into the body; acoustic meatus—opening into the ear.

**medial:** Pertaining to the middle.

**medulla:** The innermost part of an organ.

**megakaryocyte:** Giant cell of the bone marrow.

**megaloblast:** Giant precursor of abnormal red blood cells.

**-megaly:** Enlargement of an organ; splenomegaly.

**meiosis:** Sexual division of cells resulting in egg and sperm with one half the normal number of chromosomes.

**melanin:** Dark pigment of skin, hair, and various tumors; color of black moles.

**melanoma:** Tumor of melanin-pigmented cells; malignant melanoma metastasizes rapidly.

**membrane:** Thin layer of tissue covering a surface or cavity; lipoprotein covering of a single cell.

**meninges:** Three membranes covering the brain.

**meningioma:** Hard, slow growing vascular tumor along the vessels of the meninges.

**meningitis:** Inflammation of the meninges.

**menses:** Monthly flow of blood from the genital tract of women.

**meridian:** An imaginary line on the surface of a spherical body.

**mero-:** A combining form meaning part of.

**mesentery:** A membrane attaching various organs to the body wall.

**mesoderm:** Middle of three primary germinal layers of an embryo; others are ectoderm and endoderm.

**metaplasia:** A change in type of adult cells to a form not natural in that particular tissue.

**metastasis:** A growth of pathogenic organisms or abnormal cells distant from their primary site; in cancer, the spread of malignant cells to other parts of the body besides the primary tumor.

**metra:** The uterus or womb, metro, mother.

**micelle:** Unit of living matter having power of growth and division; colloid particle.

**micro-:** Meaning small.

**microbe:** A minute living organism; bacteria, protozoa, fungi.

**microtubule:** A hollow tubular protein construct in the cytoplasm, in the spindle during division, ~200Å in diameter.

**miosis:** Contraction of the pupil, alternate spelling of meiosis.

**mitochondria:** Cell organelles with double membranes that are the site of ATP synthesis.

**mitosis:** Cell division in which two daughter cells are identical.

**mole:** Fleshy mass or tumor, sometimes pigmented.

**monocyte:** A mononuclear phagocytic leukocyte.

**mononucleosis:** Presence of large numbers of monocytes in the blood.

**-morph:** Denoting form or shape, body type—endomorph, mesomorph.

**mucosa:** Membrane secreting mucus.

**muscle:** Organ which by contraction and relaxation produces movement; types: cardiac, skeletal, smooth.

**myasthenia:** Muscular debility; anomaly of muscle.

**myelin:** Lipid substance forming a sheath around certain nerve fibers.

**myeloblast:** Immature cell found in the bone marrow; myeloblastic leukemia.

**myeloma:** Tumor composed of cells normally found in the bone marrow.

**myocarditis:** Inflammation of the muscular walls of the heart.

**myocyte:** Cell of the muscle tissue.

**myofilament:** Threadlike structures in striated muscle cells; composed of actin and myosin proteins.

**myopia:** Nearsightedness.

**myosarcoma:** Malignant tumor derived from muscle.

**myxoma:** Tumor composed of primitive connective tissue cells and stroma.

**nano-:** Denoting small size.

**nares:** Nostrils; nasal openings.

**nates:** Buttocks.

**natrium:** Sodium.

**necro-:** Denoting relationship to death.

**necrosis:** Death and decomposition of a portion of tissue; may leave fluid-filled cavity.

**neoplasia:** New growth; progressive multiplication of cells unchecked by usual controls.

**nephro-:** Denoting relationship to the kidney; nephrology—study of the kidney.

**nephron:** Functional unit of the kidney.

**nerve:** Long fiber-like cells which transmit information among body parts.

**neura-:** Neuro-; pertaining to nerves.

**neuroblastoma:** Sarcoma of nervous system origin; composed of nerve precursor cells.

**neuroma:** Tumor of nerve cells and nerve fibers.

**neuron:** A single nerve cell; composed of a nerve cell body, short dendrite projections, and a long axon.

**node:** A small mass of tissue as a knot or protuberance, normal or pathological.

**noma:** Sores; nomadic—pertaining to wandering; lymphoma—a wandering tumor of lymph tissues.

**normal:** Agreeing with the regular and established type; defined set.

**nucleus:** Spheroid body in cell containing genetic information; also a group of nerve cells in the central nervous system.

**ob-:** Signifying against, in front of, towards.

**occlusion:** Act of closure; area of closure; meeting of teeth in the jaws.

**oculus:** The eye.

**odonto-:** Referring to the teeth.

**odonotoma:** Tumor of the dental tissues.

**olfaction:** The sense of smell.

**oligo-:** Denoting few, little or scanty.

**-oma:** Denoting swelling, tumor or neoplasm.

**omo-:** Relating to the shoulder.

**onco-:** Denoting relationship to a tumor or mass.

**oncogenic:** Giving rise to tumors or causing tumors.

**oocyte:** A developing egg.

**oophoro-:** Denoting relationships to the ovary.

**ophthalmo-:** Denoting relationships to the eye.

**ora-:** Referring to the mouth, oral.

**organism:** An individual living entity; unicellular or multicellular.

**orthopedics:** Branch of medicine dealing with the skeletal system.

**os:** Any orifice of the body. Also referring to bone, ossa, osseous, osteo-.

**ossification:** Formation of bone or bony substance in other tissues.

**osteoblast:** Cell which, when mature, forms bone.

**osteoclast:** Cell which absorbs and removes bone.

**osteoma:** Tumor composed of bone tissue.

**otic:** Pertaining to the ear; otor, oto.

**otitis:** Inflammation of the ear.

**ova:** Referring to eggs.

**ovario-:** Pertaining to the ovary.

**ovary:** Female gland, produces eggs.

**ovum:** The egg.

**oxy-:** Pertaining to oxygen.

**pachy-:** Meaning thick; pachyderma—thick skin.

**palate:** Top of the mouth cavity.

**paleo-:** Meaning old.

**pancreas:** Large elongated gland in the abdomen; concerned with regulation of carbohydrate metabolism; produces insulin.

**papilla:** Small nipple-shaped projections.

**papilloma:** Branched or lobulated benign tumor.

**para-:** Meaning beside, beyond, apart from, against.

**paralysis:** Loss of motor function in a part due to a lesion of the neural or muscular mechanism.

**paries:** Term for the wall of an organ or cavity.

**pars:** A division or part.

**pathogen:** Any disease-causing microorganism or material.

**pathology:** Branch of medicine which treats the nature of the disease; especially structure.

**pectoral:** Pertaining to the breast or chest.

**pedicle:** The stalk of an organ or mass; peduncle.

**pedology:** The study of the life and development of children.

**pelvis:** Lower portion of the trunk.

**peptic:** Pertaining to pepsin or digestion; peptic ulcer of the stomach.

**percutaneous:** Performed through the skin, as by needle.

**perfusion:** The passing of a fluid through the vessels of a specific organ; perfusion of the liver.

**peri-:** Meaning around.

**pericardium:** Fibrous membrane sac around the heart.

**peridontal:** Occurring around the tooth.

**peritoneum:** Membrane lining the abdominopelvic walls and connecting the viscera.

**pero-:** Meaning maimed or deformed.

**pes:** The foot.

**phago-:** Denoting eating or consumption.

**phagocyte:** A cell that ingests microbes and foreign particles.

**phalanx:** General term for any bone of the finger or toes.

**pharynx:** The throat.

**phenotype:** Entire physical, biochemical, and physiological pattern of an individual.

**-philia:** Term denoting abnormal fondness.

**phlebo-:** Denoting relationship to a vein.

**phlebotomize:** To take blood from the veins.

**phono-:** Denoting sound of voice.

**photolytic:** Decomposed by light; photo—light.

**phrenic:** Pertaining to the diaphragm; also to the mind.

**phthisis:** A wasting way of the body or a part of the body.

**physiology:** The science of the living organism and its parts; physio—pertaining to life.

**pineal:** Pertaining to the pineal body of the brain; pinealoma—a rare tumor.

**pinna:** A projecting part of the ear on the outside of the head.

**pio-:** Denoting relationship to fat; pionemia—presence of fat or oil in the blood.

**pisiform:** Resembling a pea.

**pith:** To pierce the spinal cord or brain.

**plaque:** Any patch or flat area.

**plasma:** Fluid portion of the blood.

**platelet:** Disk-shaped portion of a blood cell which participates in blood coagulation.

**-plegia:** Meaning paralysis or stroke.

**pleura:** Serous membrane investing the lung and lining the thoracic cavity.

**plexus:** A term for a network where nerves or vessels come together.

**plica:** A general term for a ridge or fold.

**pneumal:** Pertaining to the lungs.

**pneumothorax:** Accumulation of air or gas in the pleural space.

**podo-:** Denoting the foot.

**poly-:** Denoting many.

**pons:** Any slip of tissue connecting two parts of an organ; a bridge.

**post-:** Denoting after or behind.

**posterior:** Situated in back of, on the backside.

**preneoplastic:** Before the appearance of a tumor, but with potential to become malignant in a certain percentage of cases.

**prognosis:** A forecast as to the probable outcome of a disease.

**prone:** Lying face downward.

**prophylaxis:** The prevention of disease; preventative treatment.

**protoplasm:** The viscous, translucent, polyphasic water colloid that makes up the material of all plant and animal cells.

**pseudo-:** Denoting false or spurious.

**psycho-:** Denoting relationship to the psyche or mind.

**puberty:** Period of development of secondary sex organs.

**pulp:** A soft juicy animal or plant tissue; tissue contained in a bony chamber; dental pulp.

**pulse:** The rhythmic expansion of an artery which may be felt with the finger and corresponds to the beating of the heart and pulsatile blood flow.

**pupil:** The opening at the center of the iris of the eye.

**purpura:** A group of disorders characterized by purplish or brownish red spots under the skin.

**pus:** Liquid composed of white blood cells and the decomposition of tissue in a wound; pustule—pus under the skin in a bump.

**pylorus:** Distal aperture of the stomach; empties into the small intestine.

**pyro-:** Referring to fire or heat; fever; pyrogen—a fever-causing substance.

**quadrant:** One quarter of an organ.

**rachidial:** Pertaining to the spine.

**radial:** Pertaining to the radius of the forearm; or spreading outward.

**radical:** Directed to the source of a morbid process; radical surgery.

**radiology:** The science of radiant energy and radiant substances in diagnosis of disease.

**radiopacity:** Not passable by x-rays.

**radix:** A structure by which something is firmly attached.

**ramus:** A branch; term for a smaller structure given off a larger one.

**ranine:** Lower surface of the tongue; ranula—tumor beneath the tongue.

**receptor:** A chemical structure in a cell or on a cell surface which can fit together with structures of other specific chemical groups; receives information input.

**rectum:** Distal portion of the large intestine which stores feces prior to elimination.

**rectus:** Denoting a straight structure.

**reduction:** The correction of a fracture or hernia.

**reflex:** Sum total of any involuntary activity in response to particular stimuli; often a jerk of muscles.

**reflux:** A backward or return flow.

**refractory:** Not responding readily to treatment.

**regio:** General term for certain flat, defined areas of the body.

**regression:** Return to former state; for tumors, a loss in size and return to normal appearance.

**remission:** A diminution or abatement of symptoms; temporary or permanent.

**renal:** Pertaining to the kidney; reniculus—lobe of the kidney.

**resection:** Excision or removal of a portion of an organ or structure.

**respiration:** The exchange of oxygen and carbon dioxide between the air and the cells of the body.

**rete:** Anatomical term for network of blood vessels.

**reticulum:** A network of fibers or membranes inside cells.

**retina:** The back surface of the eyeball, location of photosensing cells.

**retinoblastoma:** Tumor from retinal germ cells.

**retro-:** Meaning behind or dropping back.

**retrogression:** Degeneration, deterioration, failure.

**rhabdo-:** Denoting relationship to a rod or rod-shaped.

**rhabdosarcoma:** Rhabdomyosarcoma; highly malignant tumor of striated muscle derived from primitive mesenchymal cells.

**rhinal:** Pertaining to the nose.

**rhodocyte:** Erythrocyte; rhodo—red.

**rima:** Cleft or crack; opening.

**rosette:** Any structure or formation resembling a rose; anatomical feature.

**rouleau:** A roll of red blood cells piled like coins.

**sac:** Pouch, bag-like structure; sacculus—little bag.

**sacrum:** Triangular bone at the base of the spine; sacro-.

**saliva:** Clear, alkaline secretion from the mouth.

**sangui-:** Denoting relationships to blood.

**saphena:** Two large superficial veins of the leg.

**sarcoma:** Tumor made up of cells like embryonic connective tissue; often highly malignant.

**scapho-:** Meaning boat-shaped.

**scapula:** Triangular bone in back of the shoulder.

**scatology:** Study of human feces or excrement.

**scissura:** An incision or cut.

**sclerosis:** A hardening of an interstitial surface; arterial sclerosis.

**scoliosis:** A lateral deviation of the spine.

**scoto-:** Related to darkness.

**secretion:** The process of production of a substance by a gland; the substance produced.

**sectio:** The act of cutting into parts.

**segment:** A portion of a length of an organ.

**semi-:** Denoting half; semicoma, semicolon.

**serum:** Clear portion of the blood after clotting; inoculum containing antibodies.

**sheath:** Tubular structure of membranes.

**shock:** Sudden disturbance of mental or physical equilibrium; allergic shock; blood loss shock; electric shock.

**shunt:** To turn aside, divert, bypass, alternate pathway.

**sialaden:** A salivary gland.

**sigmoid:** Terminal portion of the colon.

**sinus:** A cavity or channel in an organ; fluid and air spaces in the cranium.

**skeleton:** The hard framework of the body; the bone.

**skiagram:** X-ray picture.

**skin:** Outer covering of the body; consists of dermis, epidermis, and subcutaneous tissue layers.

**skull:** Hard bony framework of the head.

**soma:** The body as distinguished from the mind.

**soporific:** Causing or inducing profound sleep.

**sphincter:** Ring-like band of muscle constricting a passage or natural orifice; anal sphincter.

**spina:** A spine, a thornlike process; the spinal column of vertebrae.

**splanchnic:** Pertaining to the viscera.

**spleen:** A large glandlike, but ductless, organ in the upper part of the abdomen on the left side, removes dead cells from the blood.

**spongio-:** Looking like a sponge.

**sputum:** Matter ejected from the lungs upon deep coughing.

**squama:** Scale or platelike cells; squamous carcinoma.

**steal:** The diversion of blood flow from its normal course.

**stella:** Anatomical features looking like a star.

**stenosis:** Narrowing or stricture of a duct or canal.

**sternum:** Bone in middle of the chest.

**stoma:** Any minute pore or orifice between cells.

**stomach:** Muscular organ of digestion between the esophagus and duodenum.

**stratum:** A layer; sheetlike mass of uniform thickness.

**stria:** A streak or line.

**striation:** Being marked by stripes or lines.

**stroma:** The support matrix of a tissue or organ.

**sub-:** Under, near, almost.

**sulcus:** A groove, trench or furrow; often on the brain.

**suture:** A type of closely-knit joint as in the skull; or a material used in sewing up a wound.

**synapse:** Anatomical relation of the axon of one nerve to the cell body of another.

**syndrome:** A set of symptoms which occur together and have been given a name.

**tabes:** Any wasting away of the body, progressive atrophy.

**tache:** A spot or blemish.

**tacho-:** Denoting speed, tachycardia—fast heart beat.

**tapetum:** A covering layer of cells; a part of the brain.

**tarso-:** Denoting the edge of the eyelid; or the bones of the foot—tarsus.

**tegmen:** Anatomical term for covering or roof over a part of an organ.

**tela:** Thin membrane resembling a web.

**telo-:** Denoting the end.

**tendon:** Fibrous cord by which a muscle is attached to a bone.

**tenia:** A flat band or strip of soft tissue; holds organ in place.

**teres:** Long and round as a muscle.

**terminal:** End; or leading to death.

**therapy:** Treatment of disease; therapeutics.

**-thrix:** Denoting a resemblance to hair.

**thrombus:** Aggregation of blood factors causing vascular obstruction.

**thyroid:** Gland producing metabolic controller—thyroid hormone.

**tissue:** An aggregation of similarly specialized cells united to perform a special function.

**tongue:** Muscular organ on the floor of the mouth.

**toxin:** A poison; product of a microorganism.

**trachea:** Cartilagenous tube extending from the esophagus to the bronchi of the lungs.

**trauma:** A wound or injury; sometimes caused by a physical blow.

**triad:** A group of three similar elements.

**trigonum:** Term for a triangular area.

**truncus:** The main part, stem of body.

**tumor:** Swelling or new growth of tissue in which division is uncontrolled; see *cancer*.

**ulcer:** Local defect with central necrotic region.

**ultra-:** Excess or beyond.

**ureter:** Tube which conveys urine from the kidney to the bladder.

**urethra:** Tube carrying urine from the bladder to the outside.

**urine:** Filtrate of the blood excreted by the kidneys.

**uterus:** Hollow muscular organ of females in which the embryo develops until birth.

**uvula:** Little grape; pendant fleshy mass.

**vaccine:** Suspension of attenuated or dead microorganisms, injected to produce immunity to diseases.

**vagina:** Sheathed canal in females extending from the outside to the cervix.

**vagus:** Tenth cranial nerve.

**valva:** Valve; one-way passage.

**varix:** An enlarged and tortuous vein, artery or lymphatic vessel.

**vas:** A vessel, a canal for carrying fluid such as blood, lymph, or spermatozoa.

**vasoactive:** Having an effect upon the caliber of vessels.

**vein:** Vessel carrying oxygen-poor blood from the tissues to the heart.

**ventral:** Pertaining to the bellyside.

**ventricle:** A small cavity; ventricles of the brain and heart.

**vertebra:** Thirty-three bones of the spinal column.

**vertigo:** Dizziness.

**vesica:** Membranous sac for a secretion.

**villus:** Small vascular process from the surface of a membrane.

**viscera:** Referring to the internal organs.

**vitreous:** Glass-like fluid or gel inside the eyeball.

**vivi-:** Denoting relationship to life.

**vulva:** External genitalia of female at head of vagina.

**wound:** Body injury caused by physical means.

**xeno-:** Meaning foreign or strange.

**xerosis:** Abnormal dryness of the eye, skin, mouth.

**yolk:** Stored nutrients of the ovum.

**zona:** A zone; defined area of anatomy.

**zygote:** The cell resulting from the union of the egg and sperm.

## 10.4 COMMON ABBREVIATIONS

| | |
|---|---|
| AC | Alternating current |
| ADC | Analog to digital converter |
| A/D | (See ADC) |
| CT | Computed tomography |
| CPMG | Carr-Purcell-Meiboom-Gill sequence |
| cw | Continuous wave |
| DAC | Digital to analog converter |
| D/A | (See DAC) |
| DC | Direct current |
| EFG | Electric field gradient |
| DFT | Discrete Fourier transform |
| ESR | Electron spin resonance |
| FFT | Fast Fourier transform |
| FID | Free induction decay |
| FT | Fourier transform |
| FWHM | Full width at half maximum |
| IF | Intermediate frequency |
| IRSE | Inversion recovery spin–echo |
| l-He | Liquid helium |
| $l\text{-}N_2$ | Liquid nitrogen |
| NAR | Nuclear acoustic resonance |
| NMR | Nuclear magnetic resonance |
| NQR | Nuclear quadrupole resonance |
| PRFT | Partially relaxed Fourier transform |
| PSD | Phase-sensitive detector |
| QD | Quadrature detection |
| rf | Radio-frequency |
| SC | Superconducting |
| SEFT | Spin–echo Fourier transform |
| SFP (SSFP) | Steady-state free precession |
| $S/N$ | Signal-to-noise ratio |
| $T_1$ | Spin–lattice relaxation time |
| $T_{1D}$ | Dipolar spin–lattice relaxation time |
| $T_{1\rho}$ | Spin–lattice relaxation time in rotating frame |
| $T_2$ | Spin–spin relaxation time |
| VOM | Volt–ohm meter |

# APPENDIX

## A. LIST OF NMR EQUIPMENT MANUFACTURERS

### Imaging Systems

Technicare Corp.
29100 Aurora Road
Solon, Ohio 44139 USA
216-248-1800 Contact: Sheldon Shaffer

General Electric Co.
Medical Systems Operations
P.O. Box 414 NB-901
Milwaukee, Wisconsin 53201 USA
414-785-5765 Contact: Joe Vacca

FONAR Corporation
110 Marcus Drive
Melville, New York 11747 USA
516-694-2929 Contact: John Cassesse

Picker International, Inc.
595 Miner Rd.
Highland Heights, Ohio 44143 USA
216-449-3000 Contact: William Doran

USA Bruker Instruments Inc.
Manning Park
Billerica, Massachusetts 01821 USA
617-667-9580 Contact: Christian I. Tanzer

Mag Scan, Inc.
P.O. Box 540324
Houston, Texas 77254 USA
713-797-5856 Contact: Waylon House

Diasonics
1545 Barber Lane
Milpitas, California 95035 USA
408-946-9001 Contact: William Sullivan

Elscint, Inc.
930 Commonwealth Avenue
Boston, Massachusetts 02215 USA
617-739-6000

Oxford Magnet Technology
Osney Mead,
Oxford OX2 0DX ENGLAND
0865-250128 Telex 83413

Nalorac Cryogenics
1717 Solano Way, #37
Concord, California 94520 USA
415-676-9577 Contact: James Carolan

OMR Technology, Inc.
292 South LaCienaga Blvd.
Beverly Hills, California 90211 USA
213-657-5191 Contact: Jerry Gartin

Philips Medical Systems Inc.
710 Bridgeport Avenue
Shelton, Connecticut 06484 USA
203-926-7674 Contact: Richard Sano

Siemens Medical Systems, Inc.
186 Wood Avenue. South
Iselin, New Jersey 08830 USA
201-321-3400 Contact: Tom Miller

Toshiba America, Inc.
2441 Michele Drive
Tustin, California 92680 USA
816-444-7411 Contact: Steven Clark

SEIMCO
Box 51
Parnassus Station
New Kensington, Pennsylvania 15068 USA
412-339-7553

Varian
Instrument Division
611 Hansen Way
Palo Alto, California 94303 USA

### In Vitro Systems

The Praxis Corporation
8327 Potranco Road
San Antonio, Texas 78251 USA
512-684-3231 Contact: Gil Persyn

Spin–Lock Electronics, Ltd.
403-28 Helene Street N.
Port Credit L5G-3B7
Ontario, CANADA
416-278-0931

Phospho-Energetics Inc.
3401 Market Street, #320
Philadelphia, Pennsylvania 19104 USA
215-387-4429

JEOL USA
235 Birchwood Avenue
Cranford, New Jersey 07016 USA

## B. SUGGESTED READING LIST

### BOOKS

**Abragam, A.** (1961) *The Principles of Nuclear Magnetism,* Oxford University Press (Clarendon), London.

**Andrew, E.R.** (1955) *Nuclear Magnetic Resonance,* Cambridge University Press, Cambridge.

**Beall, P.T., Amtey, S.R., and Kasturi, S.R.** (1984) *NMR Data Handbook for Biomedical Applications,* Pergamon Press, New York.

**Becker, E.D.** (1980) *High Resolution NMR: Theory and Chemical Applications,* 2nd Ed., Academic Press, New York.

**Bryan, R.N., Bushong, S.C., and Willcott, M.R.** (1984) *Medical NMR Imaging,* C.V. Mosby Co., St. Louis.

**Damadian, R.** (1981) *NMR in Medicine,* Springer-Verlag, Berlin.

**Eisenberg, D., and Kauzman, D.** (1969) *Structure and Properties of Water,* Clarendon Press, Oxford.

**Farrar, T.C., and Becker, E.D.** (1971) *Pulse and Fourier Transform NMR,* Academic Press, New York.

**Foster, M.** (1984) *Magnetic Resonance in Biological Systems,* Pergamon Press, Oxford.

**Franks, F.** (1972–1979) *Water—A Comprehensive Treatise,* Vol. 1–6, Plenum Press, New York.

**Fukushima, E., Roeder, S., and Stephen, B.W.** (1981) *Experimental Pulse NMR: A Nuts and Bolts Approach,* Addison Wesley, Reading, MA (Advanced Book Program).

**Hoult, D.I.** (1980) *Magnetic Resonance in Biology* (Ed. J.S. Cohen), John Wiley, New York.

**James, T.L.** (1975) *Nuclear Magnetic Resonance in Biochemistry,* Academic Press, New York.

**Kaufman, L., Crooks, L.E., and Margulis, A.R.** (1981) *Nuclear Magnetic Resonance Imaging in Medicine,* Igaku-Shoin, New York, Tokyo.

**Kundla, E., Lippman, E., and Saluvere, T.** (1979) *Magnetic Resonance and Related Phenomena,* Proceedings of the XXth Congress AMPERE, Springer-Verlag, Berlin.

**Mansfield, P., and Morris, P.G.** (1982) *NMR Imaging in Biomedicine,* Academic Press, New York.

**Mullen, K., and Pregosin, P.S.** (1976) *Fourier Transform NMR Techniques: A Practical Approach,* Academic Press, London.

**Partain, C.L., James, A.E., Rollo, F.D., and Price, R.R. (Eds.)** (1983) *Nuclear Magnetic Resonance (NMR) Imaging,* W.B. Saunders Co., Philadelphia.

**Poole, C.P. Jr., and Farach, H.A.** (1972) *The Theory of Magnetic Resonance,* Wiley-Interscience, New York.

**Pople, J.A., Schneider, W.G., and Bernstein, H.J.** (1959) *High-Resolution Magnetic Resonance,* McGraw-Hill Book Co., New York.

**Rushworth, F.A., and Tunstall, D.P.** (1973) *Nuclear Magnetic Resonance,* Gordon and Breach Science Publishers, Inc., New York.

**Saha, A.K., and Das, T.P.** (1957) *Theory and Applications of Nuclear Induction,* Saha Institute of Nuclear Physics, Calcutta.

**Shaw, D.** (1976) *Fourier Transform NMR Spectroscopy,* Elsevier, Amsterdam.

**Slichter, C.P.** (1980) *Principles of Magnetic Resonance,* Springer-Verlag, Berlin.

**Witcofski, R., Karstaedt, N., and Partain, C.L. (Eds.)** (1981) *Proceedings Intern Symp on NMR Imaging,* Bowman Gray School of Medicine, Vanderbilt Univ. School of Medicine, and National Cancer Institute.

## REVIEWS

**Beall, P.T.** (1983) States of Water in Biological Systems, *Cryobiology* **20:**324–334.

**Bottomley, P.A.** (1982) NMR Imaging Techniques and Applications: A Review, *Rev. Sci. Instr.* **53:**1319.

**Cohen, M.H., and Reif, F.** (1957) Quadrupole Effects in Nuclear Magnetic Resonance Studies and Solids, in *Solid State Physics,* Vol. 5, ed. by Seitz and Turnbull, Academic Press, New York, 321–438.

**Das, T.P., and Hahn, E.L.** (1958) Nuclear Quadrupole Resonance Spectroscopy, in *Solid State Physics,* Supplement 1, ed. by Seitz and Turnbull, Academic Press, New York.

**Fullerton, G.D.** (1982) Basic Concepts for Nuclear Magnetic Resonance Imaging, *Mag. Res. Imaging, 1:39–55.*

**Gorter, C.J.** (1947) *Paramagnetic Relaxation,* Elsevier Publishers, New York.

**Ingram, D.J.E.** (1955) *Spectroscopy at Radio and Microwave Frequencies,* Butterworth's Scientific Publications, London.

**Ingwall, J.S.** (1982) Phosphorous Nuclear Magnetic Resonance Spectroscopy of Cardiac and Skeletal Muscles, *Amer. J. of Physiol.,* **242:**H729–H744.

**Mansfield, P., and Pykett, I.L.** (1978) Biological and Medical Imaging by NMR, *Mag. Res.* **29:**355–373.

**Mathur-DeVré, R.** (1979) The NMR Studies of Water in Biological Systems, *Prog. Biophys. Mol. Biol.,* **35:**103–134.

**O'Neill, I.K., and Richards, C.P.** (1980) Biological $^{31}$PNMR Spectroscopy, *Ann. Rep. NMR Spect.,* **10A:**133–236.

**Pake, G.E.** (1956) Nuclear Magnetic Resonance, in *Solid State Physics,* Vol. 2., ed. by Seitz and Turnbull, Academic Press, New York.

**Pykett, I.L., Newhouse, J.H., Buonanno, F.S., Brady, T.J., Goldman, M.R., Kistler, J.P., and Pohost, G.M.** (1982) Principles of Nuclear Magnetic Resonance Imaging, *Radiology,* **143:**157.

**Pykett, I.L.** (1982) NMR Imaging in Medicine, *Scientific American* **246:**78.

**Schulman, R.G.** (1983) NMR Spectroscopy of Living Cells, *Scientific American,* Jan., :86.

## CLASSICAL PAPERS

**Bloch, F.** (1946) Nuclear Induction, *Phys. Rev.* **70:**467–474.

**Bloch, F., Hansen, W.W., and Packard, M.** (1946a) Nuclear Induction, *Phys. Rev.* **69:**127.

**Bloch, F., Hansen, W.W., and Packard, M.** (1946b) The Nuclear Induction Experiment, *Phys. Rev.* **70:**474–485.

**Bloembergen, N., Purcell, E.M., and Pound, R.V.** (1948) Relaxation Effects in Nuclear Magnetic Resonance Absorption, *Phys. Rev.,* **73:**679–712.

**Carr, H.Y., and Purcell, E.M.** (1954) Effects of Diffusion on Free Precession in Nuclear Magnetic Resonance Experiments, *Phys. Rev.* **94:**630–638.

**Damadian, R.** (1971) Tumor Detection by Nuclear Magnetic Resonance, *Science* **171:**1151–1153.

**Damadian, R., Minkoff, L., Goldsmith, M., Stanford, M. and Koutcher, J.** (1976) Field Focusing Nuclear Magnetic Resonance (FONAR): Visualization of a Tumor in a Live Animal, *Science* **194:**1430–1432.

**Damadian, R., Goldsmith, M., Mindoff, L.** (1977) NMR in Cancer: XVI. FONAR Image of the Live Human Body, *Physiol. Chem. Phys.* **9:**97–100.

**Hahn, E.L.** (1950) Spin Echoes, *Phys. Rev.,* **80:**580–594.

**Lauterbur, P.C.** (1973) Image Formation by Induced Local Interactions: Examples Employing NMR, *Nature* **242:**190–191.

**Odeblad, E., Bahr, B.N., and Lindstrom, G.** (1956) Proton Magnetic Resonance of Human Red Blood Cells in Heavy-Water Exchange Experiments, *Arch. Biochem. Biophys.,* **63:**221–225.

**Purcell, E.E., Torrey, H.C., and Pound, R.V.,** (1946) Resonance Absorption by Nuclear Magnetic Moments in a Solid, *Phys. Rev.,* **69:**37.

## JOURNALS PUBLISHING BIOLOGICAL AND BIOMEDICAL NMR RESEARCH

Advances in Magnetic Resonance

American Journal of Roentgenology

British Journal of Radiology

British Medical Journal

Cancer Research

Computerized Radiology

IEEE Transactions—Biomedical Engineering

Investigative Radiology

Journal of Clinical Engineering

Journal of Computerized Tomography

Journal of Magnetic Resonance in Medicine

Journal of Physics (E): Scientific Instruments

Journal of the National Cancer Institute

Lancet

Magnetic Resonance Imaging

Medical Physics

Nature

New England Journal of Medicine

Physics in Medicine and Biology

Physiological Chemistry, Physics and Medical NMR

Proceedings of the National Academy of Sciences

Progress in NMR Spectroscopy

Radiology

Reviews of Scientific Instruments

Science

Stroke